PRENTICE-HALL INTERNATIONAL, INC., *London*
PRENTICE-HALL OF AUSTRALIA, PTY. LTD., *Sydney*
PRENTICE-HALL OF CANADA, LTD., *Toronto*
PRENTICE-HALL OF INDIA PRIVATE LIMITED, *New Delhi*
PRENTICE-HALL OF JAPAN, INC., *Tokyo*

FOUNDATIONS OF MODERN BIOCHEMISTRY SERIES

Lowell Hager and Finn Wold, editors

ORGANIC CHEMISTRY OF BIOLOGICAL COMPOUNDS*
Robert Barker

INTERMEDIARY METABOLISM AND ITS REGULATION
Joseph Larner

PHYSICAL BIOCHEMISTRY
Kensal Edward Van Holde

MACROMOLECULES: STRUCTURE AND FUNCTION
Finn Wold

SPECIAL TOPICS:

BIOCHEMICAL ENDOCRINOLOGY OF THE VERTEBRATES
Earl Frieden and Harry Lipner

* Published jointly in Prentice-Hall's *Foundations of Modern Organic Chemistry Series.*

MACROMOLECULES: STRUCTURE AND FUNCTION

FINN WOLD
Professor of Biochemistry
University of Minnesota

Prentice-Hall, Inc., Englewood Cliffs, New Jersey

Current printing (last digit)

10 9 8 7 6 5 4 3 2 1

Library of Congress
Catalog Card No. 71-173660

13-542613-8 (C)
13-542605-7 (P)

Printed in the United States of America

FOREWORD

Biochemistry has been and still is the major meeting place for biological and physical sciences, and most introductory courses in biochemistry have a rather unique and heterogeneous population of advanced students from all branches of the natural sciences. Such courses, therefore, have equally unique and heterogeneous requirements for background material and textbooks. The content of a first-year course in biochemistry based on two years of chemistry and biology would probably not be too difficult to define, but to write a single text which contains the needed material for all students becomes a much more controversial issue. As a solution to this dilemma, presenting the material in several packages offers some interesting possibilities. Without in any way compromising the basic content, such a multivolume text should above all allow for a great deal of flexibility: flexibility for the student to supplement previous experience with only the parts which represent new and unique features; flexibility for the instructor to offer a course covering an area of modern biochemistry with something less than the comprehensive text; and the flexibility to keep a text current by rewriting any outdated part without having to reject the whole text.

These thoughts were fundamental in formulating the Foundations of Modern Biochemistry series. So was the philosophy that a major purpose of a textbook is perhaps not so much to be encyclopedic as to select for the student the important principles and to illustrate and explain these in some depth. With this basic plan in mind, the next step was to divide the whole into logical subdivisions or parts. The science of biochemistry seeks the answer to three basic questions:

1. What is the nature of the molecules and structures found in living cells?
2. What is the biological function of these molecules and structures?
3. How are they synthesized (and broken down) in the cell?

These questions were adopted as the basis for the division of this text. The first question, related to the qualitative and quantitative characterization of the biochemical world and to the methods available for structural analysis, is covered in two books: *Organic Chemistry of Biological Compounds* and *Physical Biochemistry*. The second question, concerning the elucidation of the biological function of these molecules, is discussed in *Macromolecules: Structure and Function*. The third question, covering all aspects of intermediate metabolism and metabolic regulation, is considered in *Intermediary Metabolism and Its Regulation*. As the work on the individual books progressed, it became apparent that this subdivision had one unexpected advantage; namely that the physical presentation of material as diversified as the mathematical derivations in physical chemistry, the many structures of organic chemistry, and the long and complicated road maps of intermediate metabolism can be handled much more rationally in the individual format of separate books than in a single volume.

Thus, the Foundations of Modern Biochemistry series came into being consisting of four *individual* books. The books can hopefully be used either separately or in any combination to meet the requirements of the individual student according to his particular background and goals. The absence of a numbered sequence represents a deliberate effort to emphasize that the books can be used in any order. The integration of the four parts into the total program is done by extensive cross-references between the individual books. Trying to retain the individuality and utility of the separate books and at the same time aiming for an integrated program required some compromises in the selection and distribution of the material to be covered. Quite expectedly, the main price paid for this dual purpose has turned out to be some duplication, which hopefully is not extensive enough to become a serious flaw.

In arriving at the definition of the Foundations of Modern Biochemistry content and at the philosophy represented by this set of books, the thinking of the "editorial board" (the four authors and Dr. Lowell P. Hager) underwent an extensive evolution. The evolutionary pressures were graciously, patiently, and sometimes even enthusiastically provided by colleagues, by the publishers, and most importantly by students, too many to mention individually. To all of these good people we express our sincere thanks.

FINN WOLD

To BERNADINE

PREFACE

I have often had cause to feel that my hands are cleverer than my head. That is a crude way of characterizing the dialectics of experimentation. When it is going well, it is like a quiet conversation with Nature. One asks a question and gets an answer; then one asks the next question, and gets the next answer. An experiment is a device to make Nature speak intelligibly. After that one has only to listen.

G. WALD*

In this book we discuss the status of the structure–function analysis of biological macromolecules and macromolecular complexes. The ultimate goal of the analysis must be to explain all the functional properties of the molecules in question in terms of their completely defined three-dimensional structure, and the analysis thus contains three separate components: the determination of structure, the determination and quantitation of function, and final correlation of this information into the structure–function model. The first component, the structural analysis, is reviewed only briefly, and this book therefore leans heavily on Barker's and Van Holde's books in this series for proper background and documentation for this component. The second component, the analysis of functional properties, is given broader consideration (Chapters 1, 2, 5, and 9), but the main emphasis has been the step-by-step development of the structure–function models. It is hoped that this approach will clearly illustrate the typical progression of scientific model building from the first clear definition of the

*Quote taken from *Science,* **162,** 230 (1968).

problem and the statement of the hypothesis through ever-increasing refinements of experimental tests toward the final answer. It is also hoped that the statements of philosophy, principles, and scientific method that are the bases for this approach are of broad enough validity to survive even after its models have become obsolete. With this approach, it is essential to inform the reader in unequivocal terms that this book is not a summary of final conclusions and complete stories which can be submitted to memory. Each system discussed should be considered very critically, and the models should be evaluated in terms of the available evidence. The only "facts" are the experimental data; the interpretation of this data into models is only convincing to the extent that it makes logical sense to the individual examining it. Since both space and common sense prohibits a continuous reiteration of this statement throughout the book, be prepared to encounter some models and hypotheses which are based on sound experimental evidence as well as some which have no experimental basis at all. In neither case are they "facts," but in either case they represent ideas which can be subjected to further experimental tests. If the book helps to sharpen this critical evaluation of both ideas and the experimental test of the hypotheses, one of its major purposes has been fulfilled.

A few words about references and specific topics: In executing the above stated philosophy, it was decided to select and discuss a few "case histories" to illustrate the kinds of experimental tools we have available and the type of conclusions we can reach. The selection is quite arbitrary and flavored by the author's interests and ignorance. For each example, key references (mostly to the original literature) are given as footnotes in the text. These references consequently become rather specialized and do not in any way represent a cross section of the whole area of macromolecular structure and function. To remedy this shortcoming, more general references have been included at the end of each chapter. These references include mostly treatises, books, and reviews which will give the reader a start toward a more complete coverage of the field.

Finally, in looking back at the genesis of this book, one of the outstanding impressions which will linger for a long time is the generosity of the many colleagues and friends who helped along the way. To all who contributed their time and know-how in constructive criticism, far too many to name individually, I extend my most sincere thanks. I also owe a dept of gratitude to the biochemists who made their data available to me, and to the publishers who helped to catalyze the transition from an idea to a manuscript, and then effected the metamorphism from manuscript to book. Very special thanks for invaluable criticism and help go to my fellow authors and editors in the Biochemistry Series; and to my wife, Bernadine, who patiently endured the whole ordeal with a cheerful smile, and corrected, typed, and retyped, I gratefully dedicate the book.

FINN WOLD

CONTENTS

ONE CELLS, CELLULAR COMPONENTS, AND VIRUSES: DEFINITION OF THE PROBLEM

The smallest unit of all living systems is the cell, but viruses, parasitic, nonliving "incomplete cells," are also important members of the biosphere. We shall review the organization of the cell and briefly survey its main functional components. Macromolecules, macromolecular aggregates, or organelles characteristic of *all living cells* can be divided into three main groups: proteins, nucleic acids, and lipoprotein membranes. The subject of this book is the problem of correlating the structure and function of these three groups of compounds in vivo and in vitro. Hence the following chapter introduces in a general manner the objects of biochemical inquiry.

1.1 MODELS

It may be useful to start with a few words about models. Some day, as determined by need and knowledge, a Mendeleev of biology will construct a periodic table of the elements of life, and the very complex living world may then be neatly classified according to biological equivalents of valence and mass. These equivalents must necessarily be based on a precise definition of both structure and function and will require an impressive accumulation of data.

The analogy between biological taxonomy and the chemical elements is not so farfetched as it may seem and may even be of some use to a chemist looking at biology. The biological equivalent of the atom is the cell, the smallest living unit, and the equivalent of the molecule is the multicellular organism. Just as there are many different types of atoms, there are different types of cells, and as all atoms can be further broken down into common elementary particles, all cells can be defined in terms of certain structural or functional subcellular entities.

This type of analogy is a very common part of scientific communication; it represents the most obvious kind of *model*, in which familiar concepts are used to translate new and relatively unknown systems into ideas that can be readily communicated. The importance of such analogies lies primarily in communication, as they conjure up relatively precise and uniform pictures of situations for which an adequate vocabulary is not yet fully developed. Often, however, they are very restrictive and totally worthless as real scientific models. The concept of the scientific model as it will be used in this book, and as applied to biochemical systems, is much more specifically directed at the system under study. The real experimental model is synthesized from all the information available for the system, interpreted in terms of analogies, and formulated in such a way that *it predicts certain new events*; it can thus be subjected to *experimental tests.* A model that has no predictive aspects and is not subject to experimental test is no real model at all although it may still have some use in communication and in thinking about a problem. Biochemistry, in bringing together concepts from the physical and biological sciences, relies heavily on models, both of the analogy type and the real type, and the use of models will be an obvious necessity throughout this book.

1.2 UNIFYING PROPERTIES AND STRUCTURES

If living forms are defined as highly organized combinations of chemical compounds that are capable of self-directed growth and reproduction and use only the chemical starting materials available in their environment, it is reasonable to argue that all living cells must have at least two systems: the synthetic apparatus to produce the chemical compounds required for the original cell, and a means of processing information so that the synthetic product is always the necessary one. It also seems reasonable that once a set of molecules fitting these requirements is assembled, it should be separated and protected from the environment by some kind of envelope or barrier; we can thus include the plasma membrane as a third essential component of all living cells.

Chemically, we can thus propose that all living cells require at least three broad groups of functional molecules or structures: (1) those involved in synthesis (including all the enzymes involved in energy metabolism, in conversion of environmental precursors to the proper cellular starting materials, and in the biosynthetic reactions); (2) those concerned with information storage and transfer (DNA and RNA); and (3) those forming the semipermeable membrane that separates the vital processes from the nonliving environment and also provides the interphases and barriers between cellular compartments. If we examine a large number of different cells to find these types of molecules and structures, we discover that although such components are always present, their physical or morphological appearance varies greatly. Thus DNA may be found associated with basic proteins (histones) and concentrated almost exclusively in the nucleus of nucleated cells, whereas in bacteria, which have no distinct nucleus, the DNA is dispersed in the cell but perhaps concentrated at the plasma membrane. Similarly, most of the enzymes involved in the oxidative degradation of fats and carbohydrates (the energy-producing processes) are found associated with the mitochondria in the cells of higher organisms, whereas in cells without mitochondria these enzymes are distributed throughout and are again probably associated primarily with the plasma membrane. The plasma membrane and the ribosomes are perhaps the only two types of macromolecular complexes found in all living cells.

1.3 UNIQUE, SPECIALIZED PROPERTIES AND STRUCTURES

In any search for the universal structures in the different types of cells, it would become immediately apparent that each cell has several unique features. Usually, unique structures and molecules can now be associated with specific functions. The photosynthetic apparatus of green plants, the chloroplast, is a good example of such a unique cellular component with a known specific function. Other well-defined examples are the organelles responsible for motility (flagella and cilia), the cell walls of a number of unicellular organisms, and the lysosomes of higher cells. Some of these unique structures and molecules, as well as the universal ones, are briefly reviewed in Tables 1.1 and 1.1A. It should be clear that the list is far from complete.

A brief pause to recall the endless and wonderful display of different specialized expressions of life around us should adequately illustrate the complexity of the biochemical systems in living cells. Virtually every technological advance of mankind since the invention of the wheel has a natural counterpart evolved in some cells or organisms to a very high degree of efficiency. All cells carry out combustion more efficiently than our best engines; several species produce, store, and discharge electricity; light is produced by a large number of bioluminescent cells and organisms; radar, or more precisely sonar, is used by the bat; the squid moves by jet propulsion; birds not only fly but may even know elements of celestial navigation; and most cells have

TABLE 1.1 A SURVEY OF MOLECULES AND SUBCELLULAR ORGANELLES IN DIFFERENT CELLS

Subcellular Component (Major Molecular Constituents)	*Function Properties*	*Occurrence in Different Types of Cell*[a]
Cell membrane—also called plasma membrane: forming a continuous layer around the cytoplasm (lipid and protein).	Semipermeable cell envelope regulating exchange with environment	All cells
Cytosol: the soluble, nonparticulate part of the cytoplasm (proteins, nucleic acids, inorganic salts, and most metabolites.)	Solution of many components of the synthetic apparatus and the information transfer chain	All cells
Ribosomes: nucleoprotein particles (RNA and protein)	Sites of protein synthesis	All cells
Nucleus (DNA and proteins (histones) with nuclear membrane (lipoprotein)	Information storage and transfer (replication and transcription)	1 and 2
Free DNA—without nucleus and histones.	Information storage and transfer (replication and transcription)	3
Cell wall: rigid structure surrounding the cell. (Polysaccharides, glycopeptides, teichoic acids)	Protection, storage	Most of 2 and 3
Mitochondria: separate bodies made by highly convoluted membranes (lipids and proteins)	Part of the synthetic apparatus: biosynthesis and energy metabolism	1 and 2
Plastids—Chloroplasts and leucoplasts: separate bodies resembling mitochondria (lipids, including photosynthetic pigments and proteins)	Part of synthetic apparatus: photosynthesis	2 and 3 (photosynthetic)
Smooth Endoplasmic Reticulum: (golgi apparatus) cytoplasmic membrane structure, perhaps continuous with the nuclear and plasma membrane (lipids and proteins)	Provides interphase for transport and for chemical reactions and separates cell compartments (?) Secretory canals?	1 and 2
Rough Endoplasmic Reticulum: Smooth ER with ribosomes attached (microsomes)	Same as above and protein synthesis	1 and 2
Lysosomes: Separate bodies consisting of several types of hydrolytic enzymes enclosed within a membrane (lipids and proteins)	Part of synthetic (digestive) apparatus: Hydrolysis of foreign material, lysis of dead cells.	1

[a] Types of cells divided into the following 3 broad groups: 1, animals and protozoa; 2, higher plants, algae and fungi; and 3, bacteria and blue-green algae.

TABLE 1.1A ISOLATION OF SUBCELLULAR PARTICLES BY DIFFERENTIAL CENTRIFUGATION AT 4°

Preparation	*Centrifugal Force* (xg)	*Time*	*Precipitate Collected*
1. Cell homogenate in 0.3 *M* sucrose, buffer, and dilute salt	700	10 min	Nuclei
2. Supernatant from 1	7000	10 min	Lysosomes and mitochondria[a]
3. Supernatant from 2	100,000	90 min	Smooth and rough endoplasmic reticulum[a]
4. Supernatant from 3	100,000	10 hr	Free ribosomes
5. Supernatant from 4	—	—	Cytosol

[a] These can be separated by recentrifugation.

developed sophisticated molecular computers with feedback and memory circuits. Perhaps the mastery of controlled nuclear fission is man's only unique invention, not duplicated by any other living form.

Considering also that some 1.5 million different forms of animals, plants, and protista (microorganisms) have been described, and that living forms are found from high in the atmosphere to the depths of the oceans, from the tropics to the polar caps, and from the deserts to the swamps, the diversity of life as it has evolved on earth during the last 2 to 5 billion yr (years) is truly remarkable. If we take for granted that each specialized function and each adaptation to a new environment is a consequence of new molecules or molecular arrangements, the living world presents us with an endless series of exciting and challenging biochemical problems.

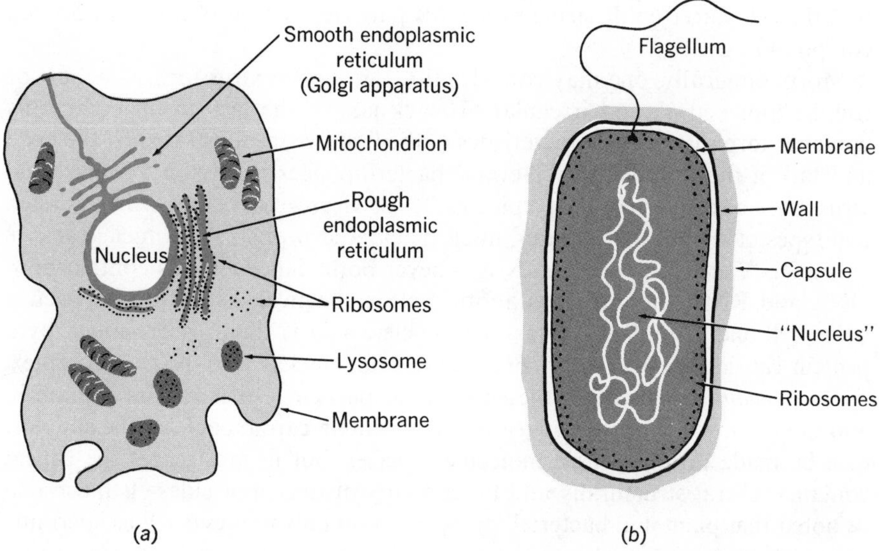

Figure 1.1 (*a*) A typical hypothetical cell of higher animals, showing the characteristic organelles. It should be noted that the membrane forming the endoplasmic reticulum and the Golgi apparatus (also referred to as smooth endoplasmic reticulum) is represented as a continuous structure that is also continuous with the plasma membrane and the nuclear membrane, implying perhaps a functional continuum as well. The ribosomes are represented both as free and as attached to the rough endoplasmic reticulum. A plant cell would be quite similar to this if one added the characteristic plastids (leucoplasts and chloroplasts) to the cytoplasmic organelles and surrounded the cell with a more rigidly defined cell wall. (*b*) The typical bacterial cell does not show any of the organelles of the higher cells. The DNA is distributed in multiple sites, the ribosomes are all free, and no membrane structure other than the cytoplasmic membrane is observed. The characteristic morphology of the bacteria (spherical or rod-shaped) is determined by the rigid cell wall. When the wall is removed, these cells all become spherical (spheroplasts), adopting the shape with the smallest surface.

It should be evident from this that no single cell can serve as a universal model for all other cells. The illustrations in Figure 1.1 are thus only meant to convey a general picture of cell structure as a basis for the subsequent discussion.

1.4 VIRUSES

The viruses are forms of molecular organization that are nonliving in that they can replicate only in a living host cell. If we refer to our definition of the minimum makeup of a living cell, we find that the viruses lack the synthetic apparatus required to produce chemical compounds; in their simplest form, viruses can be considered as informational molecules surrounded by a protective protein coat and perhaps also by a membrane. In order to replicate, a virus must thus invade a living cell (generally, each virus is very specific for its host) and substitute its information content for that of the host cell in such a manner that the cell starts synthesizing new virus particles instead of normal host-cell compounds.

Morphologically, one may consider three groups of virus forms—helical or rodlike, polyhedral, and irregular. The viruses of the last group commonly contain morphological characteristics of the first two, as illustrated by the head and tail of the tadpolelike "T-even" bacteriophages in Figure 1.2 (bacterial viruses are also referred to as "phages"). All viruses are composed of at least two types of chemical structures, nucleic acid and protein. The nucleic acid is either DNA or RNA; apparently it is never both. Since all cells contain both DNA and RNA, this provides another clear chemical distinction between a living cell and a nonliving virus. The nucleic acid is tightly surrounded by a protein capsid (coat), and for some viruses this nucleic acid–protein complex, the *nucleocapsid*, constitutes the entire virus particle. Other viruses contain in addition a loose membrane envelope. The protein capsid could in the simplest case be made up of a single molecular species, but in most cases the capsid contains several structurally and functionally distinct molecules. It must also be noted that plant and bacterial viruses contain polyamines (such as spermine or spermidine) associated with the nucleic acid as counterions to the negatively charged phosphate diester groups of the nucleic acid (see Table 5.1). Similar compounds have not yet been found in animal viruses. So far, all known plant viruses are RNA viruses, but there seems to be no other correlation between morphology and chemistry on the one hand and biological action and specificity on the other.

Some models of viruses are given in Figure 1.2. Table 1.2 lists the chemical composition and size of some viruses and demonstrates the complexity and variety among the members of this class of parasites that exists just below the level of the true biosphere.

When a virus infects a host cell, the nucleic acid is the infective principle (see Section 6.2)—it is the informational molecule, according to the model discussed before. The simplest parasite, according to this concept, should

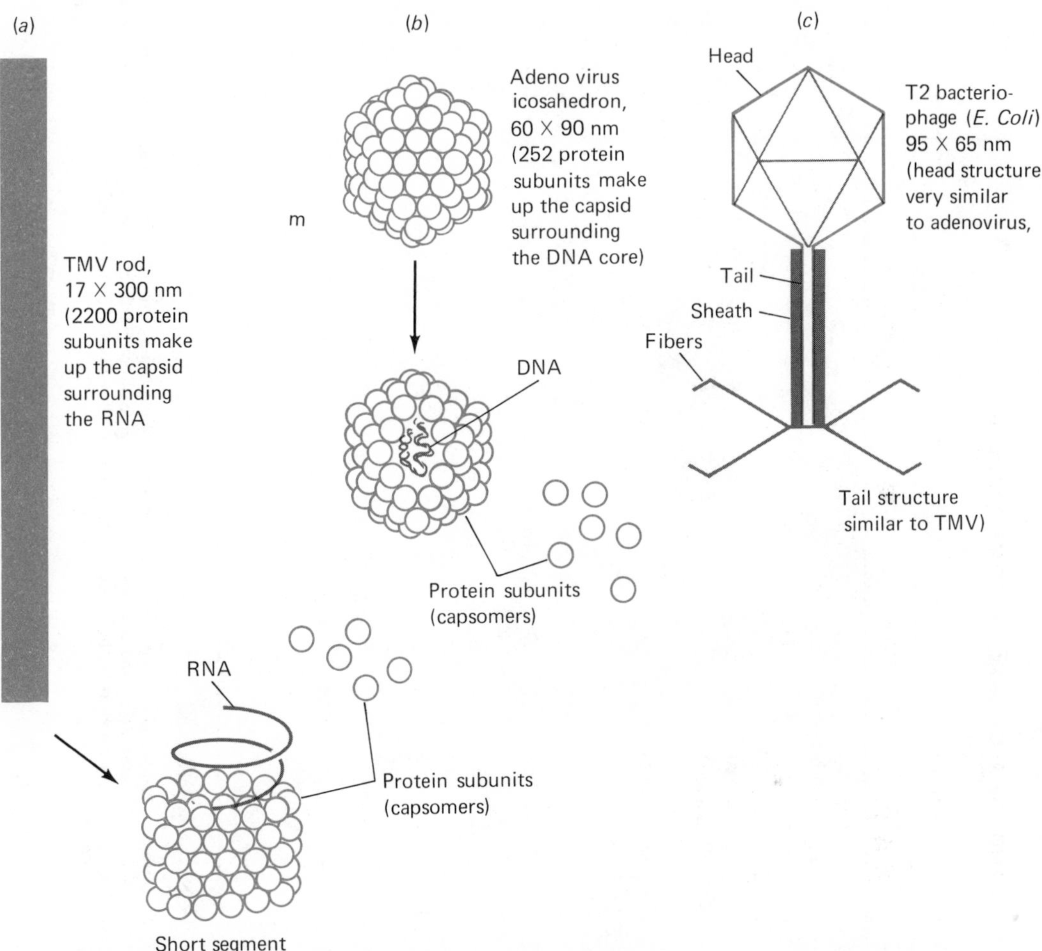

Figure 1.2 The three basic morphological forms of viruses. (*a*) Rod-shaped as illustrated by Tobacco Mosaic Virus, (*b*) Polyhedral shape as illustrated by Adeno virus, and (*c*) irregular shape as illustrated by T_2-bacteriophage. The fundamental structure of all viruses is simply nucleic acid surrounded by protein. The intact infectious virus particle is also referred to as a *virion*, the protein coat is called the *capsid*, and the protein subunits *capsomers*.

therefore be a naked nucleic acid molecule containing only the blueprints for its own replication. However, such a molecule would be extremely vulnerable to enzymatic and chemical attack and perhaps could not even enter the host cell, and no such naturally occurring, single, infective molecules are known. Some of the simple viruses, such as the satellite tobacco-necrosis virus, which contains a single RNA molecule with a capsid made up of a single molecular species of protein, come very close to this picture of the simplest parasite. Apparently the existence of viruses was predicted more than 2,000 yr ago by a Greek philosopher who, while watching a scratching dog, reasoned that in any just and orderly

TABLE 1.2 A SURVEY OF THE SIZE, SHAPE, AND COMPOSITION OF SOME VIRUSES

Virus	*Host*	*Size in 1000 Å*	*Mass (Daltons)*	*Shape*[a]	*Nucleic Acid Components*						*Others %*
					DNA[b]	%	*MW*	*RNA*[b]	%	*MW*	
Vaccinia	Animal	2.1 × 2.6	4×10^9	A, with membrane	D	7	1.6×10^8				Lipid, 5%
Herpes virus	Animal	1.8	10^9	A, with membrane	D		6.8×10^7				
Poliovirus	Animal	0.28	6×10^6	A, naked				S	30	2×10^6	
Reovirus	Animal	0.75	—	A, naked		—		D		10^7	
Arbovirus	Animal	0.45	—	?, with membrane				S			Lipids, 50%
Coliphage λ	*E. coli*	0.05	5×10^7		D	50	3×10^7				
Coliphage f_2	*E. coli*	0.2	4×10^6	A, naked				S		10^6	Glycosylated RNA
Coliphage *T-even*	*E. coli*	0.65 × 0.95		C, naked	D	60	1.6×10^8				Glycosylated DNA
Coliphage ϕx-174	*E. coli*	0.22	6×10^6	A, naked	S	26	1.7×10^7				
Tobacco mosaic	Plants	3 × 0.175	4×10^7	B, naked				S	5	22×10^6	
Turnip Yellow mosaic	Plants	0.26	5.6×10^6	A, naked				S	35	2×10^6	
Tobacco necrosis satellite	Plants	0.21	2×10^6	A, naked				S	20	0.4×10^6	

[a] Shape: A, polyhedron; B, rodlike (helical); C, irregular; Naked, no membrane.

[b] Both RNA and DNA occur as single stranded (S) or doubled stranded (D) in viruses. The selection of illustrations for this table is quite arbitrary and made for the purpose of giving at least one example of each possible structure, size, and composition. For the sake of comparison it should be recalled here that bacterial cells overlap with the larger viruses in size, most of them having diameters in the range 1–100 × 10^3 Å. A red blood cell has a diameter of 75 × 10^3 Å, and an individual hemoglobin moecule the dimensions 30 × 150 Å.

biological system, if a dog has fleas, the fleas should have fleas of their own, and those fleas should in turn have their fleas, and so on. It would appear that a nucleoprotein particle such as the satellite tobacco-necrosis virus (Table 1.2) is the ultimate "flea" that nature has produced.

It follows from the above consideration that, if the virus nucleic acid contains the information required for the synthesis of new virus particles, then the more complex the virus particle, the more information and the more nucleic acid is needed. The data in Table 1.2 in general confirm this prediction, which should have general validity and apply to cells as well.

1.5 THE PROBLEM OF STRUCTURE AND FUNCTION

The above brief review has presented some of the structures and molecules a biochemist would encounter in *all* cells, as well as some found in specialized cells and viruses. Let us now try to formulate the questions a biochemist wants to answer regarding these structures and molecules. In the simplest form, there are four basic questions: (1) What are they (the structures and molecules)? (2) What do they do? (3) How do they do it? and (4) How are they made in the cell?

These four questions, as they apply to various levels of naturally occurring compounds and structures in fact define quite precisely several broad areas of natural science: If they refer to intact cells and organisms, they cover all of classical biology; if they refer to individual molecules, they cover all of classical biochemistry; and if they refer chemically to all structures from individual molecules through molecular complexes and organelles to the cell, then they cover the area of molecular biology. The distinction between these areas, historically significant as it may be, is rapidly becoming quite meaningless. The questions are the same, the objects of study (as defined above in this chapter) are the same, and traditionally only the approach to the search for the answers has really differed. Differences in approach still exist but, rather than varying according to discipline, they follow the lines of individual interests, which range from the exploratory, descriptive work of first formulating a problem to a precise physical and chemical statement of the specific question and determination of the quantitative answer. Each approach is important and each has its own excitement and individual appeal. The impressive advances of biochemical knowledge in our time must reflect the beneficial interaction of many disciplines in solving the same problems by different but complementary approaches, each modifying its own specialized language to allow meaningful interdisciplinary communication in terms of the common goal.

Let us briefly elaborate on each of the four questions stated above. The first question represents the oldest, most prolific, and perhaps still most active area of biochemical research: the qualitative cataloging and the precise quantitative structural analysis of molecules. This area is pure chemistry with only the restriction that the molecules under examination be natural products of living cells. Some of the most exciting and ingenious research efforts of this century have been directed toward the painstaking development of precise methods for isolation, analysis, structure determination, and synthesis, allowing an increasingly refined description of relatively simple molecules and promising the future elucidation of complex ones. This area, the physical and chemical basis for structural analysis, as well as our present knowledge of structure based on these studies, is treated by Barker and Van Holde in this series.

The second and third questions are the topics of this volume, but the treatment is restricted to consideration of the macromolecules and macromolecular structures we can associate with *all living cells*: proteins, nucleic acids, and membranes.

Polysaccharides and lipids could (and perhaps should) also be considered here as individual classes of ubiquitous molecules with well-defined structural and functional properties. The only reason for discrimination is that pure polysaccharides and lipids appear to have only a very passive biological function. In pure form they are synthesized and deposited as structural material (such as cell walls in plants) or as energy reserves (such as stored glycogen and fat in animals). As this volume concentrates on the dynamic molecules, the carbohydrates and lipids will only be discussed where they appear as components of molecular complexes (such as membranes and glycoproteins) with a dynamic role in the living process.

The fourth question covers the whole area of nutrition, metabolism, and metabolic regulation. This includes the definition of the nutrients or the starting materials available in a cellular environment, the nature, thermodynamics, and kinetics of the chemical interconversions by which molecules are synthesized and degraded, and the molecular mechanisms by which these interconversions can be initiated and terminated to ensure a well-regulated and efficient synthetic apparatus. This area is treated by Larner in this series.

The topic of the present book, then, is the correlation of structure and biological function of the universal dynamic macromolecules: proteins, nucleic acids, and membranes. For each group of structures the complete problem has three facets: (1) the definition of structure as discussed in the other volumes; (2) the definition of molecular function and consideration of how functional properties (what the molecule does) can be measured; and (3) the synthesis, from these two sets of data, of molecular models that correlate and explain the biological role(s) in terms of the complete structure. In terms of models, the ideal answer to any one problem is a model that correlates perfectly all the available information and allows experimentally verifiable predictions to be made concerning the role of the system under study. The first step toward constructing such models must be a careful consideration of the levels of organization at which the data are collected and for which the models are valid. For the purpose of discussing macromolecular structure and function we will consider three levels of organization at which relevant information can be obtained: the in vivo state and, both in vitro, the native and the denatured states. The following descriptions of these states differ from other more precise—and restrictive—definitions, but it is intended that the looser operational definitions used here will serve to emphasize the care needed in interpreting and fitting experimental data into meaningful models.

The *in vivo state* represents the status of any molecule or structure in the intact, normal cell. The ultimate goal of our structure-function analyses must be to describe the total functional role and the complete structure of all cellular components in this state. Although we stress "normal cell" in this definition, there must also exist abnormal in vivo states. Pathological and artificially perturbed living systems are examples of such abnormal states, and these have been very useful for in vivo studies. The in vivo state is not readily accessible to biochemical studies of individual molecules. Structural studies in this state are generally concerned with gross morphological description (based on microscopy in conjunction with specific staining); information on molecular function can usually be obtained only by indirect methods. Because of the difficulties encountered in making detailed analyses of cellular components in intact cells, researchers are forced to isolate the different systems and study their structure and function in vitro.

The *native state* refers to isolated in vitro molecules and structures showing biological activity compatible with known in vivo properties. The native state should be defined in terms of a specific set of reference conditions of pH, ionic strength, and temperature chosen so that the original biological activity is retained and can be measured. These *native conditions* should be as similar as possible to those existing in the in vivo state in the parent cell. Virtually all the available information on macromolecules has been obtained from this state. No native state can be defined for some systems, however, as isolation leads directly to denaturation. Such systems can only be studied in vivo or in the denatured state.

The *denatured state* refers to isolated in vitro molecules and structures that have been deliberately perturbed from the native reference state. The perturbation must lead to an observable loss of or change in any of the native characteristics (biological function, physical properties, or chemical properties). The process responsible for the change from native to denatured forms may be reversible or irreversible; in the former case the perturbation changes are nearly always noncovalent, whereas in the latter they can involve covalent and/or noncovalent changes. A large amount of data on macromolecules has been collected from the denatured state (by X-ray crystallography, chemical modification for example) and from reversible transitions between the native and the denatured states.

These definitions differ from traditional usage in that here several points are explicitly understood: (1) The in vivo and the native states are different; (2) the native state can be defined for any biologically active molecule in terms of both structure and activity (presence of biological activity is required for the definition of a native state); and therefore (3) the denatured state represents any change in either native structure or native activity and thus does not require that activity is lost. (Some chemical modifications of enzymes, for example, will give increased biological activity after an irreversible covalent structural change. According to our definition, such a change is a denaturation because the native structure has been deliberately and specifically altered; however the activity of the denatured enzyme has increased.) This whole concept of three levels of organization requires a precise definition of the reference conditions for the native state; unfortunately this has not been common practice in the past, and standard test conditions for different in vitro macromolecules vary greatly, showing little relationship to what might be a reasonable approximation of in vivo conditions. [To give just one illustrative example, enzymes are generally assayed at room temperature (20 to 25°C) whether they were isolated from a mammal with an in vivo temperature of 38°C or from a fish living at 10°C.]

We are not proposing that some committee should sit down and define the "universal native reference condition" for everyone to use, but it is important to realize that caution must be exercised in interpreting in vitro experiments in terms of in vivo events. It is also often difficult to directly compare data from different laboratories, and indeed, in some instances, to actually establish the precise experimental conditions under which the data were obtained.

We have until now discussed structure and function as simple and well-defined concepts. In fact each has a very precise meaning, but when looking at a biological macromolecule, it becomes necessary to specify with what degree of precision they can actually be described. Let us first consider structure.

The isolation, criteria of purity, and structural analysis of individual macromolecules are discussed by Barker; they will not be considered here except for a statement of three important points of general validity. The first point is that all direct structural analyses must be carried out with homogeneous compounds and thus are restricted to in vitro conditions; the second (really a projection) is that once a pure compound has been obtained, it is in principle always possible to determine its covalent structure. The third and most important point is that once one has determined the covalent structure of a pure biological macromolecule in vitro, he is still far away from a proper structural description of that molecule in vivo. Most of the giant biological molecules probably do not exist free and unassociated in the cell; they certainly are not the linear polymers suggested by the covalent-structure analysis, but are twisted, bent, and folded into different three-dimensional structures (*conformations*).* Since any one of these conformations is different from the others only by its unique atomic arrangement, which is stabilized by weak, noncovalent forces, each conformation becomes an extremely elusive object for precise structural analysis.

Perhaps the most useful approach to constructing a general model for macromolecules and complex structures in dilute aqueous solution is to consider each a *dynamic mixture of rapidly interconverting structures.* The equilibrium

* "Covalent-structure analysis" means the assay of the covalent bonds that define amino acid sequence and disulfide bond location in proteins, the complete nucleotide sequence in nucleic acids, and so forth. The term "conformation" refers to the different possible structures (different arrangements of atoms) in molecules with identical covalent structure. Thus going from one conformation to another should only involve rotation of single chemical bonds, and no change in covalent structure. For a more detailed description of these terms, see Barker.

between the conformational isomers at any time is determined by all the experimental conditions and is extremely sensitive to any changes in these conditions—so sensitive, in fact, that just the performance of a measurement may be sufficient to shift the equilibrium. Based on this model, any physical, chemical, or biochemical measurement cannot in the strictest sense be interpreted in terms of a single three-dimensional structure, but rather in terms of an average of the contributing effects of all the conformations in an equilibrium mixture. One hopes, nevertheless, that these average values will represent a sufficiently restricted part of the spectrum of contributing structures to allow a meaningful interpretation of the information.

If we next look at the definition of functional properties, we should consider a similar set of precautions. In many instances the activity or function may be difficult to assess in specific and quantitative terms; however, in all instances one looks for a biological-activity assay by which the functional properties of the macromolecules can be defined quantitatively in terms of *experimental parameters.* The experimental functional parameters simply represent the numerical values obtained in measurements of biological activity. They are the "handles" by which a molecule can be pulled out for study. Any property of the in vivo system can be used as a handle. If a cell has been found to carry on a given reaction, one designs an in vitro test to determine the cellular catalyst (enzyme) for that reaction; if a cell has been found to bind a certain compound, one designs an in vitro assay that will measure the binding, to be used in the search for the molecule(s) responsible for this binding process. The assay system is in all cases designed to give quantitative results, and the values thus obtained are our experimental parameters.

An important point to stress here is that when a given compound is purified, it is generally isolated, not as a chemical entity according to known chemical properties, but as an *activity* defined by the chosen assay method. In specifying a single functional criterion as the only guideline for the isolation, one immediately imposes an arbitrary restriction on the total functional definition of a molecule. It could well have a large number of functional properties that could remain undetected because they are not specifically being sought. The particular functional criterion chosen could be a very insignificant one compared to a molecule's real total functional role in vivo. Once selected, however, a particular activity (or function) is the property that determines both the isolation procedure and the further functional characterization in vitro. The definition of the functional parameter also determines the complexity of the structural unit to be studied. It has already been pointed out that for some systems no native state exists. In other words it may not be possible to separate and study a given function in the absence of the cell. In this case the cell itself is the simplest structural unit that possesses the particular functional property under consideration.

With all these precautions and difficulties, is it then possible to obtain any relevant information about structural and functional relationships in macromolecules? Obviously it is. First, the above considerations and models are presented in the most general terms to cover a very broad spectrum of complexity. Limiting a study to the simplest systems—those of fairly rigid structural features, high purity, and well-defined activity—makes the difficulties listed above rather insignificant. Recent advances in methodology have greatly facilitated this direct approach and furthermore indirect, and yet very meaningful, approaches are also possible. Thus a compound of virtually unknown properties can be subjected to a series of physical, chemical and biochemical tests that each result in a characteristic number or characteristic parameters. This set of numbers, obtained under a standard set of reference conditions, is taken to define the "native structure and function." (The numbers may merely be indicative of molecular size and shape; on the other hand, spectral properties will indicate chromophore location and interaction, chemical reactivity will indicate interaction of reactive groups with each other and with the environment, the biological assay will quantitate the functional capabilities and so forth. (For a more detailed description of these properties see Barker and Van Holde and Chapters 2, 5 and 9 in this volume.) When one specifically changes a single variable in the total system and reexamines the compound and functions, a new set of numbers is obtained that can often be interpreted with amazing precision. In this way *changes in function* can be correlated with well-described *changes in structure*; thus a comprehensive model of the total structure-function relationship can be built step by step. Several examples of such an approach will be given throughout this book (for example, see Chapter 3).

Finally, let us return to the question of the relevance of the in vitro data to the understanding of in vivo processes in the living cell. First of all even if no direct relevance can be established, the correlation of structure and function of any biologically active molecule is of fundamental importance and is an exceedingly interesting and challenging problem in its own right. However, the extrapolation from in vitro data to in vivo interpretation may not be as dangerous as one might fear; perhaps even more important, the ingeneous methods and techniques developed for the study of macromolecules in vitro are finding more and more application in vivo, promising very exciting developments in the future.

Several years ago someone illustrated the biochemical approach to biological problems with the following analogy. Suppose we encounter a car for the first time and know absolutely nothing about this contraption except that we have observed it moving. To learn something about what makes it move, our first approach is to grind it up into small pieces and analyze the pieces. As a result of these efforts we learn that a car is made of metal, rubber, plastic, gasoline, oil, and water, which contributes little toward an explanation of what makes it move. This rather uncharitable picture is probably not a bad analogy for a certain stage of biochemical research; clearly, with *no* previous knowledge of the car, this initial approach with all its lack of finesse would even perhaps be a logical one. Our next step—perhaps the stage at which we are presently operating—which is more refined but still fraught with hazards, is to use somewhat gentler methods for fragmenting the car. As a result we isolate functionally intact pieces such as wheels, gears, carburetor, battery, and generator and we still have difficulties in explaining what makes the car move. The stage is set, however. We realize that there are many specialized auxiliary functions associated with the movement of the car; now we can formulate the questions more precisely and be more direct in our search for the answers.

REFERENCES

General (multivolume) reference sources:

Brachet, J., and A. E. Mirsky (Eds.), *The Cell*, Academic Press, New York, 1959.

Gunsalus, I. C., and R. Y. Stanier (Eds.), *The Bacteria*, Academic Press, New York, 1960.

Review of Virus Structure:

Green, M., "Chemistry of the DNA Viruses;" Schaffer, F. L., and C. E. Schwerdt, "Chemistry of the RNA Viruses," in Horsfall, F. and I. Tamm (Eds.), *Viral and Rickettsial Infections of Man*, Lippincott, Philadelphia, 1965.

Electron micrographs:

Fawcett, D. W., *An Atlas of Fine Structure, The Cell, the Organelles and Inclusions*, W. B. Saunders Co., Philadelphia, 1966.

Other general sources:

Allen, J. M. (Ed.), *Molecular Organization and Biological Function*, Harper and Row, New York, 1967.

Brachet, J., "The Living Cell," *Scientific American*, **205**, September 3, 1961. (Offprint by W. H. Freeman and Co., San Francisco). This reference is also part of a collection of 24 papers from *Scientific American* on the cell, its organelles, energetics and synthesis, division and differentiation, and special activities: *The Living Cell, Readings from Scientific American*, with Introductions by D. Kennedy, W. H. Freeman and Co., San Francisco, 1965.

Pfeiffer, J., *The Cell*, Time, Inc., New York, 1964.

TWO PROTEINS: SYSTEMS AND METHODS

Chapters 2, 3, and 4 of this book are concerned with the first of the three types of ubiquitous biological molecules, the *operation molecules* or the proteins. This chapter introduces different types of proteins and surveys the handles by which their structure and function can be studied. It is proposed that the fundamental property of all *biologically active* proteins is their ability to recognize and bind other molecules, and the methods by which the specific binding can be measured are discussed.

As was seen in Chapter 1, proteins are found either free in the cytoplasm or in association with other molecules in the different cellular organelles. Any classification of proteins is arbitrary and trivial. The traditional classification of proteins was essentially based on solubility, ionic properties, and function (see Barker in this series). For the purpose of our present discussion it is useful to classify and consider the proteins as members of the different universal operational units of living cells. Table 2.1 gives such a division, and with the exception of the two specialized proteins, γ-globulin and hemoglobin, which will be discussed in Chapter 4, we will here discuss only those found in all cells. Enzymes will be discussed in Chapter 3 and examples of contractile proteins, carrier proteins, and immune proteins in Chapter 4; proteins connected with nucleic acids are briefly considered in Chapters 6, 7, and 8 and membrane proteins in Chapters 9 and 10. It will be apparent that even the broad classification in Table 2.1 is equivocal.

TABLE 2.1 TYPES OF PROTEINS[a]

	Examples		*Classical Classification*
	Ubiquitous	*Specialized*	
Nonassociated[b] proteins (Chapters 2, 3, 4, 8)	Enzymes	Immune proteins Hormones[c]	Globular proteins: Albumins (soluble in water) Globulins (soluble in dilute salt solution)
Associated with proteins (Chapters 2, 3, 4, 8)	Enzymes Contractile proteins[d]		
Associated with nucleic acids (Chapters 5, 6, 7, 8)	Ribosomal proteins Regulatory proteins		Protamines Histones High contents of lys and arg (basic proteins) (Repressors, interferon)
Associated with membranes (Chapters 9, 10)	Lipoproteins Enzymes Transport proteins		Lipoproteins, glycoproteins, mucoproteins
(Associated with carbohydrates)[e]		(Some enzymes in glycogen metabolism)	
(Miscellaneous specialized proteins)		(Structural proteins) (Collagen, elastins keratins)	Insoluble, fibrous proteins

[a] This classification is based on a certain degree of knowledge of structural associations (still far from complete) and, therefore, an implied functional role. Even if this functional role is poorly defined at present, the above representations indicate where to look for functional roles.

[b] Association refers specifically to other macromolecules or macromolecular complexes.

[c] The protein hormones of higher animals are simple proteins, relatively easy to obtain in nonassociated forms. Practically nothing is known about their specific mode of action, but it seems clear that they must function in association with other macromolecules.

[d] It is assumed that all cells have some form of contractile apparatus [muscle, flagella, cilia, membrane, contractile systems (mitochondrion), etc.]. As illustrated by muscle contraction, this represents an excellent example of a function being undisputably linked with a multiprotein complex.

[e] A large number of proteins are associated with carbohydrate (glycoproteins, mucoproteins) but, in contrast to the other "association groups," by covalent linkage. The carbohydrate part of functional proteins presents an enigma in that no specific functional role can yet be assigned to it, nor is there any obvious pattern to the occurrence of glycoproteins. [Pancreatic ribonuclease of the cow (See Chapter 3) can be isolated in two forms, one free of carbohydrate and one containing a heptasaccharide covalently attached. The same enzyme from pig pancreas contains a covalent carbohydrate attachment of about 40 % by weight. All these three enzyme forms have very similar catalytic properties.] In the above classification, therefore, the carbohydrate is considered a part of the covalent protein structure, and glycoproteins can be found in any one of the above subdivisions. Noncovalent protein–carbohydrate association complexes appear to exist for the enzymes involved in the metabolism of glycogen.

These then, are the operation molecules—those molecules which through binding and catalysis constitute the working force, the regulatory systems, and the defence barrier of the living cell. In the next chapters we shall attempt to correlate the structural features of the proteins with these many and varied functional roles. First however, we must formulate the working hypothesis of both structure and functions in terms of the types of experimental models mentioned in Chapter 1.

2.1 SUMMARY OF STRUCTURE

The complete description of the three-dimensional structure of proteins has become possible in recent years through the very exciting advances in X-ray crystallography. Such analysis is complicated, however, and has been limited to only a few proteins; hence we are still forced to use the indirect methods for most structural definitions. The molecular weight and the chemical composition of any pure protein can be readily determined, thus providing the first level of structural information in terms of size, amino acid composition, and presence of other covalent chemical components (such as carbohydrate components in the glycoproteins). The next level, establishing the number of polypeptide chains per mole (by quantitative end-group analysis) and the number of subunits* (by dissociation of the molecule into its smallest covalent units), can also be accomplished quite readily. The third level, the description of the complete covalent structure, is more complex; however, with all the refined methods available for amino acid–sequence analysis, it should only be a question of time and effort to describe any pure protein in terms of the amino acid sequence and the disulfide bond location of all the subunits and chains that make up the protein. From this point to the desired goal of describing the folded three-dimensional structure is an extremely complex process for the reasons outlined in the previous chapter, based on the flexibility of proteins in solution. The best structural model for most proteins in solution must therefore allow for a dynamic equilibrium between different conformations as indicated in Figure 2.1. With this model in mind we must resort to the indirect methods and describe the "average structure" of the protein polymer by a number of experimental parameters. The types of experimental characteristics that can be used have been discussed by Barker and Van Holde in this series and are summarized in Table 2.2.

* Notice that "polypeptide chains" and "subunit" are not synonymous. The subunit is defined as the smallest *covalent* unit; a subunit can thus contain several polypeptide chains covalently linked together (for example, disulfide bonds). The combining of subunits to give the stable complex that is isolated and characterized as the protein molecule is accomplished by noncovalent interactions; such interactions can be disrupted by reagents that do not cause covalent-bond cleavage (Barker and Van Holde in this series).

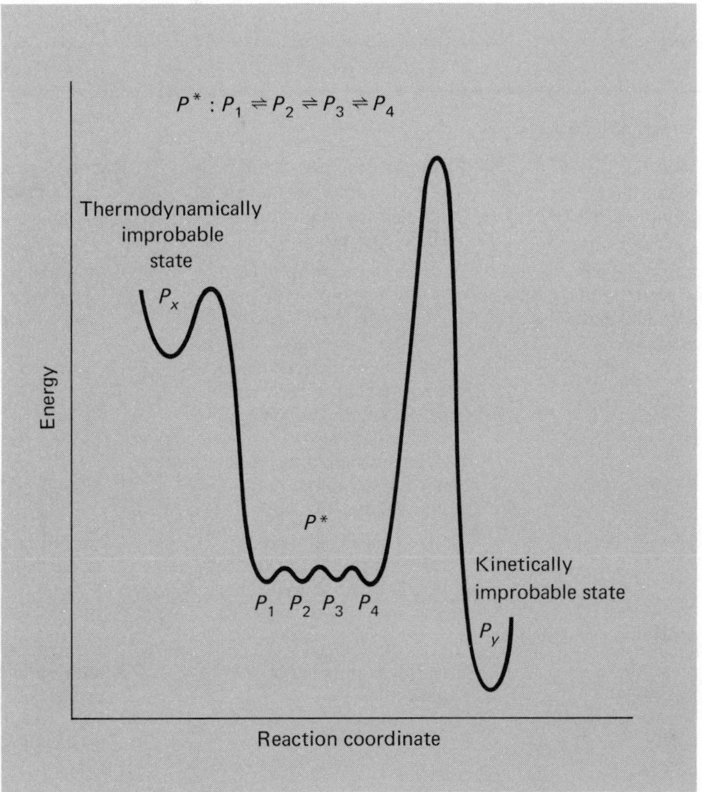

Figure 2.1 A general model of a protein as a dynamic, flexible structure. The average structure under standard conditions, P^*, is visualized as representing several distinct structures (conformers) P_1, P_2, P_3, etc. Other possible but improbable conformations are also indicated, the thermodynamically improbable state P_x and the kinetically improbable state P_y. The latter one is thermodynamically the most favorable one, but the molecules are "trapped" at the P^* levels by the high energy barrier. (Heat denaturation could in this system simply be the activation and transformation from P^* to P_y, and could well appear to be irreversible.)

2.2 GENERAL FUNCTIONAL DEFINITION

Some of the self-imposed restrictions on the discussion of function have been mentioned in connection with Table 2.1. The *passive proteins*, such as the structural proteins collagen and keratin, will not be discussed here. They are quite rigid fibrous molecules containing a high degree of natural inter- and intrapolypeptide-chain cross-links (both disulfide bonds and other bonds) and are well suited for their function of providing structural stability to connective tissue (collagen) and to hair, nails, and hoofs (keratins). Collagen and

TABLE 2.2 EXPERIMENTAL APPROACHES TO THE DEFINITION OF PROTEIN STRUCTURE

Experimental Procedure	Information Obtained	
	In the Native State	*In the Denatured State*
I. Chemical: (see Barker in this series)		
(a) Amino acid sequence; location of disulfide bonds, quantitative carboxyl and amino terminal analysis	Covalent structure. (Number of end-groups per mole gives the number of polypeptide chains per mole)	(Increase in end-groups: covalent bonds have been broken)
(b) Susceptibility to chemical reagents—chemical modification with monofunctional and polyfunctional reagents	Preferential reaction of certain residues in a protein distinguishes between "reactive" and "masked" residues. This is a function of the specific microenvironment of each residue, or the specific polymer conformation Reagents which form cross-links allow determination of interresidue distances	Changes in reaction pattern: The microenvironment (protein conformation) has changed
(c) Susceptibility to proteolytic enzyme	Native proteins are either resistant to or attacked very slowly by proteolytic enzymes	Increased rate of proteolysis: conformational change
II. Physical: (see Van Holde in this series)		
(a) Hydrodynamic properties: sedimentation, diffusion, visocity, and electrophoresis.	Size, shape and charge of the molecule	Alteration in hydrodynamic properties: change in molecular weight (association or dissociation) or in molecular shape
(b) Optical properties:		
1. Light scattering, (Raighley)	Size and shape of the molecule	Same as above
2. Absorption and fluorescence	Microenvironment of chromophores (tyr, try, phe)	Changes in absorption, fluorescence: conformational changes at or around the chromophores
3. Optical rotatory dispersion and circular dicroism	Ordered (secondary) structure (α-helix, β-structure)	Same as above (chromophores include the peptide bond)
4. X-ray diffraction		Complete three-dimensional structure of crystalline forms of the proteins

keratins have been studied extensively and represent well-documented and unique chapters in protein chemistry.* Other structural proteins will be briefly considered as membrane components in Chapters 9 and 10.

Concentrating on *active proteins*, those directly involved in the dynamics of living cells, we shall postulate one universal functional property common to all: the ability to *specifically recognize and bind other molecules* (ligands or

* For more on this matter, see W. F. Harrington and P. H. von Hippel, *Adv. Prot. Chem.*, **16**, 1 (1961); J. J. Harding, *Ibid.*, **20**, 109 (1965); W. G. Crewther *et al.*, *Ibid.*, **20**, 191 (1965).

substrates). This property is outlined for different types of active proteins in Table 2.3, which will serve also as an outline for the discussions in Chapters 3 and 4. For all of these types of active proteins we can experimentally define as the *binding site*, a minimum structural part of the total functional molecule that is directly involved in the highly specific and selective combination with the ligand. Generally the binding process can be studied directly through experimentation. After the protein-ligand complex has formed, further processes come into play: the chemical changes in the ligand that are catalyzed in enzyme-substrate complexes; the formation of a high-molecular-weight network in immunochemical precipitin reactions; and the contraction or relaxation of contractile systems.

TABLE 2.3 THE 5 MAJOR GROUPS OF ACTIVE PROTEINS, EMPHASIZING THEIR COMMON FUNCTIONAL CHARACTERISTIC: SPECIFIC LIGAND BINDING[a]

1. Catalytic proteins (enzymes): binding and chemical action on ligands (substrate)
2. Immune proteins: irreversible binding and immobilization of ligands
3. Carrier (or transport) proteins: reversible binding and transport of ligands
4. Regulatory proteins: reversible binding of ligand leading to activity changes
5. Contractile proteins: binding of ligand, leading to mechanical work

[a] It has recently been proposed to give the name *Emphore* ("that which carries in") to the groups of active proteins which are not enzymes, but which specifically bind ligands and thereby "carry the ligand into biological activity." (A. B. Pardee, in *Structural Chemistry and Molecular Biology*, A. Rich, and N. Davidson, (Eds.), W. H. Freeman and Co., San Francisco and London, 1968).

The highly specific binding site and the binding process thus provide the common functional handle in the study of all of the active proteins and the experimental approach to their functional definition must therefore be based on the quantitative description of the binding process in each individual case. The remainder of this chapter will be devoted to that end.

2.3 BINDING

The subsequent brief consideration of binding types is somewhat unusual. It is used simply to emphasize as much as possible the analogy to enzyme kinetics in the next section. For a more general treatment of binding, see Van Holde in this series.

The simple reaction, in which one molecule of protein, P, binds one molecule of ligand, A, can be written

$$P + A \rightleftharpoons PA$$

with equilibrium constant $\mathbf{K}$,

$$\mathbf{K} = \frac{[PA]}{[P][A]}\text{*}$$

It is important to note that according to conventions all equilibrium constants for ligand binding are expressed as binding constants or association (formation) constants, whereas for acid-base titration (proton binding) and for enzyme-substrate binding (see sec. 2.4), they are expressed as dissociation constants. Since $K_{\text{association}} = 1/K_{\text{dissociation}}$ (or $\mathbf{K} = 1/K$), this causes no difficulties, but the distinction must be kept in mind. Anticipating the experimental methods for evaluating $\mathbf{K}$, we next define a parameter, v,

$$v = \frac{\text{total number of moles of } A \text{ bound}}{\text{total } P} = \frac{[PA]}{[P] + [PA]}$$

which can be expressed in either of the following forms:

$$v = \frac{\mathbf{K}[A]}{1 + \mathbf{K}[A]} \qquad [2.1(a)]$$

or

$$v = \frac{1}{1 + 1/\mathbf{K}[A]} \qquad [2.1(b)]$$

This is the simplest binding equation, analogous to the Langmuir adsorption isotherm, and also to the Michaelis–Menten equation, which will be developed below for enzyme kinetics.

Let us next examine the case when the protein has two equivalent binding sites, which do not interact but are completely independent:

$$\begin{array}{ccccc} & \mathbf{K}'_1 & -PA & \mathbf{K}'_3 & \\ -P- & \overset{\mathbf{K}_1}{\rightleftharpoons} & (PA) & \overset{\mathbf{K}_2}{\rightleftharpoons} & APA \\ & \mathbf{K}'_2 & AP- & \mathbf{K}'_4 & \end{array}$$

* Molar concentration will be used in all the treatments in this chapter. (Brackets indicate concentration.) This is a common approximation in most experimental work, valid for small molecules in dilute aqueous solution but not for protein studies in general. The binding constants have not been corrected for activity coefficients and are thus only apparent constants.

$—PA$ and $AP—$ indicate the two sites, $\mathbf{K}'_1$, $\mathbf{K}'_2$, $\mathbf{K}'_3$, and $\mathbf{K}'_4$ describe the four microscopic equilibria involved in going from P to APA (that is, PA_2), and $\mathbf{K}_1$ and $\mathbf{K}_2$ are the experimentally observable equilibrium constants. Note here that in distinguishing between the two sites ($—PA$ and $AP—$), the intermediate, (PA), becomes the sum of the two binding forms. Since we have stipulated that the two sites are equivalent and independent (noninteracting), it follows that $\mathbf{K}'_1 = \mathbf{K}'_2 = \mathbf{K}'_3 = \mathbf{K}'_4 = \mathbf{K}'$ where $\mathbf{K}'$ is the *intrinsic binding constant* for each site, the constant describing the intrinsic affinity of each site for A.

The first step in this reaction is

$$P + A \rightleftharpoons (PA)$$

$$\mathbf{K}_1 = \frac{[PA]}{[P][A]} = \frac{[—PA] + [AP—]}{[P][A]} = \frac{(\mathbf{K}'_1 + \mathbf{K}'_2)[P][A]}{[P][A]}$$

giving $\mathbf{K}'_1 = \mathbf{K}'_1 + \mathbf{K}'_2 = 2\mathbf{K}'$.

In the same manner the next step,

$$(PA) + A \rightleftharpoons PA_2$$

$$\mathbf{K}_2 = \frac{[PA_2]}{[PA][A]} = \frac{[PA_2]}{([—PA] + [AP—])[A]} = \frac{1}{1/\mathbf{K}'_3 + 1/\mathbf{K}'_4}$$

gives $\mathbf{K}_2 = \mathbf{K}'/2$.

These relationships will be used below, but it should also be pointed out that they have a general significance to two-site systems. It is clear from this that the experimentally determined binding constants ($\mathbf{K}_1$ and $\mathbf{K}_2$) for two identical and noninteracting sites are related in the following way:

$$\frac{\mathbf{K}_1}{\mathbf{K}_2} = \frac{2\mathbf{K}'}{\mathbf{K}'/2} = 4$$

where 4 is called the *statistical factor* for a two-site compound. It predicts that simply on the basis of probability, the first association constant should be 4 times as large as the second in a system of two identical, noninteracting sites.

Now, let us develop an expression for v for this system. According to the definition,

$$v = \frac{[PA] + 2[PA_2]}{[P] + [PA] + [PA_2]}$$

(Note that since the numerator is the total number of moles of A bound, PA_2, containing two moles of A, per mole of complex must be multiplied by 2.) Substituting the two experimentally observable binding constants, $\mathbf{K}_1$ and $\mathbf{K}_2$, this can be expressed as

$$v = \frac{\mathbf{K}_1[P][A] + 2\mathbf{K}_1\mathbf{K}_2[P][A]^2}{[P] + \mathbf{K}_1[P][A] + \mathbf{K}_1\mathbf{K}_2[P][A]^2} = \frac{\mathbf{K}_1[A] + 2\mathbf{K}_1\mathbf{K}_2[A]^2}{1 + \mathbf{K}_1[A] + \mathbf{K}_1\mathbf{K}_2[A]^2}$$

Since for identical and noninteracting sites $\mathbf{K}_1 = 2\mathbf{K}'$ and $\mathbf{K}_2 = \mathbf{K}'/2$ we have in terms of the intrinsic binding constant, $\mathbf{K}'$,

$$v = \frac{2\mathbf{K}'[A] + 2\mathbf{K}'^2[A]^2}{1 + 2\mathbf{K}'[A] + 2\mathbf{K}'^2[A]^2} = \frac{2\mathbf{K}'[A](1 + \mathbf{K}'[A])}{(1 + \mathbf{K}'[A])^2}$$

or

$$v = \frac{2\mathbf{K}'[A]}{1 + \mathbf{K}'[A]} \qquad [2.2(a)]$$

or

$$v = \frac{2}{1 + 1/\mathbf{K}'[A]} \qquad [2.2(b)]$$

By exactly the same manipulations it can be shown that the corresponding equations for n equivalent and independent binding sites gives

$$v = \frac{n\mathbf{K}'[A](1 + \mathbf{K}'[A])^{n-1}}{(1 + \mathbf{K}'[A])^n} \,*$$

where $\mathbf{K}'$ again is the intrinsic constant and thus

$$v = \frac{n\mathbf{K}'[A]}{1 + \mathbf{K}'[A]} \qquad [2.3(a)]$$

or

$$v = \frac{n}{1 + 1/\mathbf{K}'[A]} \qquad [2.3(b)]$$

This is the general equation for multiple binding sites with no interaction.

All the previous discussion has been limited to binding at a single site or binding at noninteracting multiple sites (since noninteracting sites are essentially equivalent to several single sites, the two cases are completely analogous), and these represent good approximations of several naturally occurring binding systems. However, the cases where the sites are interacting and binding at one site affects binding at a second site, either by increasing or decreasing its affinity for the ligand, are much more common. Let us briefly look at some simple examples of interacting sites.

* The general form of this equation for a system with n binding sites is

$$v = \frac{\sum_{i=1}^{i=n} i\mathbf{K}_i'[A]^i}{1 + \sum_{i=1}^{i=n} i\mathbf{K}_i'[A]^i}$$

It is known as the Adair equation, and describes all systems of binding involving from 1 to n binding sites without interaction ($\mathbf{K}_1' = \mathbf{K}_i' = \mathbf{K}_n' = \mathbf{K}'$) or with interaction ($\mathbf{K}_i' \neq \mathbf{K}_j'$). [Adair, G. S., *J. Biol. Chem.* **63**, 529 (1929); Weber, G. and Anderson, S. R., *Biochem.* **4**, 1942 (1965).]

Consider first the case where binding of a second ligand, B, will either activate or inhibit the binding of A.

Inhibition:

$$P + A \rightleftharpoons PA \qquad \mathbf{K}' = \frac{[PA]}{[P][A]}$$

$$P + B \rightleftharpoons PB \qquad \mathbf{L}' = \frac{[PB]}{[P][B]}$$

Activation:

$$P + B \rightleftharpoons PB \qquad \mathbf{L}' = \frac{[PB]}{[P][B]}$$

$$PB + A \rightleftharpoons PBA \qquad \mathbf{K}' = \frac{[PBA]}{[PB][A]}$$

For the binding of A we have already defined v as follows:

$$v = \frac{[PA]}{[P] + [PB] + [PA]} \quad \text{and} \quad v = \frac{[PBA]}{[P] + [PB] + [PBA]}$$

Thus

$$v = \frac{\mathbf{K}'[A]}{1 + \mathbf{K}'[A] + \mathbf{L}'[B]} \quad \text{and} \quad v = \frac{\mathbf{K}'\mathbf{L}'[A][B]}{1 + \mathbf{K}'\mathbf{L}'[A][B] + \mathbf{L}'[B]}$$

or

$$v = \frac{1}{1 + (1/\mathbf{K}'[A])(1 + \mathbf{L}'[B])} \quad \text{and} \quad v = \frac{1}{1 + (1/\mathbf{K}'[A])(1 + 1/\mathbf{L}'[B])} \qquad \text{(2.4 and 2.5)}$$

These can both be represented in the usual form,

$$v = \frac{1}{1 + 1/\mathbf{K}''[A]}$$

where $\mathbf{K}''$ is an apparent intrinsic binding constant dependent on B as determined at any one concentration of B and related to the real $\mathbf{K}'$ (the binding for the uninhibited or fully activated system) in the following way,

Inhibition:

$$\mathbf{K}'' = \frac{\mathbf{K}'}{1 + \mathbf{L}'[B]}$$

when $[B] = 0, \mathbf{K}'' = \mathbf{K}'$

and when $[B] \longrightarrow \infty, \mathbf{K}'' \longrightarrow 0$

Activation:

$$\mathbf{K}'' = \frac{\mathbf{K}'}{1 + 1/\mathbf{L}'[B]}$$

when $[B] = 0, \mathbf{K}'' = 0$

and when $[B] \longrightarrow \infty, \mathbf{K}'' \longrightarrow \mathbf{K}'$

just as the equations for the binding reaction stated.

This relationship has general application to a number of different systems. A very common one is the case where B is a proton and binding of A is thus dependent on pH. The equations for that case should actually use the acid-dissociation constant, K_a, rather than the binding constant $\mathbf{L}'$, but the equations would otherwise be identical to those above. (It may be useful for the student to work out the proper form of the equation.)

Without getting involved in details of the mechanism we can in the broadest sense define B in the cases discussed above as an *effector* of the binding of A and, since it is different from A, it is referred to as a *heterotropic effector*. It is perhaps implicit in this type of nomenclature that A can affect its own binding and thus be a *homotropic effector*; this very important case will be treated next.

Consider the simplest case of two equivalent sites for the binding of A, interacting in such a way that when A binds to one site it modifies the second site, increasing the affinity for the second molecule of A. This *cooperative binding* effect has been observed in a number of systems and is an important property of regulatory proteins (see hemoglobin in Chapter 4 and allosteric systems in Chapters 3 and 8.) For this case Equation (2.5) becomes

$$v = \frac{1}{1 + 1/\mathbf{K}'[A] + 1/\mathbf{K}'\mathbf{L}'[A]^2} \tag{2.6}$$

This equation containing a square term of $[A]$ gives a sigmoidal binding curve (a plot of v vs. $[A]$) and represents a simple case of cooperative binding (interacting binding sites). A more general expression is given below.

It is generally extremely difficult to evaluate the individual binding constants for any multisite binding process. This is practical only when binding constants are intrinsically very different for each site or where a negative interaction between identical sites causes a large difference in binding constants so that each binding step can be determined as a separate process. (The binding of protons to the triple-site PO_4^{3-} anion represents a simple case of such negative interaction with binding constants of about 10^{12}, 10^7, and 10^2.) In view of this difficulty the degree of interaction between sites is generally evaluated by an empirical method. The empirical equation

$$v = \frac{n\mathbf{K}'[A]^c}{1 + \mathbf{K}'[A]^c}$$

rewritten in the more common form as the *Hill equation*,*

$$y = \frac{v}{n} = \frac{\mathbf{K}'[A]^c}{1 + \mathbf{K}'[A]^c} \tag{2.7}$$

* A. V. Hill, *J. Physiol.* (London) **40**, 4 (1910).

Figure 2.2 (*a*) Direct titration of protein with ligand
(*b*) Determination of binding in two-compartment systems.
1. Equilibrium dialysis: Determine A_{total} (inside), A_{total} (outside), and P_{total}. $A_{free} = A_t$ (outside); $A_{bound} = A_t$ (inside) $-$ A_t (outside); and $v = A_{bound}/P_t$ at the observed concentration of A_{free}.
2. Gel filtration: The compartmentalization is achieved on a gel whose structure is such that small molecules can penetrate into the gel where larger molecules cannot enter. The excluded volume corresponds to the inside of the dialysis bag (above) and the included volume to the outside, and the

(*a*) Direct titration

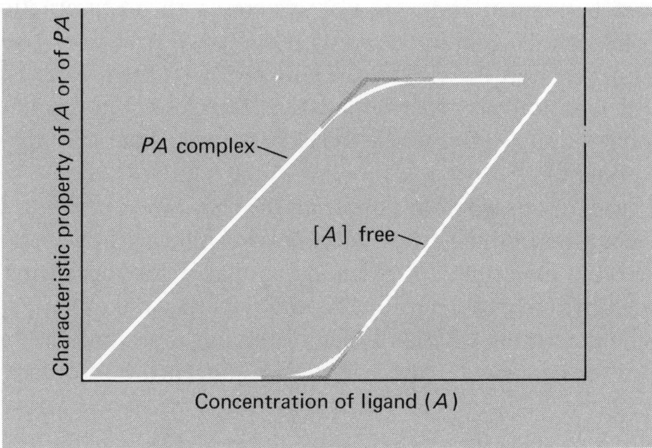

(*b*) Two-compartment systems.

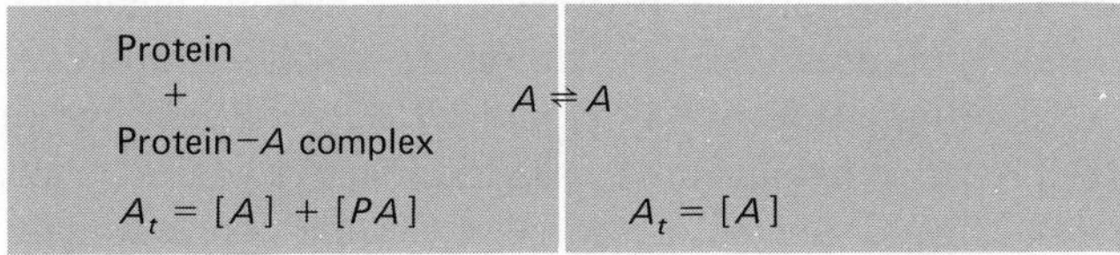

1. Equilibrium dialysis

Outside solution of A

$P + PA$

$[A]$

$[A]$

Protein inside dialysis bag

2. Gel filtration

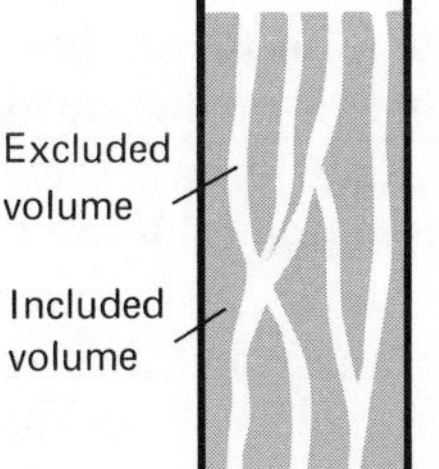

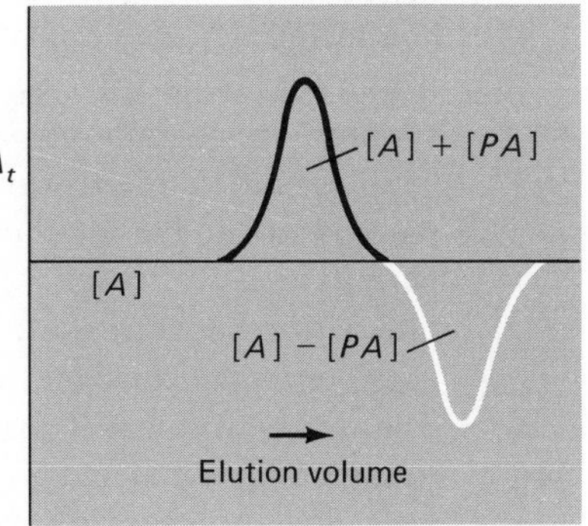

---

treatment is analogous to equilibrium dialysis. The type of data obtained is illustrated in the figure. The column is equilibrated with a solution of known concentration of A (A_{free}). A known quantity of protein is dissolved in the same solution and part of A is bound and the concentration of free A is less than the starting concentration. The protein will now pass down the column and continuously re-equilibrate with A_{free}, till it elutes. The A-deficient solution from the start of the experiment will appear as a trough in the plot.

where y is the degree of saturation, describes all binding processes. The Hill coefficient, c, is an arbitrary measure of the interaction between the sites and can have values between 0 and $+n$. For systems with no interaction, $c = 1$, and a normal hyperbolic binding curve is obtained. For cooperative binding, c is a positive number and the greater the cooperativity, the greater is the value of c (In the Hill equation n is generally used instead of c but, to avoid confusion with the number of sites (n, above), c has been used here for the empirical interaction exponent.)

The simple cases discussed above cover all the specific examples to be treated in subsequent chapters of this book. Before leaving binding, however, we should briefly consider the experimental methods available for measuring binding. The first goal is obviously to determine $\mathbf{K}'$ and n for any unknown PA pair and at the same time determine whether different binding sites interact. In order to do this we need to determine v and $[A]_{\text{free}}$, the concentration of free ligand in equilibrium with the protein-ligand complex. Since we always know the total concentrations of protein ($[P]_t$) and ligand ($[A]_t$) in the system, v and $[A]_{\text{free}}$ can be evaluated if we can determine $[A]_{\text{bound}}$, the concentration of bound ligand ($v = [A]_{\text{bound}}/[P]_t$ and $[A]_{\text{free}} = [A]_t - [A]_{\text{bound}}$). Different methods that are used for this purpose have been outlined in Figure 2.2.

Any method which distinguishes between free and bound ligand in a mixture (spectrophotometric (fluorescence), potentiometric, polarographic, etc.) can be used for direct determination of $[A]_{\text{free}}$ and $[A]_{\text{bound}}$. Where free and bound ligand cannot be readily distinguished (using radioactive labeled A for example), so that only *total* ligand can be measured, a double compartment system is used such that the protein is present in only one compartment, and ligand can pass freely between the compartments. At equilibrium the experimentally determined total ligand in the protein compartment is the sum of free ligand and bound ligand; in the protein free compartment, total ligand is equal to free ligand (the same value as free ligand in the protein compartment). Two such systems are commonly used and are illustrated in Figure 2.2(*b*). When $[A]_{\text{bound}}$, and therefore v and $[A]_{\text{free}}$, are known, there are a number of graphical methods available for the determination of $\mathbf{K}'$ and n; the Hill equation then permits the evaluation of c, the empirical quantitative measurement of the interaction between sites. Some of these methods are illustrated in Figure 2.3.

Let us finally consider the question of what our knowledge of $\mathbf{K}'$ and n tells us. The answer, which will be restated in connection with the kinetic constants in the next section, is that it tells us very little. These constants simply represent numerical values for what we have defined as the functional property of the protein and do not give any more information than a melting point does. They do give some clue to biological function but, for our purpose, that of correlating the structure and function of proteins, the *changes* in the two characteristic constants $\mathbf{K}'$ and n are probably more important. By correlating changes in $\mathbf{K}'$ and n with well-defined changes in structure, we can start building up an inventory of the components of the polymer that are involved in the binding process; it is primarily for this purpose we are interested in $\mathbf{K}'$ and n.

1. $\nu = \dfrac{n}{1 + \dfrac{1}{\mathbf{K}[A]}}$

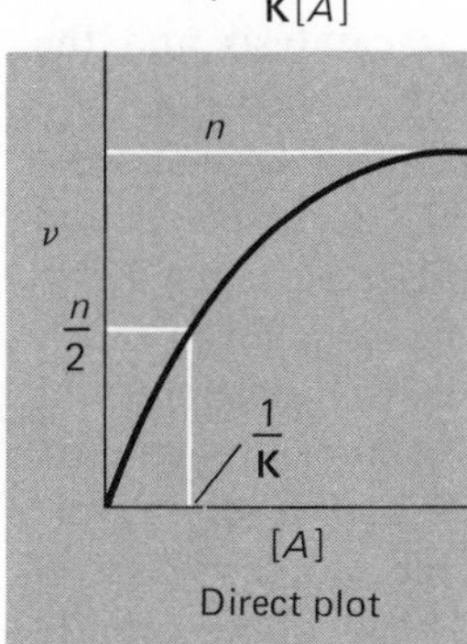

2. $\dfrac{1}{\nu} = \dfrac{1}{n} + \dfrac{1}{n\mathbf{K}}\dfrac{1}{[A]}$

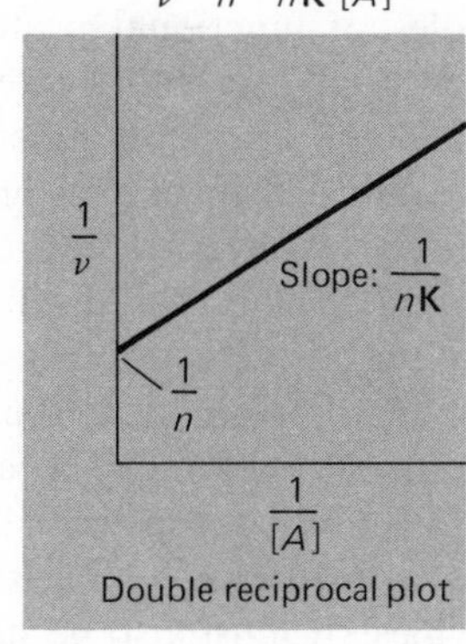

3. $\dfrac{\nu}{[A]} = n\mathbf{K} - \nu\mathbf{K}$

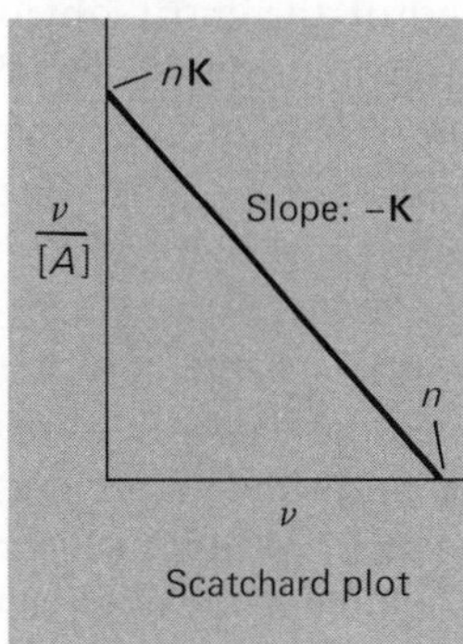

(a)

1. $\dfrac{\nu}{n} = \dfrac{\mathbf{K}[A]^c}{1 + \mathbf{K}[A]^c}$

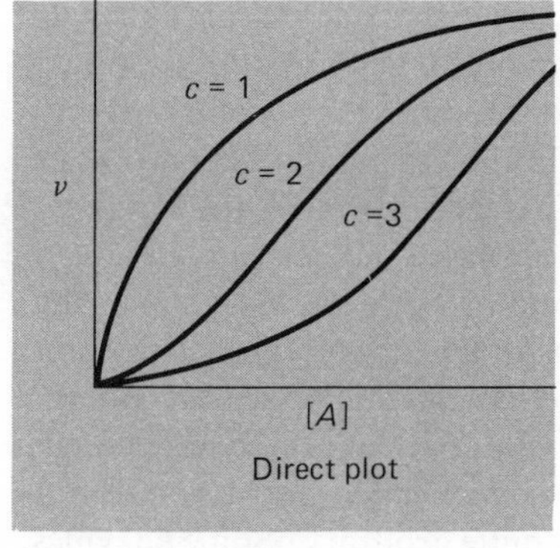

2. $\log \dfrac{\nu}{n-\nu} = c \log[A] + \log\mathbf{K}$

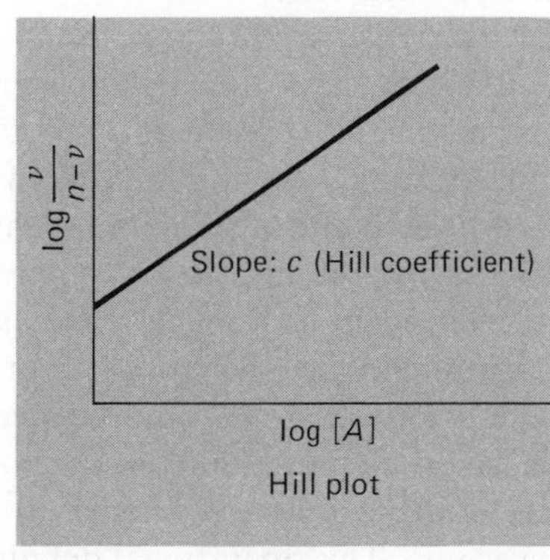

(b)

Figure 2.3 (a) Evaluation of **K** and **n** for noninteracting sites ($c = 1$ in equation 2.7).
1. Direct plot (ν vs. $[A]$).
2. Double reciprocal plot ($1/\nu$ vs. $1/[A]$).
3. Scratchard plot (single reciprocal plot) ($\nu/[A]$ vs. ν). [From Scratchard, G., J. S. Coleman, and A. L. Shen, *J. Am. Chem. Soc.*, **56**, 658 (1934)].

(b) Evaluation of **K**, n and c for interacting sites (plots (a)2 and (a)3 will not be linear; $1 < c \leqslant n$ in Equation 2.7).
1. Direct plot (ν vs. $[A]$) is sigmoidal.
2. Hill plot, based on rearranging the Hill equation to

$$\frac{\dfrac{\nu}{n}}{\dfrac{1-\nu}{n}} = \mathbf{K}[A]^c$$

$$\log \frac{\nu}{n-\nu} = c \log [A] + \log \mathbf{K}$$

It should be noted that in practice good linear plots are only obtained between 10 percent and 90 percent saturation.

2.4 ENZYME KINETICS

One of the most obvious roles of functional proteins is catalysis and the catalytic proteins, the enzymes, have been more extensively studied than any other group of functional molecules. The development of enzymology and the kinetic analysis of enzyme-catalyzed reactions probably represented the "coming of age" of biochemistry as an exact and quantitative science. One should therefore not be surprised to find that enzyme kinetics is given much emphasis in most biochemistry courses. However, based on the philosophy that kinetic analysis is just one of the tools available to the biochemist in his studies, the following treatment is limited to the most general principles of enzyme kinetics. Although the equations are developed for only the simplest cases, it should be made clear that this simple treatment has surprisingly broad validity when applied to isolated enzymes. It must also be stated that the information gained from kinetic analyses increases significantly with greater theoretical and experimental refinement. With an understanding of the first principles, however, the further steps into the theory and practice at more advanced levels are not complicated ones.

All known enzymes are proteins. Some carry out their catalytic role as *simple enzymes*, in which case only the protein is involved in the catalysis (these simple enzymes could presumably contain nonprotein components—carbohydrate, for example—having no known role in the catalytic process). Other enzymes contain nonprotein components as integral parts of the catalytic apparatus and are referred to as *conjugated enzymes*. The active conjugated enzyme is referred to as a *holoenzyme* and contains both the *apoenzyme* (the protein component) and the *coenzyme* or the *prosthetic group*. The distinction between the prosthetic group and the coenzyme has essentially been an operational one, based on the stability of the holoenzyme complex: If the nonprotein component is dialyzable it is referred to as a coenzyme; if not it is called a prosthetic group. This operational definition is quite arbitrary. Some enzymes are synthesized in the cell as inactive *proenzymes* or *zymogens*; these can be converted to active enzymes by different processes involving covalent structural changes. Enzymes may act as single agents or as molecular complexes in association with other proteins or lipoproteins. A ligand that is bound and then chemically altered by the enzyme is a *substrate*. Each enzyme is generally very specific in its interaction with its substrate and is a very efficient catalyst. In accord with our general definition of protein function, the first step in enzyme catalysis is the binding of the substrate, but the binding is followed by chemical alteration of the substrate to produce a particular product. As in the previous section on general binding, in this section we are interested in describing the enzyme-substrate binding in terms of the affinity of the protein for the ligand. In principle it should be possible to do this by the direct-binding studies outlined in Section 2.3. The main difficulty with this approach is that the ligand is undergoing chemical changes during the binding studies. It is therefore generally simpler to determine the affinity from the more natural measurements of the rate of the enzyme-catalyzed reaction. This, then, is the realm of kinetics. At the same time we obtain a measurement of the catalytic efficiency of the enzyme by determining the maximum rate at which it converts substrate to product. This is expressed as V_{max} or more precisely as the *turnover number*, the number of moles of substrate converted to product per min (minute) per mg (milligram) of enzyme when the enzyme is saturated with substrate. Just as general binding was described in terms of **K** and n, the special kind of binding (as well as the accompanying chemical alteration) observed in the case of the enzyme, E, and the substrate, S, are described in terms of a "dissociation constant" referred to as the Michaelis constant, K_m, and the catalytic efficiency, V_{max}. It is the purpose of this section to develop the simplest equations and methods by which these constants can be determined, to review the information they give about the system, and to show how this information can be used in the study of structure-function relationships. The basic enzyme-catalyzed reaction can be written

$$E + S \underset{k_{-1}}{\overset{k_1}{\rightleftharpoons}} ES \overset{k_2}{\overline{\rightleftharpoons x \rightleftharpoons EP \rightleftharpoons}} E + P$$

This represents a mechanistic proposal that the enzyme combines with the substrate to give an enzyme-substrate complex. In a series of subsequent steps (x) the transformation of the substrate to product—still enzyme bound (EP)—takes place and in the last step the enzyme-product complex dissociates to yield the product and regenerate the active enzyme.

A graphic representation of enzyme catalysis is given in Figure 2.4. It should be emphasized that all enzyme-catalyzed reactions are formally reversible and that an enzyme thus has at least two substrates (S is the substrate in the conversion of S to P and P is the substrate in the reversed conversion of P to S). It is also very important to keep in mind that the enzyme only affects the *rate* of the reaction and the equilibrium is determined solely by the properties of substrate and products. This holds true only under the normal in vitro conditions of enzyme catalysis (low enzyme concentration), where the concentration of substrate (or product) bound to the enzyme is very low compared to the total substrate (or product) concentration. If substratelike quantities of enzyme are added the concentration of ES and EP complex becomes a significant component in the equilibrium mixture; this could obviously alter the proportions of free S and P at equilibrium. In vivo it is quite possible that the enzyme and substrate concentrations are more equal, especially when the substrate is another macromolecule.

Our first task, then, is to go from the general mechanism to a rate equation that will permit us to evaluate the characteristic constants K_m and V_{max} in terms of the known quantities: (1) total substrate concentration, S_0; (2) total enzyme concentration, E_0; and (3) an experimentally observed rate, v_0. The basic mechanism and the assumptions and experimental restrictions used to arrive at the rate equation are listed below.

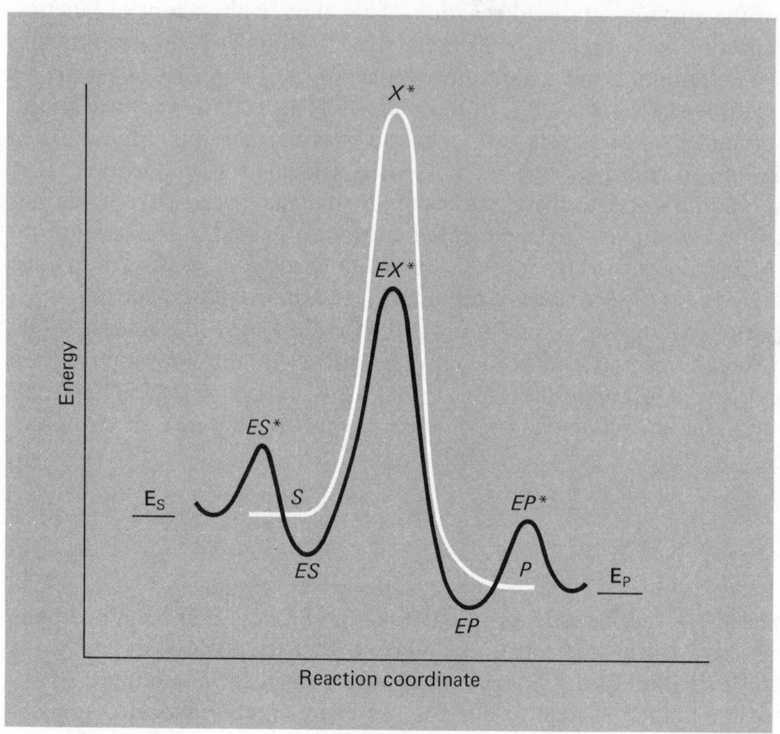

$$S \rightleftharpoons X^* \rightleftharpoons P$$

$$E + S \rightleftharpoons ES^* \rightleftharpoons ES \rightleftharpoons EX^* \rightleftharpoons EP \rightleftharpoons EP^* \rightleftharpoons P + E$$

(*a*)

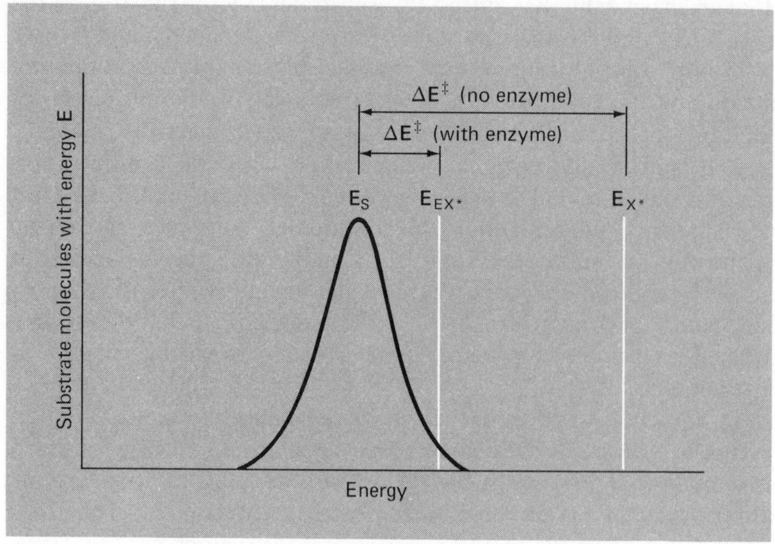

(*b*)

$$E + S \underset{k_{-1}}{\overset{k_1}{\rightleftharpoons}} ES \underset{k_{-2}}{\overset{k_2}{\rightleftharpoons}} E + P$$

1 Only initial rates are measured. This means that the reverse reaction (k_{-2}) can be disregarded and that we may assume the substrate concentration at the time of measurement is equal to the initial (total) substrate concentration, S_0.

2* Beyond the *ES* formation is a rate-limiting step (k_2). In other words the overall rate of product formation is given as $v_0 = k_2\,[ES]$.

3* The total enzyme concentration, E_0, can be expressed in terms of free and bound enzyme, or $[E]$ and $[ES]$ ($[EP]$ and other intermediate forms, indicated as x in the mechanism above, are insignificant during the initial stages of the reaction): $E_0 = [E] + [ES]$. Equations such as this are termed conservation equations.

4 The concentration of free enzyme, $[E]$, and the concentration of enzyme-substrate complex, $[ES]$, as unknown quantities can be related to each other and to a characteristic *kinetic constant* through either of two assumptions: (1) the rapid-equilibrium assumption,

$$E + S \underset{k_{-1}}{\overset{k_1}{\rightleftharpoons}} ES$$

$$K_s = \frac{k_{-1}}{k_1} = \frac{[E][S]}{[ES]}$$

where K_s is the dissociation constant; or (2) the steady-state assumption,

$$E + S \underset{k_{-1}}{\overset{k_1}{\rightleftharpoons}} ES \xrightarrow{k_2} E + P$$

$$k_1[E][S] = (k_{-1} + k_2)[ES]$$

$$K_m = \frac{k_{-1} + k_2}{k_1} \equiv \frac{[E][S]}{[ES]}$$

* These restrictions (2 and 3) are only imposed in order to define k_2 precisely. If these restrictions are not obeyed, the same rate equation is obtained but k_2 is a much more complex function of several rate constants.

Figure 2.4 (*a*) An oversimplified energy diagram for the conversion of substrate, *S*, to product, *P*, through some transition state, X^*. The white line indicates the path for reaction in the absence of enzyme, and the black line indicates the path for the enzyme catalyzed reaction. The main role of the enzyme or any other catalyst is to reduce the energy of activation required to reach the transition state, and thus increase the rate of the reaction. The equilibrium of the reaction by which *S* and *P* are interconverted is independent of the path, and is only dependent upon the energies of *S* and *P*. The equilibrium is therefore not affected by the enzyme. (*b*) Another way of looking at the "energy barrier" for a reaction. The distribution of energy content of a population of substrate molecules is shown relative to the energy of the transition state in the absence of enzyme (reaction rate zero) and in the presence of the enzyme (finite reaction rate). [For a detailed and very useful discussion of enzyme catalysis see R. Lovrien, *J. Theoret. Biol.*, **24**, 247 (1969.]

Both of these assumptions are consistent with assumption 2 above. The rapid-equilibrium assumption says that k_2 is so low compared to k_{-1} that it can be disregarded altogether and that the concentration of the *ES* complex is determined by the equilibrium constant K_s and $[E]$ and $[S]$. The steady-state assumption does consider the k_2 step as part of the overall *ES*-determining system but says that at some stage of the initial reaction the concentration of the *ES* complex remains constant (in a steady state), that is, its rate of formation, $k_1[E][S]$, is equal to its rate of breakdown, $(k_2 + k_{-1})\,[ES]$; the steady-state kinetic constant, K_m (or Michaelis constant), is thus not a true equilibrium constant, though in its interpretation it is generally considered simply a measure of the affinity between enzyme and substrate. Although this may be a good approximation for many systems, for others it often leads to unacceptable interpretations. The critical feature is the relative values of the rate constants. Obviously as k_2 becomes smaller and smaller compared to k_{-1}, K_m approaches the true dissociation constant K_s.* These restrictions and assumptions (the steady-state assumption will be used in the following) makes it possible to derive a rate equation from the proposed mechanism and to express it in terms of the known quantities E_0 and S_0 and the characteristic constants.

The basic rate equation for the mechanism

$$E + S \underset{k_{-1}}{\overset{k_1}{\rightleftharpoons}} ES \xrightarrow{k_2} E + P$$

is

$$v_0 = k_2[ES]$$

where v_0 is the initial observed rate. We now use our assumptions to rewrite this equation with known quantities. From the steady-state assumption we get

$$(1)\ K_m = \frac{[E][S]}{[ES]} \quad \text{or} \quad [E] = K_m\frac{[ES]}{[S]}$$

and from the conservation laws,

$$(2)\ E_0 = [E] + [ES]$$

and

$$(3)\ S_0 = [S] + [ES]$$

* Michaelis and Menten based their treatment on the rapid-equilibrium assumption and the original Michaelis constant was thus in a strict sense the equilibrium constant K_s. Briggs–Haldane later introduced the steady-state treatment. However, it has become customary to use the term Michaelis constant for both K_s and K_m. [L. Michaelis, and M. L. Menten, *Biochem. Z.*, **49**, 333 (1913), and G. E. Briggs, and J. B. S. Haldane, *Biochem. J.* **19**, 338 (1925).]

since $[S] \gg [ES]$, $[S] \equiv S_0$. Substituting (1) into (2) and substituting S_0 for $[S]$ according to (3) gives

$$E_0 = [E] + [ES] = \frac{K_m[ES]}{S_0} + [ES] = [ES]\left(1 + \frac{K_m}{S_0}\right)$$

Rearranging this to

$$[ES] = \frac{E_0}{1 + K_m/S_0}$$

we can now substitute this directly into the basic rate equation:

$$v_0 = \frac{k_2 E_0}{1 + K_m/S_0}. \tag{2.8}$$

This is the well-known Michaelis–Menten–Briggs–Haldane equation. Since $v_0 = k_2[ES]$ we can see that v_0 approaches a limiting value of V_{max} when $[ES]$ approaches E_0. Thus $V_{max} = k_2 E_0$ and Equation 2.8 can also be written in the more commonly used form

$$v_0 = \frac{V_{max}}{1 + K_m/S_0}. \tag{2.9}$$

This equation relating the experimentally obtained rate v_0 with the known quantity S_0 allows the evaluation of the constants K_m and V_{max} for an enzyme-catalyzed reaction. It should be noted that the equation is identical to the equation for general binding [2.3(*b*)] if one allows for the use of dissociation constants in enzyme kinetics and association constants in binding. (Equation 2.3(*b*) can be written

$$\frac{v}{n} = \frac{1}{1 + 1/\mathbf{K}'[A]}$$

where v/n is the degree of saturation of the protein with ligand. Since v_0/V_{max} from equation 2.9 is the ratio of the rate at some partial saturation of enzyme with substrate, v_0, to the rate at complete saturation, V_{max}, it also expresses degree of saturation.)

Let us note here one interesting feature of Equation 2.9, namely the relationship of the kinetic constants for a reversible reaction to each other and to the equilibrium constant for the $S \rightleftharpoons P$ interconversion.
For the complete mechanism

$$E + S \rightleftharpoons (ES \rightleftharpoons EP) \rightleftharpoons E + P$$

the rates of the forward (left to right) reaction, $v_{0(f)}$, and the reverse (right to left) reaction, $v_{0(r)}$, can be described by

$$v_{0(f)} = \frac{V_{\max(f)}}{1 + K_{m(f)}/S_0}$$

and

$$v_{0(r)} = \frac{V_{\max(r)}}{1 + K_{m(r)}/P_0}$$

and it can be shown that

$$K_{eq} = \frac{[P]_{eq}}{[S]_{eq}} = \frac{K_{m(r)}V_{\max(f)}}{K_{m(f)}V_{\max(r)}} \tag{2.10}$$

(It may be useful for the student to work out this relationship.) This equation is known as the Haldane relationship* and relates the equilibrium constant to the K_m and $V_{\max}$ of the forward and the reverse reactions.

As in the case of binding, the evaluation of K_m and $V_{\max}$ gives some information about the biochemical properties of a particular enzyme but the two constants have little value beyond that. Again we are primarily interested in these constants as functional "indicators" and wish to study how they change when physical and chemical "effectors" are introduced into the system. In the following paragraphs a simple general form of the Michaelis–Menten equation is derived that will account for the most common effects. This equation has no fundamental value but provides a convenient way of obtaining rate equations for relatively complicated mechanisms.

Consider the cases where changes in reaction conditions (such as pH, inhibitors, or activators) will give rise to different inactive (nonproductive) forms of the enzyme, the substrate, or the enzyme-substrate complex. The conservation equations must account for all of these forms and the general Michaelis-Menten equation must be modified accordingly. For the general mechanism

$$\begin{array}{ccccc} E & + & S & \underset{k_{-1}}{\overset{k_1}{\rightleftharpoons}} & ES \xrightarrow{k_2} E + P \\ \upharpoonleft\downharpoonright & & \downharpoonleft\upharpoonright & & \upharpoonleft\downharpoonright \\ eE & & sS & & esES \end{array}$$

sS, eE, and $esES$ represent any number of *inactive* species in equilibrium with active S, E, and ES, respectively. The equilibrium relationships are described by the appropriate equilibrium constants and the concentration of the corresponding derivatives sS, eE, and $esES$ (we also assume that all the interconversions between active and inactive species are rapid compared to the k_2 step); the rate equation is still

$$v_0 = k_2[ES]$$

* J. B. S. Haldane, *Enzymes*, Longmans, Green and Co., London, 1930 (Reprinted MIT Press, 1965).

and the steady-state law is also unaltered:

$$(1)\ K_m = \frac{[E][S]}{[ES]} \quad \text{or} \quad [E] = K_m \frac{[ES]}{[S]}$$

The conservation laws, however, now become

$$(2)\ E_0 = [E] + e[E] + [ES] + es[ES]$$

$$= [E](1 + e) + [ES](1 + es)$$

and

$$(3)\ S_0 = [S] + s[S] = [S](1 + s)$$

and the resulting form of the Michaelis–Menten equation is

$$v_0 = \frac{k_2E_0/(1 + es)}{1 + (K_m/S_0)[(1 + e)(1 + s)/(1 + es)]} \tag{2.11}$$

This general equation tells immediately that K_m varies with variations in E, S, and ES while V_{max} ($= k_2E_0$) only depends on variations in the ES complex; this follows from the definition of V_{max}, the reaction rate when all the enzyme has been converted to ES complex (V_{max}, by operational definition, is attained when enough active S has been added to saturate the active E; under these conditions there is no free active E present, and the observed rate is independent of $[S]$).

Equation 2.11 has general application to all systems where only one active form each of S, E, and ES is involved; it can be used to develop rate equations for different types of special conditions.

Inhibition

Consider first the effect of inhibitors: Like the binding of a second ligand to a protein (equation 2.4) the net effect of an inhibitor, I, is a decrease in the observed enzyme activity. We can visualize the inhibitor as operating by several different mechanisms. Three of these are given in Table 2.4 with the corresponding rate equations. All three have been observed for a number of different enzymes and inhibitors. They are readily distinguished experimentally and in fact the assignment of an inhibitor as competitive (see the illustration of competitive inhibition in Fig. 2.5), noncompetitive (note that the special case of simple noncompetitive inhibition, where $K_I = K_I'$, is illustrated; the more complex case where $K_I \neq K_I'$ is more common), or uncompetitive is based entirely on the experimental data (see Figure 2.6 below).

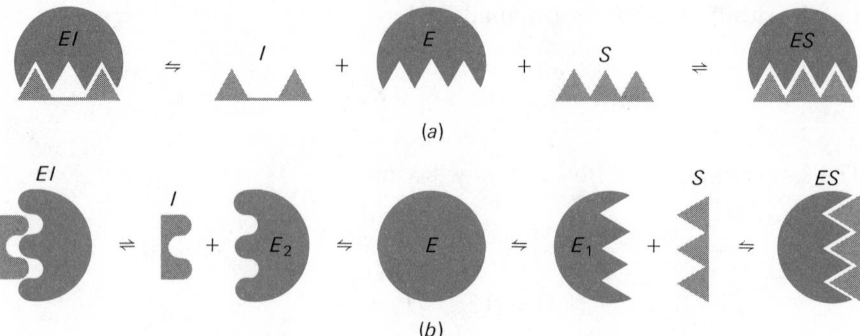

Figure 2.5 Illustration of competitive type inhibition. The key to this type of inhibition is that *the binding of substrate and the inhibitor are mutually exclusive*; that is, they cannot bind simultaneously to the enzyme, and therefore *compete* for the enzyme. (*a*) Classical competitive inhibition: the inhibitor is a structural analog of the substrate and competes with the substrate for the active site. (*b*) "Allosteric" or "different site" inhibition (see Chapters 3 and 8): the substrate and the inhibitor bind at different sites which are realized by different and *mutually exclusive* conformations of the enzyme. In both cases the competitive nature of the inhibition is recognized by the fact that V_{max} (the rate at infinitely high substrate concentration) is independent of the presence of inhibitor.

TABLE 2.4 THREE TYPES OF INHIBITION OF ENZYMES

General Mechanism and Rate Equation for Inhibition:

$$\begin{array}{ccc} E + S & \rightleftharpoons & ES \rightarrow E + P \\ \downarrow\uparrow & & \downarrow\uparrow \\ eE & & esES \end{array}$$

$$v_0 = \frac{V_{max}/(1 + es)}{1 + K_m/S_0[(1 + e)/(1 + es)]}$$

Conservation Law: $E_0 = E + eE + ES + esES,\ (E_0 = E + EI + ES + ESI)$

Competitive Inhibition	*Noncompetitive Inhibition*	*Uncompetitive Inhibition*
$E + S \rightleftharpoons ES \rightarrow E + P$ $E + I \rightleftharpoons EI$	$E + S \rightleftharpoons ES \rightarrow E + P$ $E + I \rightleftharpoons EI$ $ES + I \rightleftharpoons ESI$	$E + S \rightleftharpoons ES \rightarrow E + P$ $ES + I \rightleftharpoons ESI$
$\left(K_I = \frac{[E][I]}{[EI]}\right)$	$\left(K_I = \frac{[E][I]}{[EI]},\ K'_I = \frac{[ES][I]}{[ESI]}\right)$	$\left(K_I = \frac{[ES][I]}{[ESI]}\right)$
($es = 0$, $e = [I]/K_I$ in general rate equation)	($es = [I]/K'_I$, $e = [I]/K_I$ in general rate equation)	($es = [I]/K_I$, $e = 0$ in general rate equation)
Rate Equations:		
$v_0 = \frac{V_{max}}{1 + K_m/S_0(1 + [I]/K_I)}$	Complex: $K_I \neq K'_I$ $v_0 = \frac{V_{max}\frac{1}{1 + [I]/K'_I}}{1 + K_m/S_0\frac{1 + [I]/K_I}{1 + [I]/K'_I}}$	$v_0 = \frac{V_{max}\frac{1}{1 + [I]/K_I}}{1 + K_m/S_0\frac{1}{1 + [I]/K_I}}$
	Simple: $K_I = K'_I$ $v_0 = \frac{V_{max}}{(1 + K_m/S_0)(1 + [I]/K_I)}$	$v_0 = \frac{V_{max}}{1 + [I]/K_I + K_m/S_0}$

Inhibition by substrate itself is a fairly common occurrence in enzyme-catalyzed reactions. A simple mechanism for substrate inhibition can be written

$$E + S \underset{k_{-1}}{\overset{K_1}{\rightleftharpoons}} ES \xrightarrow{k_2} E + P$$

$$E + S \overset{K_I}{\rightleftharpoons} SE$$

$$ES + S \overset{K'_I}{\rightleftharpoons} SES$$

The two nonproductive steps are

$$K_I = \frac{[S][E]}{[SE]}$$

and

$$K'_I = \frac{[S][ES]}{[SES]}$$

The inactive species are

$$eE = [SE] = \frac{S_0[E]}{K_I}$$

and

$$esES = [SES] = \frac{S_0[ES]}{K'_I}$$

and substituting for e and es in Equation 2.11 we can thus write

$$e = \frac{S_0}{K_I}$$

$$es = \frac{S_0}{K'_I}$$

For the special case of $K_I = K'_I$, the rate equation becomes

$$v_0 = \frac{V_{max}}{(1 + K_m/S_0)(1 + S_0/K_I)}$$

Note that at low substrate concentration the first term is large and the second one insignificant, but at high substrate concentration the second term becomes the predominant one. Substrate inhibition is most generally observed at high substrate concentrations, as predicted by this mechanism.

Activation

Consider next the case where the second ligand activates instead of inhibiting the enzyme. The most common example of this is activation by divalent metal ions, A. In complete analogy to the treatment in Table 2.4 we can write for this case

$$E + A \overset{K_A}{\rightleftharpoons} EA$$

$$K_A = \frac{[E][A]}{[EA]}$$

$$EA + S \underset{k_{-1}}{\overset{k_1}{\rightleftharpoons}} EAS \xrightarrow{k_2} EA + P$$

The active enzyme is EA, $es = 0$ and $e = K_A/[A]$; the rate equation thus becomes

$$v_0 = \frac{V_{max}}{1 + (K_m/S_0)(1 + K_A/[A])}$$

(Notice that in all the inhibitor cases an increase in I causes a decrease in v_0 whereas an increase in A causes an increase in v_0). It is interesting to note that the mechanism

$$S + A \overset{K_A}{\rightleftharpoons} SA$$

$$K_A = \frac{[S][A]}{[SA]}$$

$$E + SA \underset{k_{-1}}{\overset{k_1}{\rightleftharpoons}} ESA \xrightarrow{k_2} E + PA$$

would give the same rate equation [since s and e effect K_m in the same manner, this can be readily predicted from equation (2.11)].

Simple kinetic analysis will thus not distinguish between enzyme activation and substrate activation and must be supplemented with other data. One possible method would be to determine K_A for both substrate and enzyme by direct-binding studies, comparing these values with the kinetically determined constant.

A special case of activation of great importance occurs when the substrate itself is the activator. This is of course analogous to the homotropic-binding effect (cooperative binding) discussed above. The mechanism would be the same as for general activation,

$$E + S \overset{K_A}{\rightleftharpoons} ES$$

$$K_A = \frac{[E][S]}{[ES]}$$

$$ES + S \underset{k_{-1}}{\overset{k_1}{\rightleftharpoons}} ESS \xrightarrow{k_2} ES + P$$

and the rate equation would also be the same as before,

$$v_0 = \frac{V_{max}}{1 + (K_m/S_0)(1 + K_A/S_0)} = \frac{V_{max}}{1 + K_m/S_0 + K_mK_A/S_0^2} \tag{2.12}$$

Again, in analogy to the binding equation (2.6), the dependence of rate on the square of substrate concentration yields the characteristic sigmoidal v_0 versus S_0 plot, which is characteristic of *regulatory* systems, as will be discussed later.

Simultaneous Effect of Inhibitors and Activators

The best illustration of this case is the effect of pH on an enzyme-catalyzed reaction. For the complex mechanism

$$\begin{array}{ccccc} E & & ES & & \\ K_b^E \upharpoonleft\downharpoonright & & k_1 K_b^{ES} \upharpoonleft\downharpoonright & & \\ EH + S & \underset{k_{-1}}{\rightleftharpoons} & EHS & \xrightarrow{k_2} & EH + P \\ K_a^E \upharpoonleft\downharpoonright \; K_a^S \upharpoonleft\downharpoonright & & K_a^{ES} \upharpoonleft\downharpoonright & & \\ EH_2 \quad SH & & EH_2S & & \end{array}$$

the rate equation is derived by substituting into Equation 2.11 the proper e, es, and s factors from $s = [H^+]/K_a^S$, $e = [H^+]/K_a^E + K_b^E/[H^+]$, and $es = [H^+]/K_a^{ES} + K_b^{ES}/[H^+]$ to give the rate equation

$$v_0 = \frac{V_{\max}\dfrac{1}{1 + [H^+]/K_a^{ES} + K_b^{ES}/[H^+]}}{1 + \dfrac{K_m(1 + [H^+]/K_a^S)(1 + [H^+]/K_a^E + K_b^E/[H^+])}{S_0(1 + [H^+]/K_a^{ES} + K_b^{ES}/[H^+])}} \tag{2.13}$$

This last treatment of the effect of pH on the rate of enzyme-catalyzed reactions illustrates well the type of information that can be obtained from enzyme kinetics. The mechanism and rate equation (2.13) contain, in addition to the normal constants $V_{\max}$ and K_m, acid-dissociation constants for both the free enzyme and for the enzyme-substrate complex. Thus the complete kinetic analysis of this system, leading to an evaluation of the acid-dissociation constants in question, will give information about the identity of the amino acid residues responsible for the enzymatic activity. (The treatment of the data to permit such an analysis will be illustrated below in Figure 2.7.)

Reactions with More Than One Substrate

For the enzyme-catalyzed reaction $A + B \rightleftharpoons C + D$ the simple rules for the development of a rate equation no longer apply in a strict sense. The steady-state treatment becomes more complex and again we make assumptions in order to arrive at a good approximate rate law. One way of looking at the above reaction is to consider it as two activation steps (the formation of EA or EB) followed by two substrate binding steps (the formation of the active enzyme–two-substrate complex EAB). Thus we can write the general reaction

$$
\begin{array}{lll}
 & & EA + B \\
 & + A \overset{K_A}{\rightleftharpoons} & \quad \rightleftharpoons K_m^B \\
E & & \quad EAB \xrightarrow{k_2} ECD \longrightarrow E + C + D \\
 & + B \underset{K_B}{\rightleftharpoons} & \quad \rightleftharpoons K_m^A \\
 & & EB + A
\end{array}
$$

If we make the rapid equilibrium assumption, all four constants above are simple dissociation constants:

$$K_A = \frac{[E][A]}{[EA]} \qquad K_B = \frac{[E][B]}{[EB]}$$

$$K_m^A = \frac{[EB][A]}{[EAB]} \qquad K_m^B = \frac{[EA][B]}{[EAB]}$$

Since the rate of product formation again is directly proportional to $[EAB]$ we have, as before,

$$v_0 = k_2[EAB]$$

and the conservation law is

$$E_0 = [E] + [EA] + [EB] + [EAB]$$

Since we have four equations by which to convert the three unknowns $[EA]$, $[EB]$, and $[EAB]$, one equation is redundant (the four constants are interrelated by $K_A K_m^B = K_B K_m^A$). In developing the rate equation the K_B equation will arbitrarily be left out (this is common practice). From substitutions we get

$$E_0 = [EAB]\left(1 + \frac{K_m^A}{[A]} + \frac{K_m^B}{[B]} + \frac{K_A K_m^B}{[A][B]}\right)$$

and thus the rate equation

$$v_0 = \frac{V_{\max}}{1 + K_m^A/[A] + K_m^B/[B] + K_A K_m^B/[A][B]} \tag{2.14}$$

Equation 2.14 is the most general rate equation for two-substrate kinetics and has been found to describe adequately a number of enzyme-catalyzed reactions. The equation actually corresponds to an *ordered* two-substrate reaction since either path of the mechanism specifies an ordered addition of substrates to the enzyme; as a consequence, $K_m^A \neq K_A$ and $K_m^B \neq K_B$. However, we must also consider other possibilities. One of these is a *random reaction* where $K_m^A = K_A$ (binding of A is uneffected by B) and $K_m^B = K_B$ (binding of B is uneffected by A). For this case Equation 2.14 becomes

$$v_0 = \frac{V_{\max}}{(1 + K_A/[A])(1 + K_B/[B])} \tag{2.15}$$

Again this case fits a number of enzyme-catalyzed reactions.

Another special case is the so called *ping-pong* mechanism, evolved to explain experimental observations that are best described by a rate equation like 2.14 where the $[A][B]$ product term has dropped out. One mechanism that will yield such a rate law is

$$E + A \rightleftharpoons EA \longrightarrow E' + C$$

$$E' + B \rightleftharpoons E'B \longrightarrow E + D$$

and the rate equation for this is

$$v_0 = \frac{V_{\max}}{1 + K_m^A/[A] + K_m^B/[B]} \tag{2.16}$$

These three cases cover virtually all known systems. They can obviously be distinguished experimentally (otherwise they would have no significance) as indicated below in Figure 2.7.

There are several methods of treating the kinetic data in order to evaluate the kinetic constants. The initial rates are determined at different substrate concentrations and at a fixed set of conditions; the v_0–S_0 data can then be plotted by several methods given in Figure 2.6, any of which will provide a graphic determination of K_m and $V_{\max}$. The most common method is the double-reciprocal plot of Lineweaver and Burke. Figure 2.7 gives the Lineweaver–Burke plots for the different rate equations developed above. From the data in this figure one should be able to see the significance of the K_m and $V_{\max}$ evaluation to the general study of enzyme structure and function.

It was stated at the start of this section that only the simplest cases of kinetic analysis would be discussed in this book. Thus the generalization that only one form each of the enzyme, the substrate, and the enzyme-substrate complex is active has been made throughout. In many cases this simplification will not accommodate the experimental data; hence more complex mechanisms and correspondingly more complicated rate equations must be used. The principle is the same, however.

Of the aspects of kinetic analysis that have been omitted in the above discussion, two must be briefly mentioned before closing: (1) the effect of temperature on enzyme-catalyzed reactions; and (2) the use of rapid-kinetic analysis.

The effect of temperature on the rate of enzyme-catalyzed reactions is very complex in that many components of the total system are influenced by temperature. If one looks at the mechanism proposed for the pH effect above and the corresponding rate equation (2.13), it should be clear that any effect of temperature on v_0 can have many components. Thus the three rate constants k_1, k_{-1}, and k_2 will vary with temperature; in addition the five acid-dissociation constants are probably also temperature dependent. In addition to these effects the enzyme itself, according to our general model, exists as a dynamic-equilibrium mixture of conformational isomers sensitive to changes in temperature. At sufficiently high temperatures heat denaturation of the enzyme will occur. A complete analysis of temperature effects must therefore be complex but will also yield significant new information. Any experimental observation of a temperature effect on an equilibrium constant such as the acid-dissociation constants in Equation (2.13) can be treated quantitatively by the van't Hoff equation,

$$\Delta H^0 = RT^2 \frac{d \ln K_{eq}}{dT}$$

and the effects of temperature on rate constants can similarly be described by the empirical Arrhenius equation,

$$E_a = RT^2 \frac{d \ln k}{dT}$$

where E_a is activation energy.

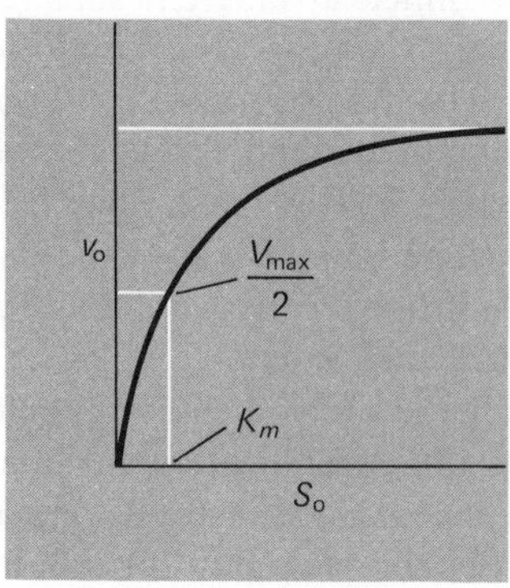

1. Direct plot

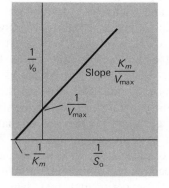

2. Double reciprocal plot

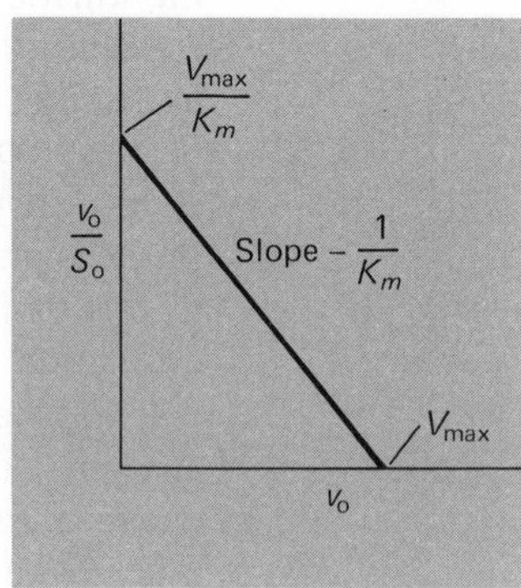

3. Single reciprocal plot

Figure 2.6(*a*) Different methods of treating kinetic data to evaluate K_m and V_{max} for one-substrate systems (compare to Figure 2.3):

$$v_0 = \frac{V_{max}}{1 + K_m/S_0}$$

1. Direct plot, v_0 vs S_0.
2. Double reciprocal plot, $1/v_0$ vs $1/S_0$. [From Lineweaver, H. and D. Burk, *J. Am. Chem. Soc.*, **56**, 658 (1934).]

$$\frac{1}{v_0} = \frac{1}{V_{max}} + \frac{K_m}{V_{max}} \frac{1}{S_0}$$

3. Single reciprocal plot, v_0/S_0 vs v_0. [From Eadie, G. S., *J. Biol. Chem.*, **146**, 85 (1942); B. H. J. Hofstee, *Science*, **116**, 329 (1952).]

$$\frac{v_0}{S_0} = -\frac{1}{K_m} v_0 + \frac{V_{max}}{K_m}$$

A. General solution (double reciprocal plots)

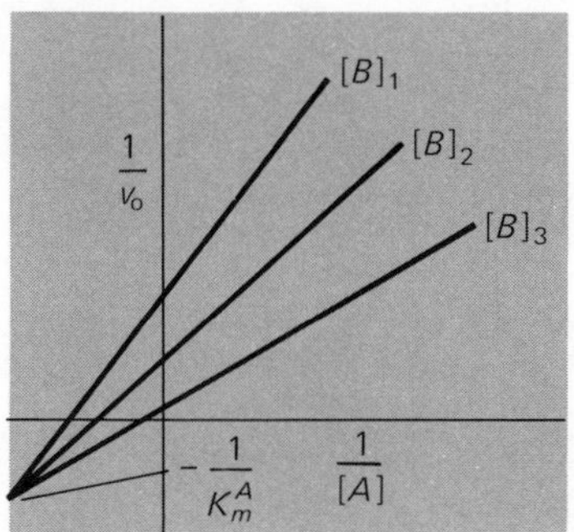

1. $[B]$ constant, $[A]$ variable

$$\frac{1}{v_0} = \underbrace{\frac{1}{V_{max}}\left(1 + \frac{K_m^B}{[B]}\right)}_{\text{Intercept}} + \underbrace{\frac{K_m^A}{V_{max}}\left(1 + \frac{K_A K_m^B}{K_m^A [B]}\right)}_{\text{Slope}} \frac{1}{[A]}$$

$([B]_1 < [B]_2 < [B]_3)$

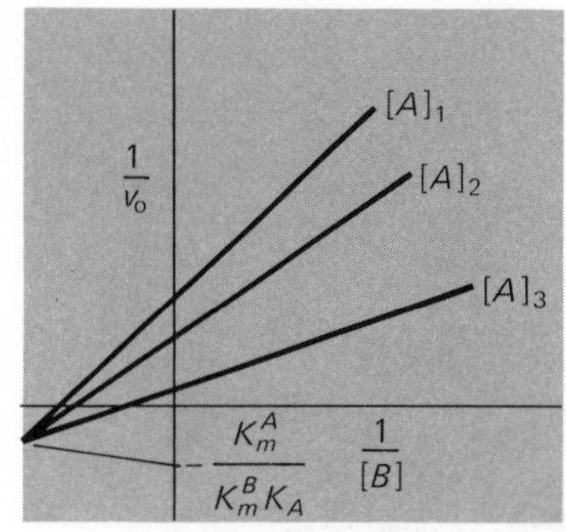

2. $[A]$ constant, $[B]$ variable

$$\frac{1}{v_0} = \underbrace{\frac{1}{V_{max}}\left(1 + \frac{K_m^A}{[A]}\right)}_{\text{Intercept}} + \underbrace{\frac{K_m^B}{V_{max}}\left(1 + \frac{K_A}{[A]}\right)}_{\text{Slope}} \frac{1}{[B]}$$

$([A]_1 < [A]_2 < [A]_3)$

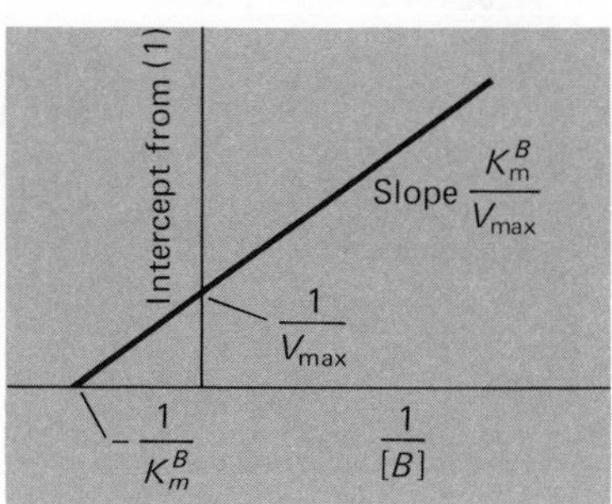

3. $\text{Intercept} = \frac{1}{V_{max}} + \frac{K_m^B}{V_{max}} \frac{1}{[B]}$

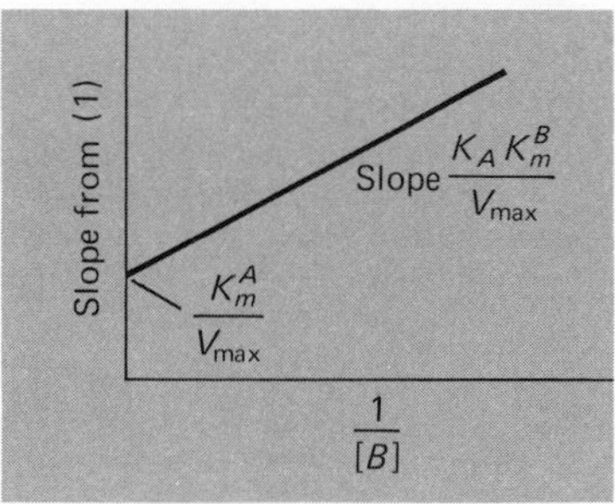

4. $\text{Slope} = \frac{K_m^A}{V_{max}} + \frac{K_A K_m^B}{V_{max}} \frac{1}{[B]}$

Figure 2.6(*b*) Different methods of treating kinetic data to evaluate K_m and V_{max} for two-substrate systems:

$$v_0 = \frac{V_{max}}{1 + K_m^A/[A] + K_m^B/[B] + K_A K_m^B/[A][B]}$$

Using only the double reciprocal plot:

$$\frac{1}{v_0} = \frac{1 + K_m^A/[A] + K_m^B/[B] + K_A K_m^B/[A][B]}{V_{max}}$$

and rearranging according to experimental design.

Part A. General Solution:

1. Hold $[B]$ constant, vary $[A]$, plot $1/v_0$ vs $1/[A]$.
2. Hold $[A]$ constant, vary $[B]$, plot $1/v_0$ vs $1/[B]$. To obtain values for V_{max} and K_m, replot (using 1 as illustration).
3. Intercept vs $1/[B]$.
4. Slope vs $1/[B]$.

B. Special solutions ($[B]_1 < [B]_2 < [B]_3$)

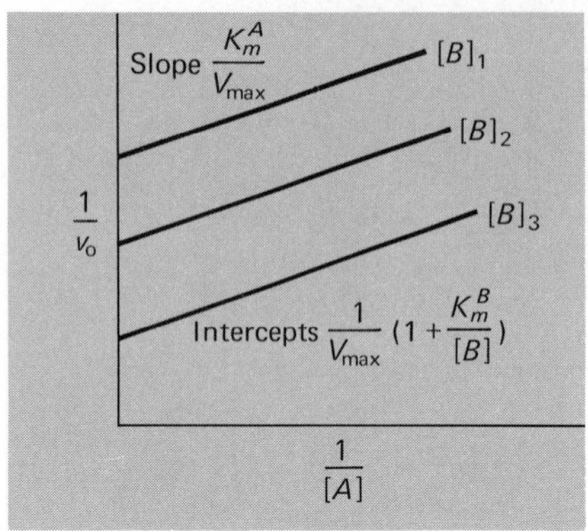

1. Ping pong mechanism

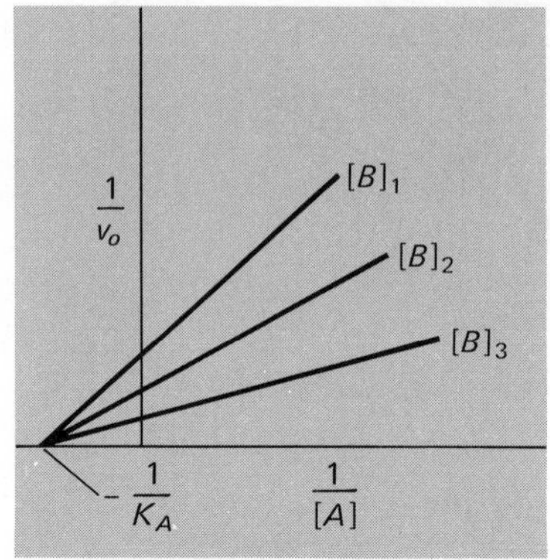

2. Random order

$$\frac{1}{V_o} = \underbrace{\frac{1}{V_{max}}(1 + \frac{K_B}{[B]})}_{\text{Intercept}} + \underbrace{\frac{K_A}{V_{max}}(1 + \frac{K_B}{[B]})}_{\text{Slope}}\frac{1}{[A]}$$

Figure 2.6(*b*) Part B. Special Solutions:

1. If B is not inhibitory at high concentration, saturate with B, and the equation reduces to that for a one-substrate system. (There are only two forms of the enzyme EB and EBA.) This represents the limiting case ($[B] = \infty$) of the general case above.
2. Ping-pong mechanism:

$$\frac{1}{v_0} = \frac{1 + K_m^B/[B]}{V_{max}} + \frac{K_m^A}{V_{max}}\frac{1}{[A]}$$

A plot of $1/v_0$ vs $1/[A]$ gives a series of parallel lines at different concentration of B (a slope independent of $[B]$ is diagnostic for the ping-pong mechanism).

3. Ordered reactions ($K_m^A \neq K_A$ and $K_m^B \neq K_B$): This is the general case discussed under (*b*)A. Diagnostic picture: The lines intersect in the 1st or 3rd quadrant.
4. Random order ($K_m^A = K_A$ and $K_m^B = K_B$): the sites for A and B do not interact.

$$\frac{1}{v_0} = \frac{(1 + K_A/[A])(1 + K_B/[B])}{V^{max}}$$

Diagnostic picture: The lines intersect on the x-axis.

Figure 2.7 Graphic representations and solutions for different cases of activation and inhibition. The evaluation of K_I, K_A and pK_a values from double reciprocal plots. (*a*) Different types of inhibition (see Table 2.4): evaluation of K_I from $v_0 = V'_{max}/1 + (K'_m/S_0)$, where V'_{max} and K'_m are the experimentally observed constants in the presence of inhibitor I.

Curve A: no inhibitor ($V'_{max} = V_{max}$; $K'_m = K_m$)
Curve B: competitive inhibition ($V'_{max} = V_{max}$; $K'_m = K_m/(1 + [I]/K_I)$
Curve C: Simple noncompetitive inhibition ($V'_{max} = V_{max}/1 + [I]/K_I$; $K'_m = K_m$)

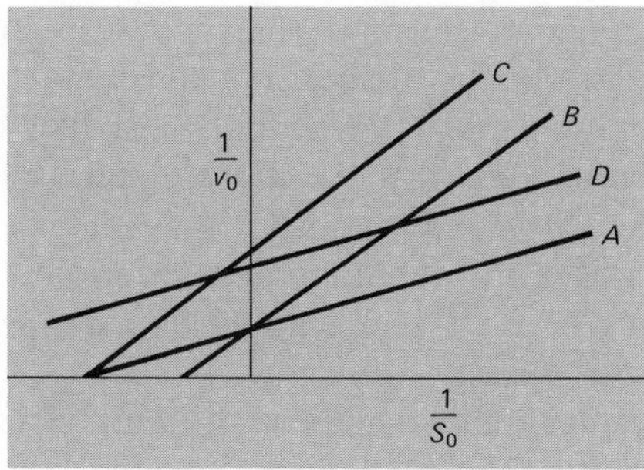

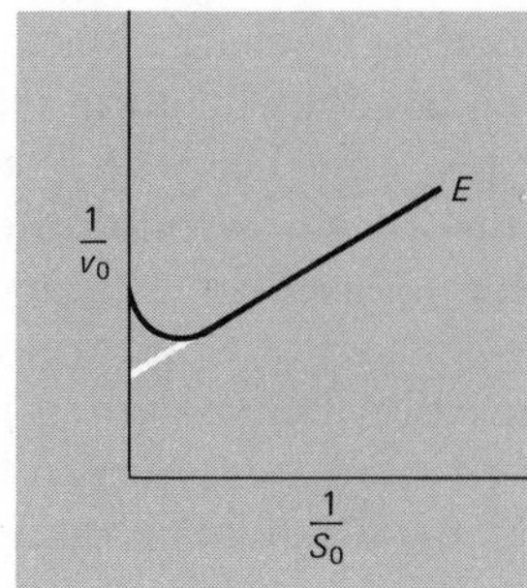

(a) Different types of inhibition

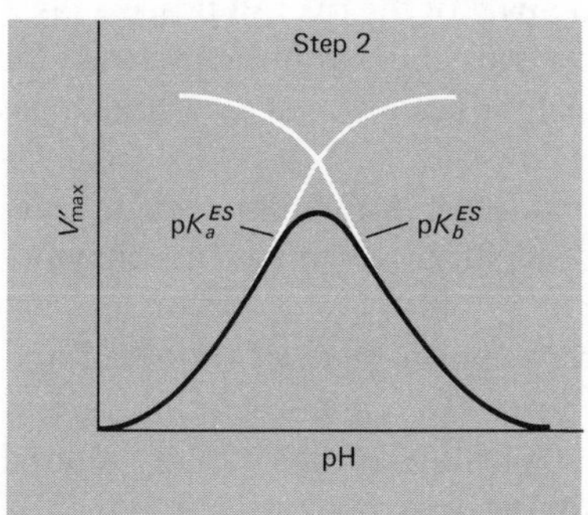

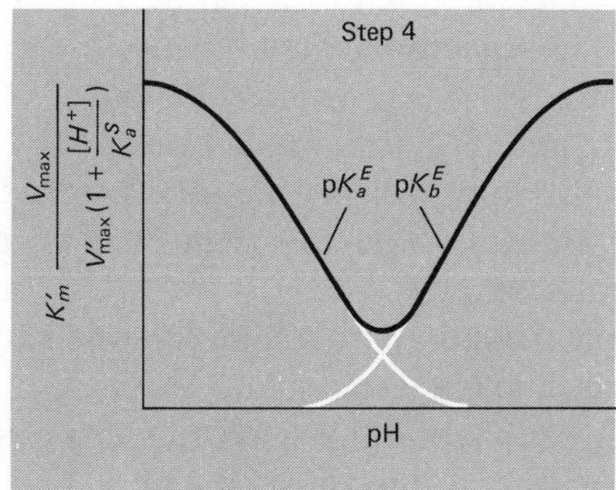

(b) Effect of pH

Curve D: uncompetitive inhibition ($V'_{max} = V_{max}/1 + [I]/K_I$; $K'_m = K_m/1 + [I]/K_I$)
Curve E: substrate inhibition ($V'_{max} = V_{max}/1 + S_0/K_I^S$; $K'_m = K_m$)
(b) Effect of pH. Evaluation of the acid dissociation constants in equation 3.13: At any pH. $v_0 = V'_{max}/1 + K'_m/S_0$ where

$$V'_{max} = \frac{V_{max}}{1 + [H^+]/K_a^{ES} + K_b^{ES}/[H^+]}$$

and

$$K'_m = \frac{K_m(1 + [H^+]/K_a^E + K_b^E/[H^+])(1 + [H^+]/K_a^S)}{1 + [H^+]/K_a^{ES} + K_b^{ES}/[H^+]}$$

Step 1: Get $1/v_0$ vs $1/S_0$ plots at many different pH, and thus V'_{max} and K'_m at many pH values.

Step 2: Plot V'_{max} vs pH and determine the shape of each arm of the bell-shaped curve by curve fitting. This gives pK_a^{ES} and pK_b^{ES} and V_{max} (the true V_{max}, independent of pH).

Step 3: Titrate the substrate to obtain pK_a^S (this should be done whenever possible).

Step 4: Reduce the complex pH function for K'_m by the constants determined in steps 2 and 3:

$$K'_m = \frac{K_m(1 + [H^+]/K_a^E + K_b^E/[H^+])(1 + [H^+]/K_a^S)}{1 + [H^+]/K_a^{ES}/[H^+]}$$

$$x = K'_m \frac{V_{max}}{V'_{max}(1 + [H^+]/K_a^S)} = K_m(1 + [H^+]/K_a^E + K_b^E/[H^+])$$

and plot the experimental parameter x vs pH. Again the bell-shaped curve can be analyzed to give pK_a^E, pK_b^E, and K_m.

There is much useful information to be obtained from such analyses. ΔH^0, for example is a characteristic thermodynamic constant that can help in identifying the component in question. Thus for the dissociation constants in 2.13 the pK_a together with the ΔH^0 for each ionization would most likely identify the ionizing group unequivocally, whereas the pK_a alone could well be subject to ambiguous interpretations. Most importantly, however, the complete analysis of temperature effects aids in the quantitative description of all the reaction parameters, a prerequisite to the understanding of how the total system functions (see Van Holde in this series).

Rapid kinetic analysis is designed to evaluate individual rate constants. The more detailed the analysis, the more information can be obtained from the data. The only way in which to ascertain just how complex the enzyme-catalyzed reaction is must be based on sorting out and studying each individual step rather than on looking only at the rate-limiting part of the overall process (as is done in Michaelis-Menten kinetics). There are two major approaches used for this purpose: (1) *presteady-state kinetics* by techniques of *rapid flow* and *rapid mixing*; and (2) *relaxation kinetics.*

The first approach is primarily designed to study the initial bimolecular reaction (presteady-state) between enzyme and substrate forming the enzyme-substrate complex.

In the second approach all intermediate steps can in theory be studied and the principle is quite simple. The system is allowed to come to equilibrium at a given set of conditions. When some external variable is then rapidly altered (temperature is the most common one, and the approach is consequently referred to as the *temperature-jump method*), a new equilibrium has to be reached and the change in any component of the system that can be observed (spectrophotometrically, for example) is recorded as the system "relaxes" to the new equilibrium condition. In this way it is possible to determine rate constants for individual steps whose presence could not even have been predicted by analysis through steady-state kinetics.

It is clear that the kinetic analysis is not merely an exercise to determine a couple of constants but a powerful tool to test a proposed mechanism and to obtain specific information about the active-site components. The information that becomes available through kinetic analysis is summarized in Table 2.5. It may be worthwhile to stress that the kinetic analysis never proves a

Figure 2.8 Exploration of the chemical architecture of an enzyme by chemical modification. (*a*) Preferential reaction in the "active site": reaction at a single specific amino acid residue leads to inactivation. Presence of substrate "protects" the enzyme against reaction. (*b*) Nonspecific reaction: When several sites must react before inactivation results, interpretation of the results becomes very complex. The sequential reaction first in the presence of substrate (to react all the "unimportant" residues) and then in the absence of substrate (to react the crucial residue-using radioactive reagent to distinguish from the first step) often gives clean results. (*c*) Affinity labeling: inducing the preferential reaction with the active site by using a substrate analog as the reagent.

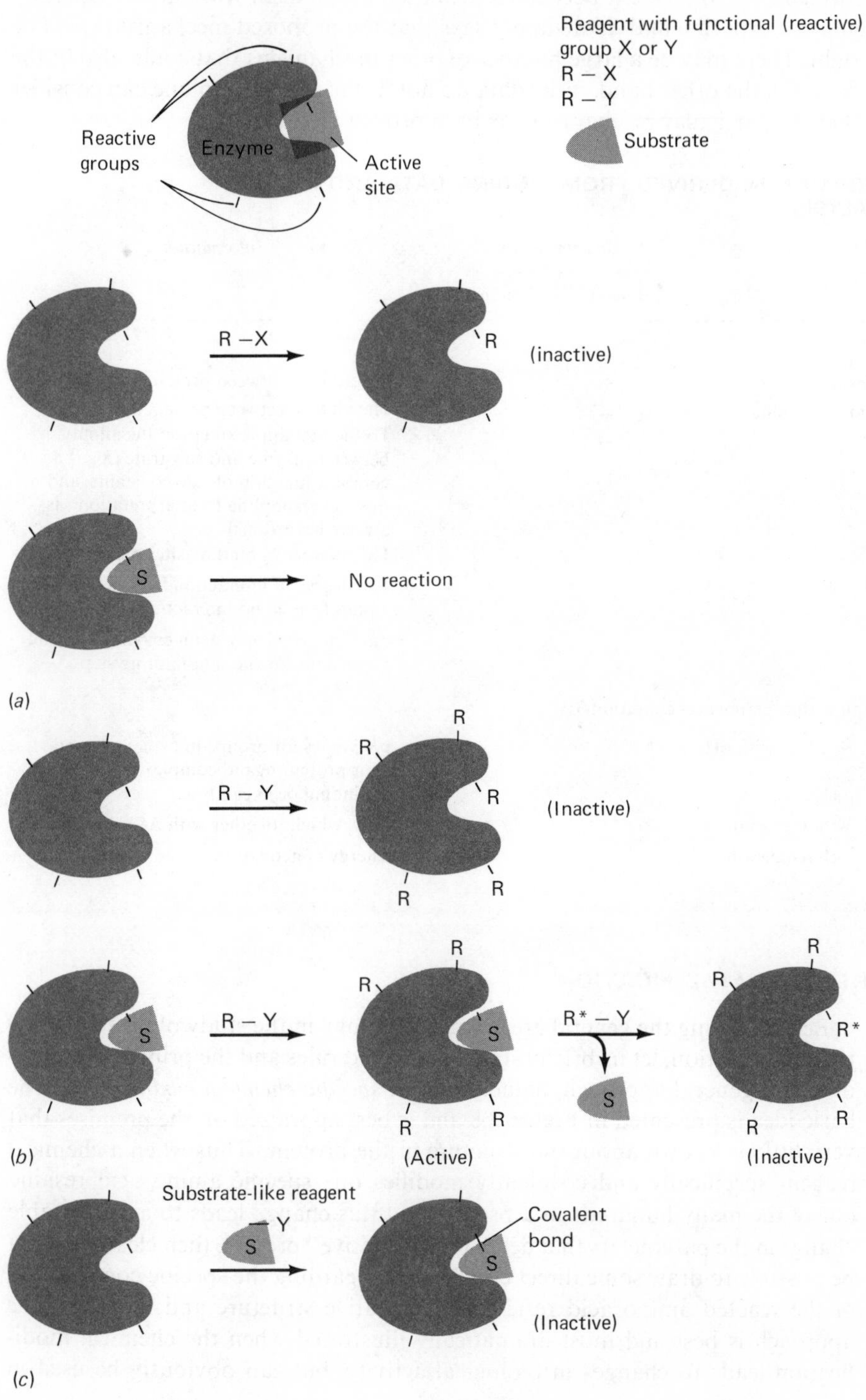
Reagent with functional (reactive) group X or Y
R – X
R – Y
Substrate
Reactive groups
Enzyme
Active site
R –X
R
(inactive)
S
No reaction
(a)
R – Y
R
(Inactive)
R – Y
S
(Active)
R* – Y
R*
(Inactive)
(b)
Substrate-like reagent
Y
Covalent bond
(Inactive)
(c)

mechanism. Agreement between a proposed mechanism with its rate equation and the experimental data simply says that the proposed mechanism *could* be right. There may be a large number of other mechanisms that could also fit the data. On the other hand, if the data do not fit the mechanism one can consider that the particular mechanism has been proved incorrect.

TABLE 2.5 INFORMATION DERIVED FROM BINDING DATA AND KINETIC ANALYSIS

Experimental Constant	*Determined by*		*Information*
	Binding	*Kinetics*	
First level:			
K (Association constants)	+		The affinity between protein and ligand
K_I, K_A (Dissociation constants)	+	+	The affinity between protein and ligand
K_m	−	+	To the first approximation the affinity between enzyme and substrate (K_m is a complex function of rate constants, and this oversimplified interpretation is always hazardous)
n	+	−	The number of binding sites per mole
c (Hill coefficient)	+	+	The degree of interaction between multiple sites ($c = 1$: no interaction)
V_{max}	−	+	Catalytic efficiency of an enzyme: the rate constant for the rate-limiting step.
Second level (variation with experimental conditions):			
Variation of **K**, K, K_m, V_{max} with pH			pK values for groups in protein, ligand, or protein-ligand complex involved in binding or in catalysis.
Variations of **K**, K with temperature			ΔH^0, which together with ΔG^0 gives ΔS^0
Variations in V_{max} with temperature			Energy of activation

2.5 CHEMICAL MODIFICATION

Before leaving the general area of methodology in the study of protein structure and function, let us briefly state the ground rules and the principal features of a very general approach, namely that of *specific chemical modification.* The basic idea is presented in Figure 2.8 and is best appraised on the premises that very little is known about the structure of the protein. Thus when a chemical reagent specifically and covalently modifies one specific amino acid residue out of the many hundreds in a protein and this change leads to a measurable change in the parameters that describe the "native" protein, then clearly it must be possible to draw some direct conclusions regarding the specific contribution of the reacted amino acid residue to the native structure and function. The approach is best and most dramatically illustrated when the chemical modification leads to changes in biological activity, but can obviously be used in evaluating the effect of specific derivativization on any parameter. The types of reagents and treatments that can be used for chemical modification have been reviewed by Barker in this series and include a wide variety of different residue specificities. Special reagents have been designed to improve the specificity of reaction with particular proteins. An example of this approach, referred to as *affinity labeling*, is the use of a substrate analogue containing a reactive group as an enzyme reagent that is "active-site specific." Polyfunctional reagents have also been used, introducing bridges into the folded structure by covalently cross-linking suitably positioned amino acid residues and thus giving specific information about interresidue distances in the folded polypeptide chain.

With these methods in hand let us in the next two chapters summarize the status of knowledge of some proteins, illustrating with specific examples both the type of information that has been obtained and the specific methods by which the information has been uncovered.

REFERENCES

General (multivolume) reference sources in general protein chemistry:

Advances in Protein Chemistry, Academic Press, New York.

Neurath, H. (Ed.), *The Proteins 2nd Edition*, Academic Press, New York, 1963.

Experimental Techniques:

Hirs, C. H. W. (Ed.), *Enzyme Structure. Methods in Enzymology, Vol. XI*, Academic Press, New York, 1967, and Hirs, C. H. W., and S. N. Timasheff (Eds.), *Enzyme Structure II. Methods in Enzymology*, Academic Press, New York, 1972 (in press).

Leggett Bailey, J., *Techniques in Protein Chemistry, 2nd Edition*, Elsevier, Amsterdam, 1967.

Up-to-date status of protein structure: Sequence

Dayhoff, M. O. (Ed.), *Atlas of Protein Sequence and Structure 1969*, National Biomedical Research Foundation, Silver Springs, Md., 1969.

Three-dimensional structure (stereo-views):

Blow, D. M. and T. A. Steitz, "X-ray Diffraction Studies of Enzymes," *Ann. Rev. Biochem.*, **39**, 63 (1970).

Dickerson, R. E. and I. Geis, *The Structure and Action of Proteins* (with a separate stereo supplement), Harper and Row, New York, 1969.

Subunit structure of a number of proteins: a tabulation:

Klotz, I. M., and D. W. Darnell, "Protein Subunits," *Science*, **166**, 126 (1969).

Klotz, I. M., *et al.*, "Quarternary Structure of Proteins," *Ann. Rev. Biochem.*, **39**, 25 (1970).

Articles on Ligand Binding:

Klotz, I. M., "Analysis of Metal-Protein Complexes," in *Methods of Biochemical Analysis*, D. Glick (Ed.), Vol. III, Interscience Publishers, Inc., New York, 1956.

Steinhardt, J., and S. Beychok, "Interaction of Proteins with Hydrogen Ions and Other Small Ions and Molecules," in *The Proteins*, H. Neurath (Ed.), Vol. II, 139, Academic Press, New York, 1964.

Wyman, J. Jr., "Linked Functions and Reciprocal Effects in Hemoglobin: A Second Look," *Adv. Prot. Chem.*, **19**, 223 (1964).

Books and articles on Kinetics:

Amdur, I, and G. Hammes, *Chemical Kinetics: Principles and Selected Topics*, McGraw-Hill, New York, 1966.

Boyer, P. D., H. Lardy, and K. Myrbäck (Eds.), *The Enzymes, 2nd Edition*, Vol. I, Academic Press, New York, 1958. Hearon, J. Z., *et al.*, "Enzyme Kinetics," p. 49; Alberty, R. A., "The Rate Equation for an Enzymic Reaction," p. 143; Lumry, R., "Some Aspects of the Thermodynamics and Mechanisms of Enzymic Catalysis," p. 157.)

Christensen, H. N., and G. A. Palmer, *Enzyme Kinetics: A Programmed Text*, W. B. Saunders Co., Philadelphia, 1967.

Hammes, G., "Relaxation Spectrometry of Biological Systems," *Adv. Prot. Chem.*, **23**, 1 (1968).

Books and articles on chemical modification:

Baker, B. R., *Design of Active-Site-Directed Irreversible Enzyme Inhibitors*, John Wiley and Sons, Inc., New York, 1967.

Glazer, A. N., "Specific Chemical Modification of Proteins," *Ann. Rev. Biochem.*, **39**, 101 (1970).

Means, G. E., and R. E. Feeney, *Chemical Modification of Proteins*, Holden-Day, Inc., New York, 1971 (in press).

Spande, T. F., *et al.*, "Selective Cleavage and Modification of Peptides and Proteins," *Adv. Prot. Chem.*, **24**, 97 (1970).

Vallee, B. L., and J. F. Riordan, "Chemical Approaches to the Properties of Active Sites of Enzymes," *Ann. Rev. Biochem.*, **38**, 733 (1969).

THREE | ENZYMES

In this chapter several enzymes will be discussed in some detail to illustrate basic properties of catalytic proteins. Emphasis will be placed on experimental results, their interpretation, and model building from these results. The enzymes have been selected both to illustrate important concepts in enzymology and to survey a broad spectrum of experimental techniques. No attempt has been made to present the discussion of the experimental findings in chronological order.

3.1 RIBONUCLEASE, A "SIMPLE" ENZYME

We shall first discuss bovine pancreatic ribonuclease as it is one of the best defined enzymes in terms of both structure and function. The search for the active site of ribonuclease (RNase) by kinetic analysis and by chemical modification will be reviewed. The use of chemical modification to study the contributions of different parts of the molecule to the stabilization of the enzymatically active conformation(s) will also be surveyed. Most important to our discussion is the fact that the structural and functional model of RNase arrived at by these indirect studies now can be compared to the complete three-dimensional structure determined from X-ray diffraction.

General Description

Pancreatic RNase is found in large amounts only in ruminants and its metabolic significance is not obvious.* (The enzyme hydrolyzes RNA in the intestine and produces 3′-nucleotides, which must be hydrolyzed to nucleosides before being taken up from the intestine. Since nucleosides or free bases are not intermediates in the biosynthesis of nucleotides and nucleic acids, the enzyme may function only in the degradation of dietary nucleic acid, which is used in energy metabolism.) The enzyme is specific for RNA and attacks only the pyrimidine components. Two separate steps are catalyzed by the enzyme, cyclization and hydrolysis:

Pu Py Py Pu Pu
2′→ –OH –OH –OH –OH –OH (cyclization →)
3′→ P P P P
5′→

Pu Py: –OH, –O–P(=O)O⁻–O– (cyclic), P + Py: cyclic –O–P(=O)O⁻–O– + Pu Pu: P

↓ hydrolysis ↓

Pu Py: P, –OH, –OP(=O)(O⁻)O⁻ Py: –OH, –OP(=O)(O⁻)O⁻

(Purine is shown by Pu, and pyrimidine by Py; shorthand notation explained in Chapter 5.) It is possible to assay for the overall process (cyclization and hydrolysis) using RNA as substrate and measuring as a function of time either production of acid-soluble nucleotides or cleavage of the diester bond. The former assay method is not entirely reliable in that any bond cleavage leading to polynucleotides with 6-10 monomer units that are not acid soluble will not be detected. The latter assay method, measuring the acid produced each time a diester is broken, is more precise. It is also possible to assay the second step (hydrolysis) using the cleavage of cytidine *cyclic*-2′,3′ monophosphate, 2′,3′-CMP or of uridine-*cyclic*-2′,3′ monophosphate, 2′,3′-UMP by a direct spectrophotometric method based on a difference between the spectra of *cyclic* monophosphates and monoesters.†

The enzyme can be isolated and crystallized from the pancreatic juice of different species. The crystalline enzyme from the bovine pancreas contains at least two catalytically indistinguishable forms, RNase *A* and RNase *B*.

* E. Zendzian, and E. A. Barnard, *Arch. Biochem. Biophys.*, **122**, 699, 714 (1967).

† E. M. Crook, A. P. Mathias, and B. R. Rabin, *Biochem. J.*, **74**, 234 (1960).

Ribonuclease *B* is a glycoprotein form of RNase *A* with five mannose and two glucosamine residues attached to the protein at residue 34 (asparagine).* Bovine pancreatic RNase *A*, the form that has been most extensively studied, is a single polypeptide chain free of carbohydrate with a molecular weight of 13,600; the complete sequence of its 124 component amino acids is given in Figure 3.1. By careful analysis of tryptic peptides with intact disulfide (S—S) bonds, it has been possible to determine how the eight half-cystine residues are paired in the enzyme; Figure 3.1 thus represents the complete covalent structure of the biologically active enzyme.

Bearing in mind that we are interested in three-dimensional structures, it will be relevant for the ensuing discussion to ask the following question: Does this specific covalent structure predetermine a narrow range of three-dimensional folded conformations under physiological conditions, or is the unique active conformation only one of many possible conformations? An approach to answering this question is illustrated by the following experiment, which asks another question: If the four S—S bridges are cleaved by reduction under conditions in which the normal native conformation is vastly distorted and the enzyme is then returned to physiological conditions and allowed to reoxidize,

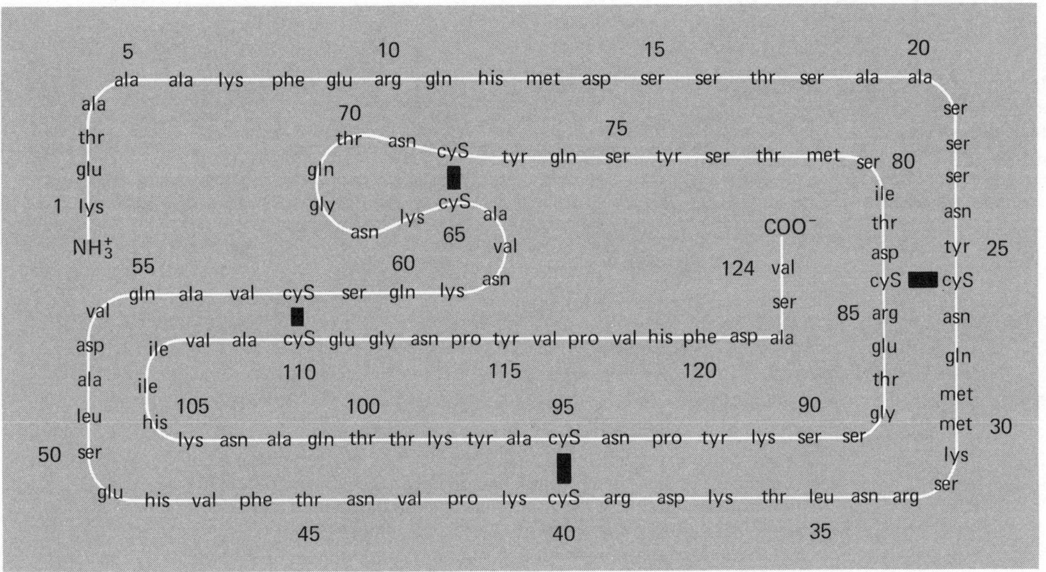

Figure 3.1 The primary structure of bovine pancreatic ribonuclease A (RNAse A) [Smythe, D. G., W. H. Stein, and S. Moore, *J. Biol. Chem.*, **238**, 227 (1963)].

* T. H. Plummer, Jr., and C. H. W. Hirs, *J. Biol. Chem.*, **239**, 2530 (1964).

will it return to only the original structure or to many different structures? The original structure can be recognized by catalytic activity, absence of sulfhydryl (—SH) groups, and formation of particular S—S pairs. The results obtained in these experiments are given in Figure 3.2 and demonstrate that (1) the features of the conformation that are uniquely required for catalytic activity must be thermodynamically favored in this particular covalent structure and (2) other less stable *intermediates* are also formed, recognizable by the presence of oxidized forms of the enzyme (no —SH groups) with no activity (Figure 3.2). Apparently when conditions for S—S interchange exist, wrong combinations

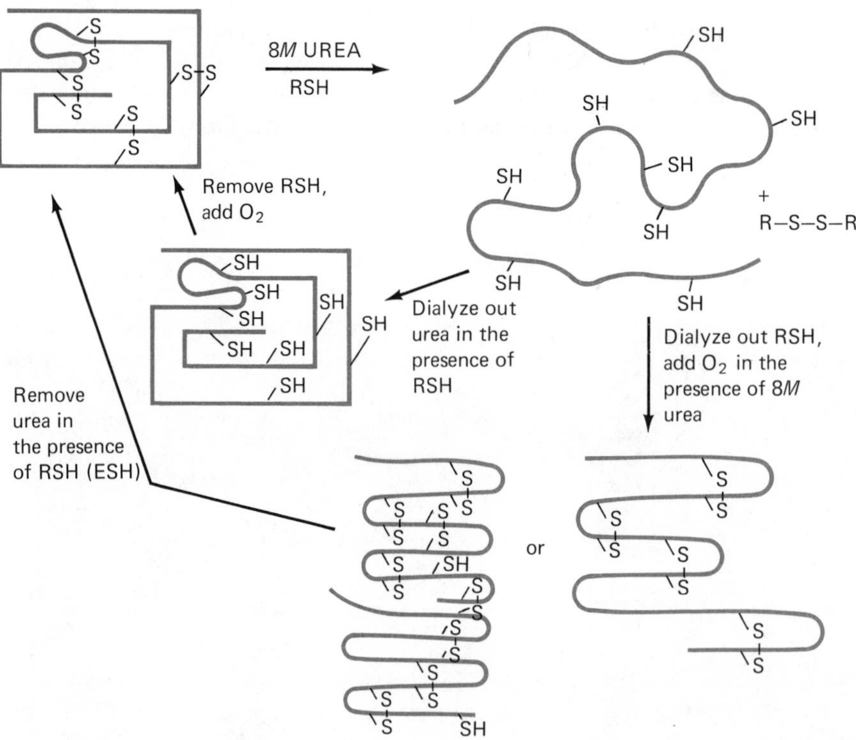

Figure 3.2 The reduction and reoxidation of RNAse. The cleavage of the S–S bonds is carried out with RSH, such as mercaptoethanol, in 8*M* urea to cause "unfolding" of the polypeptide chain and thus insure that the reduction goes to completion. Reoxidation (removal of RSH and addition of O_2) and refolding (removal of the urea) can then be studied as separate or simultaneous processes to give "right" or "wrong" S–S bonds. The "wrong" S–S bonds can be reshuffled to give the "right" S-S bonds in the presence of catalytic amount of RSH. An enzyme (ESH) which will catalyze the process has also been isolated. [From the data of White, F. H., Jr., *J. Biol. Chem.*, **236**, 1353 (1961) and Anfinsen, C. B. and E. Haber, *J. Biol. Chem.*, **236**, 1361 (1961).]

of S—S pairs are interchanged until the thermodynamically most stable combination is reached; this is also the catalytically active (native) form.

The reaction can be described as

RNase (S—S, S—S) native $\xrightarrow[\text{RSH}]{8M\text{ urea}}$ RNase (HS, HS, SH, SH) $\xrightarrow[\text{(no urea)}]{O_2}$ RNase (S—S, S—S) renatured, active enzyme; RNase (S—S, S—S) incorrect structure $\xrightarrow[\text{slow}]{\text{RSH}}$ renatured, active enzyme

It is significant that an enzyme that will catalyze the last slow S—S interchange reaction has recently been isolated: this process going from "incorrect" to "correct" disulfide matching thus may well be a natural one. The importance of this concept to the problems of protein synthesis will be discussed later.

Two items should be pointed out in connection with this experiment. First, although similar results have been obtained with other proteins, these cases must be considered exceptions rather than rules. Conceptually the conclusions could have general validity, but the experimental difficulties in demonstrating the same sequence of events in other proteins could be formidable. Thus if any of the initial incorrect forms of the reoxidized protein should be insoluble, for example, the conversion to the correct form would be very unfavorable and difficult to demonstrate. The other point is related to the implied conclusion that the reoxidized, active enzyme is identical to the native starting material. As presented above, only catalytic activity and absence of —SH groups were used as the criteria of a renatured enzyme; thus any completely reoxidized conformation with full activity would be counted as native. Much more extensive characterization has actually been carried out with reoxidized enzymes, however, and all the parameters studied (such as absorption spectra, sedimentation, and chemical reactivity) are identical to those of the native starting material with one exception. It has recently been found that the circular dichroic spectrum of the reduced and reoxidized enzyme shows small but significant differences from the spectrum of the native enzyme. This suggests that some three-dimensional order has been lost upon reduction and reoxidation, but not enough to be revealed by the other parameters. It is a very general law in macromolecular biochemistry, as in all experimental sciences, that the validity of any argument or conclusion is directly proportional to both the quantity and quality of the experimental evidence on which it is founded. It is always tempting to overlook omissions (though not necessarily flaws) in a scientific proof when a certain point is to be made. It is often impractical and nearly always tedious to thoroughly plug all holes in every argument, but the necessary "howevers" must be included in a "leaky" proposition.

Kinetic Analysis

When applied to careful kinetic analysis, the activity assay for RNase is somewhat complicated. The assay using high-molecular-weight RNA can be ambiguous in terms of K_m and V_{max} determination as the products of the initial cleavages are themselves substrates. Since a titrimetric assay of a simple dinucleotide substrate would measure the combined rates of cyclization and hydrolysis, the best kinetic assay is that using pyrimidine *cyclic*-2′,3′-monophosphates as substrates. It has already been mentioned that cytidine-3′-monophosphate (3′-CMP) and 2′3′-CMP have slightly different UV spectra.* The maximal positive difference (ΔOD for 3′-CMP − 2′,3′-CMP) is at a wavelength of 284 nm (nanometers); a direct measurement of the rate of hydrolysis of the substrate 2′,3′-CMP can be obtained at this wavelength. Since this substrate has a high extinction coefficient at 284 nm, the assay concentration is limited by the total optical density of the sample and the capability of the spectrophotometer. For this reason the kinetic assay is normally carried out at substrate concentrations well below the K_m of the enzyme.

With these difficulties in mind let us consider what kind of information can be obtained by the kinetic analysis of the RNase–2′,3′-CMP system.

The first experiment measuring the rates at different substrate concentrations under well-defined conditions of pH and ionic strength allows the determination of K_m and V_{max}, which is satisfying but highly uninformative. When the pH is varied in the range from 4 to 10, however, both K_m and V_{max} are affected and an analysis of their variation gives valuable information. The data fit well with the mechanism below and, using the treatment discussed in Chapter 2, it is possible to evaluate the four relevant ionization constants.

$$\begin{array}{ccccc} E & & ES & & \\ \uparrow\downarrow K_b^E & & \uparrow\downarrow K_b^{ES} & & \\ EH + S & \underset{k_{-1}}{\overset{k_1}{\rightleftharpoons}} & EHS & \xrightarrow{k_2} & EH + P \\ \uparrow\downarrow K_a^E & & \uparrow\downarrow K_a^{ES} & & \\ EH_2 & & EH_2S & & \end{array}$$

(Note that the substrate pK values are 2.0, 4.5, and 13.1 for primary —PO_4, and the —NH_2 and —OH of pyrimidine, respectively, and thus do not affect the kinetics significantly in the pH range studied.)

The experimental values for the four pK values are $pK_a^E = 5.22 \pm 0.2$, $pK_b^E = 6.78 \pm 0.2$, $pK_a^{ES} = 6.3 \pm 0.009$, and $pK_b^{ES} = 8.0 \pm 0.09$. Ionization constants in the free enzyme are represented by pK_a^E and pK_b^E, and of the known amino acid side-chain ionizations, only that of the imidazole group of histidine ($pK_a = 6$ to 7) is close to both values. However, it is possible that an abnormally basic carboxyl group could be responsible for pK_a^E and an abnormally acidic amino group for pK_b^E. Therefore it is necessary to establish the exact nature of

* E. M. Crook, A. P. Mathias, and B. R. Rabin, *Biochem. J.*, **74**, 234 (1960).

the group by further experiments. One possible approach is to repeat the whole set of experiments involving the effect of pH on K_m and V_{max} at several different temperatures, evaluate the ionization constants at each temperature, and from a van't Hoff's plot (log K versus $1/T$) obtain the heat of ionization (ΔH_i^0) for each ionization. (The ΔH_i^0 of carboxyl groups is near 0, that of histidine is near 6 Kcal (kilocalories)/mole, and that of amino groups is near 10 Kcal/mole.) Another approach was actually used with RNase. The effect of different solvents such as water, dioxan-water, (1:1) and dimethylformamide-water (1:1) on the pK_a values of a number of buffer systems was determined and compared with the effect of the same solvent systems on the pH-activity curves for the RNase. The enzyme was found to exhibit exactly the same dielectric dependence as the cationic model systems (which was different from that of the anionic ones), and carboxyl groups were thus eliminated. The conclusion drawn from these experiments was that the active site of RNase contains two cationic groups, one of which must be protonated and the other, deprotonated, in the active enzyme. Based primarily on the pK values, the two groups are tentatively identified as histidines.*

By kinetic methods a number of substrate analogues were tested as competitive inhibitors of the hydrolysis of *cyclic* cytidylic acid. The resulting inhibitor constants will be used in the next part of this discussion.

Chemical Modification: Carboxymethylation Experiments

When RNase was exposed to the action of bromoacetate or iodoacetate at a pH of 5.5, rapid inactivation occurred (Figure 3.3). When the proper conditions were effected, the only significant chemical change in the completely inactive enzyme was the carboxymethylation† of one histidine residue per mole of RNase. The rate of carboxymethylation of the enzyme is 5,000 times greater than the rate of carboxymethylation of free histidine under the same conditions. If iodoacetamide is substituted for iodoacetate in carboxymethylation of the enzyme, very slow inactivation is observed. Careful ion-exchange chromatography of the carboxymethylated enzyme showed that there were two main inactive products in a ratio of 85:15 (Figure 3.3). Enzymatic digestion, separation of peptides, and final characterization of the carboxymethylated peptides demonstrated that the major component was a derivative in which the 1 position of his-119 (histidine) had been carboxymethylated whereas the minor component was a derivative in which the 3 position of his-12 had been carboxymethylated.‡

It is significant that *no* derivative was found in which both his-119 and his-12 had reacted. The carboxymethylated his-119 derivative was found to be com-

* See B. R. Rabin, *et al.*, *Biochem. J.* **85**, 127, 134, 139, 145, 152 (1962).

† E. A. Barnard, and W. D. Stein, *J. Mol. Biol.*, **1**, 339, 350 (1959).

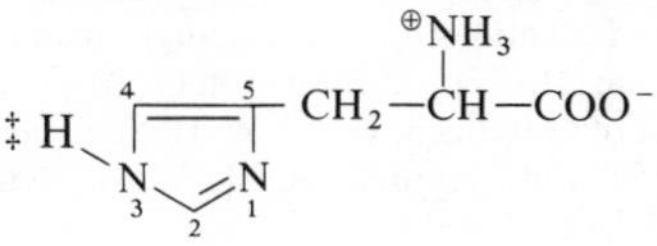

‡ $$H{-}\underset{3}{N}\;(4,\,5,\,2)\;\underset{1}{N}{-}CH_2{-}\overset{\overset{\oplus}{N}H_3}{\overset{|}{C}H}{-}COO^{-}$$

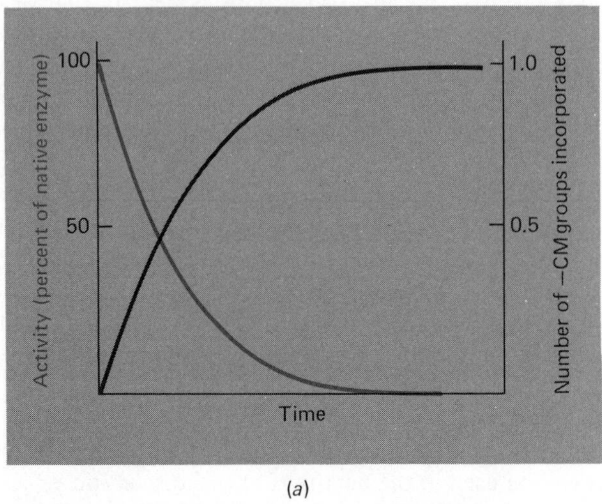

(*a*)

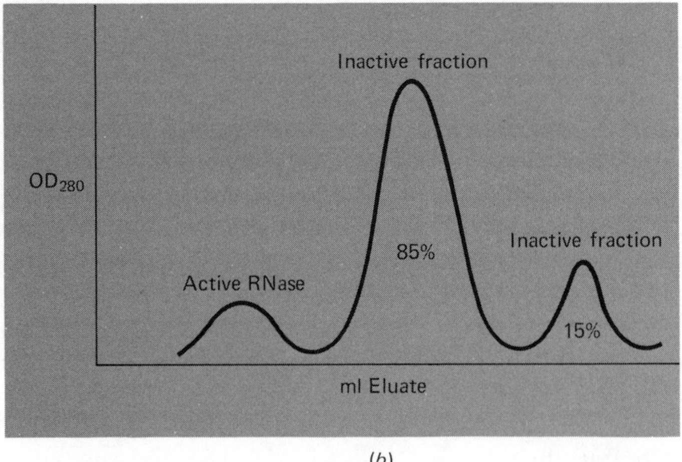

(*b*)

Figure 3.3 Carboxymethylation of RNAse. (*a*) The rate of inactivation of the enzyme and the incorporation of carboxymethyl groups upon exposure of RNAse to bromoacetate at pH 5.5. Note that the carboxymethylation and the inactivation are almost exactly parallel, and that essentially all the activity is lost when one carboxymethyl group has been taken up. (*b*) The separation of unreacted RNAse A and two inactive CM-RNAse derivatives by ion-exchange chromatography on a weak cation-exchange resin (IRC 50), eluting with 0.2 *M* phosphate buffer at pH 6.47. [From the data of Gundlach, H. G., W. H. Stein, and S. Moore, *J. Biol. Chem.*, **234**, 1754 (1959) and Crestfield, A. M., W. H. Stein, and S. Moore, *J. Biol. Chem.*, **238**, 2413, 2421 (1963).]

pletely inactive while the carboxymethylated his-12 derivative was found to exhibit only 15 per cent of normal enzyme activity. It was concluded from these observations that the enzyme contains two histidines (in positions 119 and 12), both of which are essential for full enzyme activity. The two histidines interact in such a way that only one of them reacts rapidly with the alkylating reagent.

If these are the same two histidines as those implicated by the kinetic experiments, one should be able to predict some aspects of their behavior. For example, there should be a correlation between the affinities of competitive inhibitors for the enzyme (obtained by kinetic methods) and the value of these inhibitors as protective agents against carboxymethylation. The results for several compounds tested as inhibitors of 2′,3′-CMP hydrolysis and carboxymethylation at a pH of 5.2 are given in Table 3.1; they clearly support the

TABLE 3.1 COMPARISON OF THE EFFECT OF INHIBITORS ON TWO PROPERTIES OF RNase[a]

Inhibitor	$K_I \times 10^6 (M)$	
	Catalysis[b]	*Alkylation*[c]
Cytidine-2′-monophosphate	1.9 ± 0.6	0.6 ± 0.3
Cytidine-3′-monophosphate	15.9 ± 2.4	7.3 ± 4.1
Pyrophosphate	7.6 ± 3.6	4.9 ± 2.5
Deoxythymidine-3′-monophosphate benzyl ester	1.01 ± 0.8	3.8 ± 0.7

[a] From A. P. Mathias, A. Deavin, and B. R. Rabin, "Studies on the Active Site and Mechanism of Action of Bovine Pancreatic Ribonuclease," in T. W. Goodwin, J. I. Harris, and B. S. Hartley, (eds.), *Structure and Activity of Enzymes*, p. 19, Academic Press Inc., New York, 1964.

[b] The catalysis of 2′,3′-CMP hydrolysis.

[c] The rate of alkylation (and inactivation) of the enzyme by iodoacetate.

hypothesis that his-12 and his-119 are those involved in the active site. Another prediction is that the carboxymethylation reaction should show a pH-rate profile similar to that of the catalytic reaction. This has also been tested and found to be the case. The carboxymethylation follows the general mechanism

$$\begin{array}{ccccc} EH_2 & \overset{K_a^E}{\rightleftharpoons} & EH & \overset{K_b^E}{\rightleftharpoons} & E \\ & & \Big\downarrow k_1 \quad \text{BrCH}_2\text{—COO}^- & & \\ & & EH\text{—CM} & & \end{array}$$

and from the pH dependence of the rate constant k_1, K_a^E and K_b^E were found to be 4.85 and 5.5 respectively, in reasonable agreement with the values determined by substrate kinetics, yet different enough to leave some question as to their identity.*

* See reference in Table 3.1.

Based on these findings a model of the active site of RNase has been formulated as illustrated in Figure 3.4.

If this model is correct, only a certain fraction of *E*H (the active protonated form of the enzyme) in solution (either 15 percent or 85 percent) can be active enzyme in a formal sense. The high reactivity of the two histidines must be related to their low p*K* compared to other protein histidines and to free histidine in solution. It is important to note that in denatured RNase no preferential reaction of his-12 and his-119 is observed.

Enzyme Digestions

The proteolytic enzyme subtilisin catalyzes the hydrolysis of a single peptide bond between residues 20 and 21 in RNase.* The product, designated RNase *S* is still active. The two component peptides are strongly associated and can only be separated by acid treatment (Figure 3.5). Gel filtration with 50 percent acetic acid as the solvent gives the elution pattern shown in Figure 3.5. The first peak represents a peptide that contains residues 21 through 124 of the original and is designated *S*-protein; the second, smaller component (which is retarded on the gel column) is the *N*-terminal peptide containing residues 1 through 20, designated *S*-peptide. Neither of the separated components shows any biological activity. However, when *S*-peptide is added back to *S*-protein there is complete recovery of the activity expected of a corresponding amount of intact RNase *A* and the titration curve shown in Figure 3.5 is obtained, showing a sharp end point at an *S*-peptide:*S*-protein ratio of 1:1. The reconstituted active enzyme, obtained by mixing equimolar amounts of pure *S*-protein and *S*-peptide, is designated RNase *S'*; it offers exciting possibilities for further studies of the structural requirements of the active site of the enzyme. First, the above results indirectly support the postulate that both his-12 and his-119 are required to form an active site. Second, the RNase *S* system makes it possible to ask some new questions as well as to obtain more refined answers to some old questions.

The new feature added to this system is the strong noncovalent interaction between the two peptide components. Clearly this is the type of interaction that maintains the intact protein in its proper conformation. From the titration curve two features of this interaction can be quantitated: The initial slope gives the affinity between the components and the end point gives a measure of the intactness of the active site. Systematic chemical modification of the *S*-peptide (the simplest of the two components) allows estimation of each residue's

* See reference in Figure 3.5.

Figure 3.4 A model of the active site of RNAse A. (*a*) A proposed mechanism for the carboxymethylation which explains the high reactivity and specificity of bromoacetate (the negative charge is required to align the reagent) in comparison to iodoacetamide (no negative charge). [Taken from Findlay, D., *et al.*, *Biochem. J.*, **85**, 152 (1968).] (*b*) A proposed mechanism of catalysis [reproduced with permission from Roberts *et al.*, *Proc. Nat. Acad. Sci.*, **62**, 1151 (1969).]

(*a*) Carboxymethylation

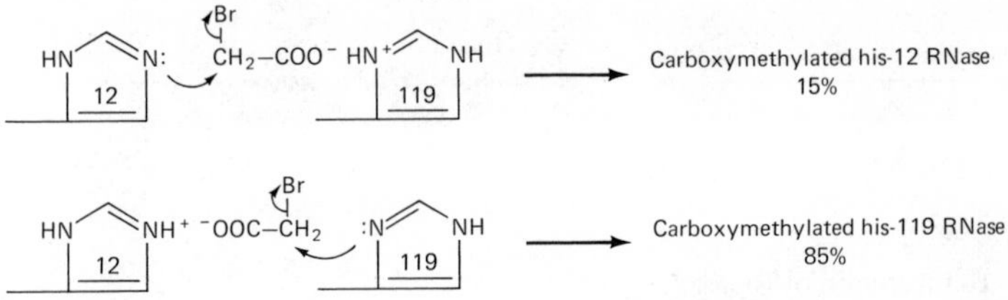

Carboxymethylated his-12 RNase 15%

Carboxymethylated his-119 RNase 85%

(*b*) Catalysis

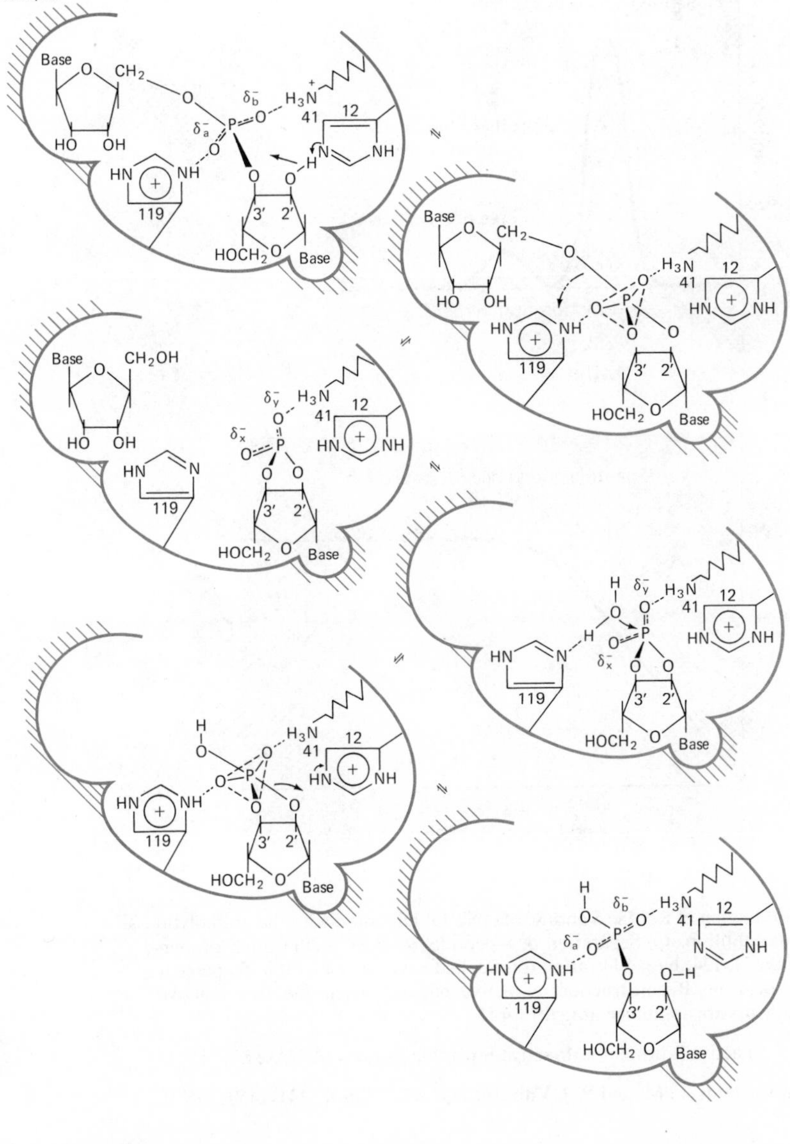

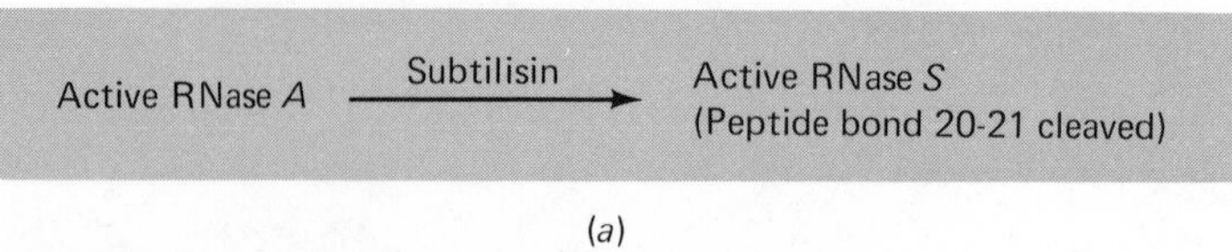

(a)

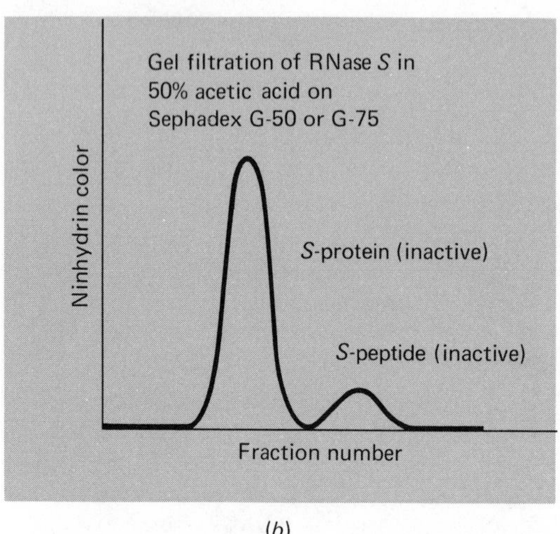

(b)

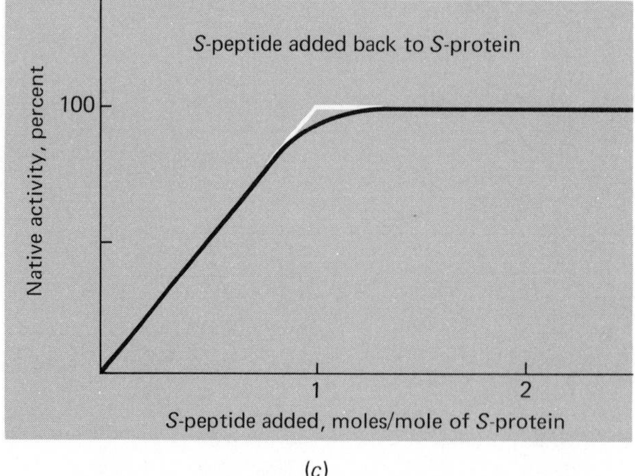

(c)

Figure 3.5 The production of RNAse S and RNAse S′. (*a*) Digestion with the proteolytic enzyme subtilisin. (*b*) Separation of S-peptide (residues 1–20) from S-protein (residues 21–124) by gel filtration (on Sephadex G-50 or G-75) in 50 percent acetic acid. (*c*) Reconstruction of active enzyme when the two inactive components are combined at neutral pH:

$$\text{Inactive S-protein} + \text{Inactive S-peptide} \rightarrow \text{Active RNAse S}'$$

[From Richards, F. M. and P. J. Vithayathil, *J. Biol. Chem.*, **234**, 1459 (1959)].

contribution to the peptide-protein interaction and to the activity. Table 3.2 shows the results of such experiments from a number of laboratories. Polypeptides corresponding to segments of *S*-peptide (and variations thereof) have also been synthesized; this synthesis represents the most exciting and precise manner in which a systematic survey of the role of each residue can be carried out (Table 3.3).*

TABLE 3.2 EFFECT OF CHEMICAL MODIFICATION OF DIFFERENT POSITIONS IN *S*-PEPTIDE[a]

S-peptide

lys-glu-thr-ala-ala(5)-ala-lys-phe-glu-arg(10)-gln-his-met-asp-ser(15)-ser-thr-ser-ala-ala(20)

Peptide Derivative	*Maximal Activity (percent) with 2′,3′-CMP as Substrate*	*Modified S-Peptide per S-Protein to Give Half of Maximal Activity*
S-Peptide, intact	100	0.5
1α-OH (deaminated)	100	0.5
1ε, 7ε-diguanidinated	100	0.5
1α, 1ε, 7ε-triguanidinated	100	0.5
1α-acetylated-1ε, 7ε-diguanidinated	100	0.5
1α, 1ε, 7ε-triacetylated	100	1.0
13-methionine sulfonated	100	20
13-S-carboxymethylated	100	75
13-S-carbaminomethylated	100	75

[a] From F. M. Richards, and P. J. Vithayathil, *Brookhaven Symp. in Biol.*, **13**, 115 (1960).

The conclusions from all the data on the interaction of *S*-protein with *S*-peptide and *S*-peptide derivatives are that his-12 is definitely required for biological activity and that certain residues in *S*-peptide are important for proper binding to *S*-protein. Residues 15 through 20 do not contribute to binding at all; however, asp-14 (aspartic acid), met-13 (methionine), and lys-1 (lysine) all have some role in binding and glu-2 (glutamic acid) appears to be very important. Until the binding partners in *S*-protein have been determined it is impossible to draw any more specific information from the data.

* It must be emphasized that all the data given above were obtained using activity as the criterion of binding. Clearly what is measured, therefore, is the *S*-protein–*S*-peptide affinity in the presence of a third binding component—namely the substrate—and some caution must be exercised in the interpretation of the data. Full activity is obtained with a 1:1 ratio of a modified *S*-protein to *S*-peptide at concentrations around $10^{-9}\,M$, and the dissociation constant must therefore be less than $10^{-9}M$. Whether at this concentration the probability of binding will be affected by the third component is questionable. Statistically the three-body collision under these conditions would be very unlikely. However the substrate concentration is high and if there is any tendency for *S*-protein or *S*-peptide to bind substrate, then the dissociation constants for the system *S*-protein–2′,3′-CMP and *S*-peptide on the one hand and the system *S*-protein and *S*-peptide on the other, for example, could be very different. That this may be so is indicated by a recent study of the direct binding of *S*-protein and *S*-peptide by spectrophotometric titration; this gives a dissociation constant of $10^{-5}M$ [B. M. Woodfin, and V. Massey, *J. Biol. Chem.* **243**, 889 (1968)].

TABLE 3.3 ACTIVITY OF SYNTHETIC SEQUENCES OF S-PEPTIDE

Peptide	*Activity Ratio*[a]
Residues 1–14	0.6
Residues 1–14 (13-methionine sulfoxide)	1.1
Residues 1–13	3.0
Residues 1–13 (13-methionine sulfoxide)	32
Residues 1–13 (13 α-amino butyric acid)	9
Residues 1–12	88
Residues 1–11	8000 (no activity)
Residues 2–13	41
Residues 3–13	2000

[a] Molar ratio of *S*-peptide derivative to *S*-protein required to give 50 percent activity. [From F. M. Finn, and K. Hofman, *J. Am. Chem. Soc.*, **87**, 645 (1965). A more extensive survey of all the replacements that have been tested can be found in E. A. Barnard, *Ann. Rev. Biochem.*, **38**, 716 (1969).]

Dimerization and Hybridization

An interesting phenomenon related to the *S*-peptide–S-protein binding was discovered a few years ago when it was found that lyophilization of RNase *A* from 50 percent acetic acid resulted in a 25 percent yield of a stable, enzymatically active dimer (as well as about a 5 percent yield of higher polymers).* Lyophilization from neutral buffers will not yield dimers, so it appears that the process in which dimer formation is favored can somehow be related to the process by which *S*-peptide could be separated from *S*-protein (in the presence of strong acid). The dimer can be isolated from the polymers and the unmodified monomer by gel filtration and studied separately. It has the same specific activity as the monomer—that is, there are two active sites per dimer (assuming one active site for the monomer)—and it is completely stable at room temperature in neutral buffers. Heating at 65 degrees C for 10 min will dissociate the dimer into active monomers and, as might be expected, strong acetic acid will also affect dissociation. The model proposed for the dimer is given in Figure 3.6 and is based on the fact that the *N*-terminal peptide is liberated from its strong association with the rest of the protein on exposure to 50 percent acetic acid. Two clean experimental tests of this model have been performed. If the model is correct, then hybrid dimers with partially restored activity should form from inactive RNase *A* carboxymethylated at his-119 and virtually

* A. M. Crestfield, W. H. Stein, and S. Moore, *Arch. Biochem. Biophys.*, Suppl. 1, 217 (1961).

Figure 3.6 Molecular Models of RNAse A, RNAse S, and RNAse S′ based on the strong interaction between S-peptide (containing his-12) and S-protein (containing his-119) at neutral pH, and the disruption of this interaction in strong acid. *x*, his-12; ●, his-119; *x-CM*, ●*-CM*, carboxymethylated histidines 12 and 119. (*a*) The RNAse S → RNAse S′ reaction. (*b*) Dimerization: formation of active dimers by lyophilization of acid-treated RNAse A. (*c*) Hybridization

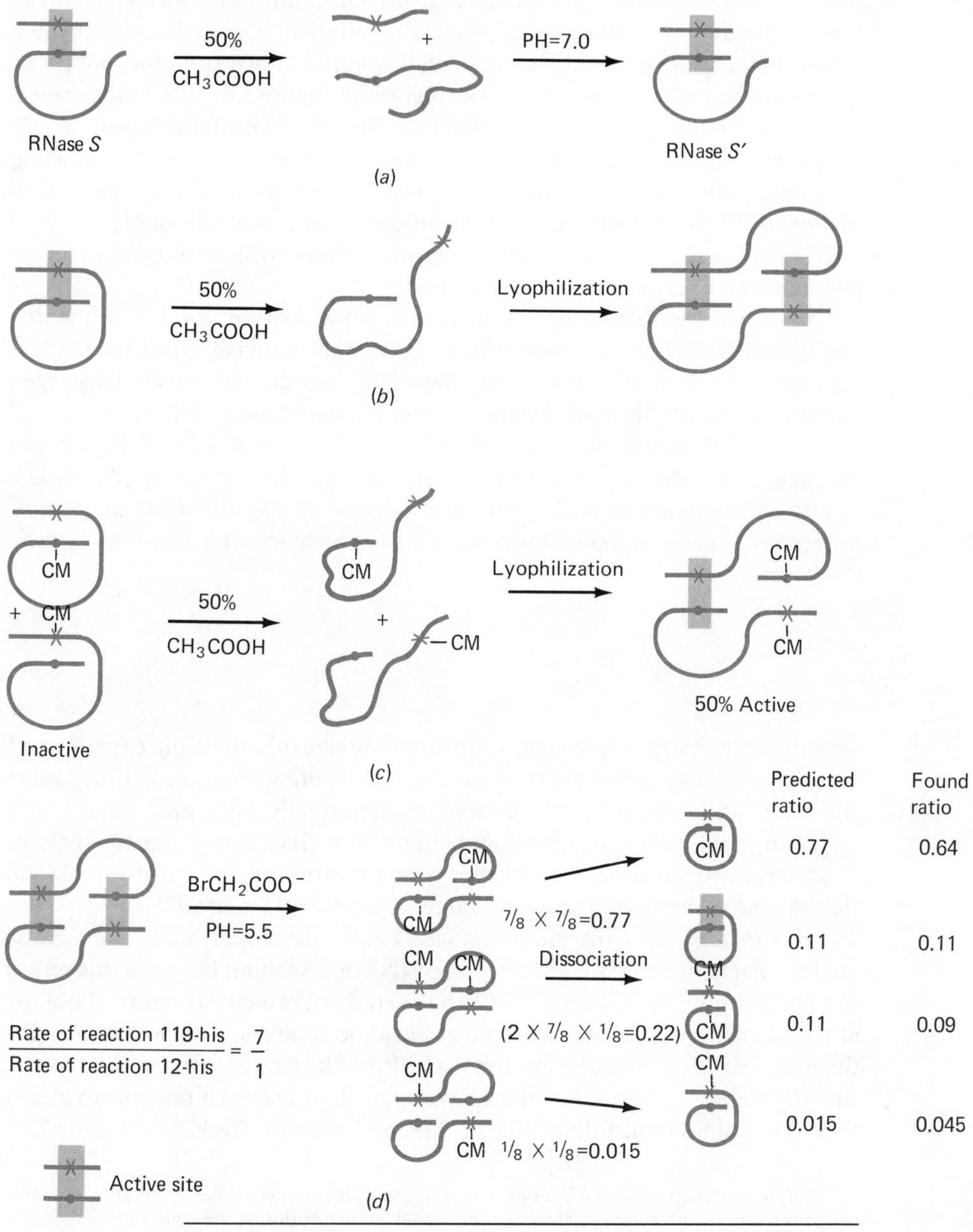

of inactive *CM*-his-12 RNAse and inactive *CM*-his-119 RNAse to give dimer with a predicted 50 percent of normal activity. The activity actually found was 45 percent. [From Crestfield, A. M., *et al.*, *J. Biol. Chem.*, **238**, 2421 (1963).] (*d*) Carboxymethylation of RNAse dimer and the demonstration of the formation of the his-12, his-119 dicarboxymethyl derivative which is never formed in the carboxymethylation of monomer. [From Fruchter, R. and A. M. Crestfield, *J. Biol. Chem.*, **240**, 3875 (1965).]

inactive RNase *A* carboxymethylated at his-12. Conversely, carboxymethylation of the dimer should give a predictable amount of dicarboxymethylated RNase upon dissociation, a derivative that is not obtained from the monomer. The rationale for this experiment is also given in Figure 3.6. When the experiment was carried out, about a 30 percent yield of a hybrid dimer with nearly 50 percent of native activity was obtained from the two inactive starting materials. Similarly, the amount of dicarboxymethylated monomer from carboxymethylated dimer is in excellent agreement with the model. It is of interest to note that the model represents an excellent artificial illustration of the phenomenon referred to as complementation.*

The above discussion has summarized some features of the structure–function model of ribonuclease, arrived at by what we have called the indirect approach. There are several other pieces of information which have been incorporated into this model, some of these are summarized below.

Abnormal tyrosines. Residues 25, 73, 76, 92, 97, and 115 of RNase are tyrosines. The phenolic side chain of tyrosine can be recognized by its UV absorbance (maximum at 276 nm), and because of the difference in spectral properties of the protonated form and the phenolate ion, it is possible to study the process

$$\text{prot}-C_6H_4-OH \rightleftharpoons \text{prot}-C_6H_4-O^- + H^+$$

directly by a spectrophotometric titration. When this titration is performed with RNase, three of the six tyrosine residues titrate with a normal pK_a value (10–10.2), while the other three have an abnormally high pK_a value (>11) indicating a unique environment and interaction. Based on extensive analyses of spectral properties and on chemical modification, it has been established that the three abnormal tyrosine residues are tyr-25, 92, and 97.†

Direct titration of active site histidines. Recent developments in the area of nuclear magnetic resonance spectroscopy (NMR—see Van Holde in this series) has made it possible to apply NMR to the study of protein structure. Looking at the characteristic proton spectra of histidine residues, it is possible to distinguish 4 different histidine protons; and from the effect of pH on the spectra, the pK_a values 6.7, 6.2, 5.8, and 6.4 are obtained. It has been possible to assign these pK_a values to histidines 105, 12, 119, and 48 respectively.‡

* Complementation has been proposed as a mechanism to explain the occurrence of active enzymes from inactive sources. Thus two mutants that have both lost the ability to synthesize a given multichain enzyme (but still synthesize proteins, which are immunochemically cross-reacting) may upon mixing produce an active enzyme. It can be demonstrated that in some cases the repair mechanism does not take place at the genetic level (see Chapter 7) but rather at the level of the finished gene product, the protein. The RNase case demonstrates that this type of repair is possible.

† C. Tanford, J. D. Hauenstein, and D. G. Rands, *J. Am. Chem. Soc.*, **17**, 6409 (1955). L. Li, J. P. Riehm, and H. A. Scheraga, *Biochem.*, **5**, 2043 (1966).

‡ D. H. Meadows, *et al.*, PNAS, **60**, 776 (1968).

Chemical modification with fluorodinitrobenzene (FDNB) at pH 8.5. RNase is rapidly inactivated by FDNB, and at pH 8.5 there is a distinct sequence of reaction preference for different lysine residues. Lys-41 reacts first, followed by lys-7 and lys-1. By degradation of inactivated DNP-RNase (dinitrophenyl-RNase) and isolation of DNP-peptides, it was shown that the inactivation by FDNB is due to the reaction with lys-41. The models of RNase must thus include the role of this residue in the active site of the enzyme.*

X-ray Diffraction Models of Ribonuclease

One of the most exciting events in enzymology and protein chemistry occurred recently when the complete three-dimensional structures as determined by x-ray crystallography of both ribonuclease A† and ribonuclease S'‡ were published, and it became possible to compare the conclusions from the indirect experiments to one real conformation. Not only was the ribonuclease model tested, but the investment of years of hard work and the validity of ingenious methodology and reasoning could be evaluated. Consequently, similar conclusions regarding enzymes for which x-ray diffraction data were unavailable would also be strengthened or weakened by the outcome of the comparison.

The three-dimensional structure of RNase S' is given in Figure 3.7, and a close examination of this structure reveals a most gratifying agreement with the model. Histidines 12 and 119 are found in the proper juxtaposition and lys-41 is superimposed over the "active site pocket" in such a manner that the inactivation by its dinitrophenylation is not difficult to rationalize. The three "buried" tyrosines (residues numbered 25, 92, and 97) are buried indeed, and the *N*-terminal peptide is found well lodged in the folds of *S*-protein.

In spite of the excellent agreement, the inevitable question is raised: How can we establish validity for the comparison between a crystalline protein immobilized in the single conformation (presumably) in the solid state and an active flexible enzyme in dilute aqueous solution? Two types of experiments were conducted in an attempt to answer this question.

One approach involves a reaction between the protein in solution and bifunctional reagents. These reagents form covalent cross-links in the three-dimensional structure; the interresidue distances, deduced from the locations of the bridges, can be compared with the distances in the crystal. The bridges formed with three types of reagents are given in Table 3.4. Note that when lys-41 reacts low activity results—but not complete inactivation, as in the case of inactivation by FDNB. There is no obvious explanation of this finding. All these cross-links are in agreement with interresidue distances derived from the crystal structure, supporting the supposition that the crystal structure is the same as the structure in solution.

* C. H. W. Hirs, *et al.*, *Arch. Biochem. Biophys.*, **111**, 209 (1965).

† G. Kartha, *et al.*, *Nature*, **213**, 862 (1967).

‡ H. W. Wyckoff, *et al.*, *J. Biol. Chem.*, **242**, 3984 (1967).

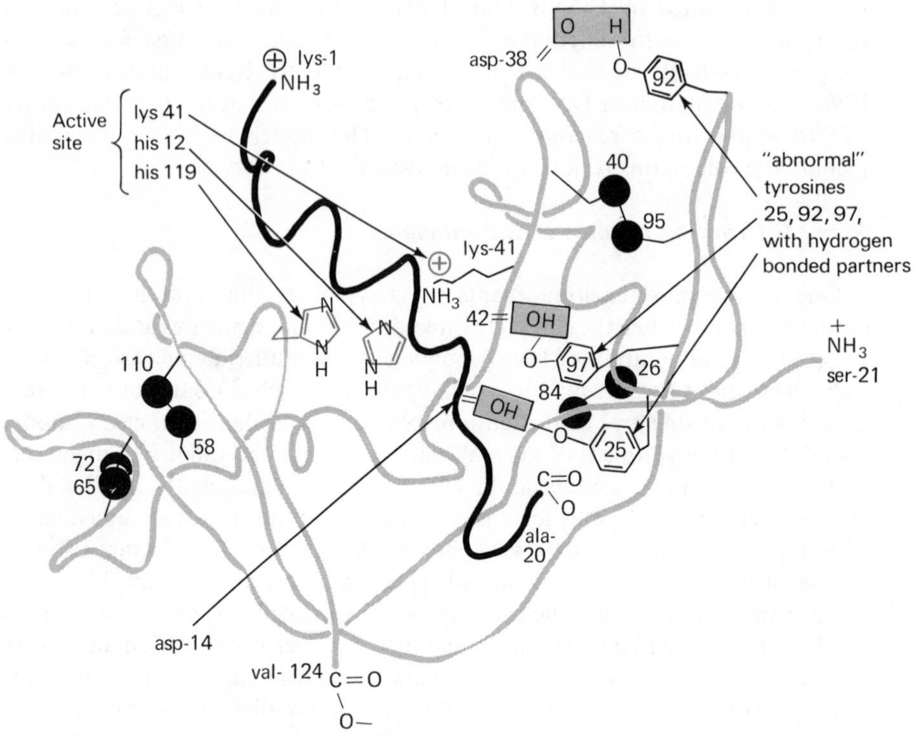

Figure 3.7 Tracing of the three dimensional structure of RNase S′. [Reproduced with permission from Spande, T. F. *et al.*, *Adv. Prot. Chem.*, **24**, 97 (1970).]

A much more direct and significant answer was obtained by allowing substrate to flow through the RNase crystal while it was mounted in an X-ray machine.* From the crystal dimensions, the number of molecules of enzyme in the crystal was estimated and the rate of substrate hydrolysis was determined and corrected empirically for substrate-diffusion time into the crystal and product-diffusion time out of the crystal. The results showed that, within the errors of the method, the enzyme was as active in the crystal as in solution. Thus it appears that we do have a complete three-dimensional structure of a biologically active enzyme and that we should be able to eventually formulate the complete mechanism for the catalytic work carried out by this structural unit.

* M. S. Doscher, and F. M. Richards, *J. Biol. Chem.*, **238**, 2399 (1963).

TABLE 3.4 ARTIFICIAL CROSSLINKS IN RNase[a]

Reagents Used	*Product*	*Activity of Derivative (percent)*	*Bridge*
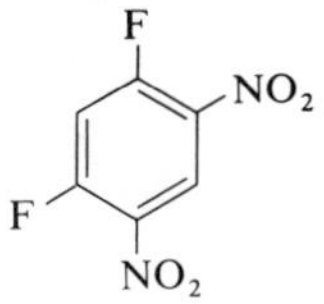 difluoro-dinitro benzene	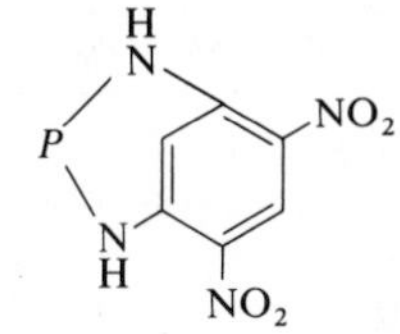	30%	lys-7–lys-41
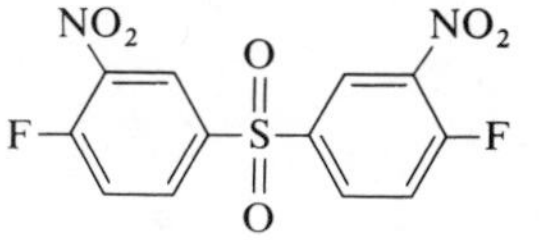 p,p′-difluoro-m,m′-dinitro-diphenylsulfone		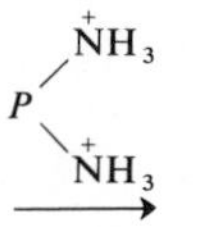 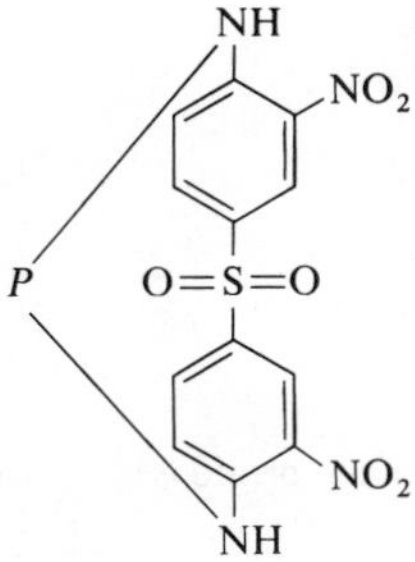 100%	lys-37–lys-98
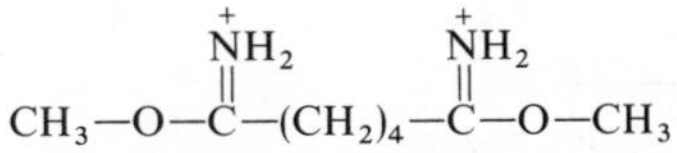 Dimethyladipimidate $CH_3-O-\overset{+}{C}(=NH_2)-(CH_2)_4-\overset{+}{C}(=NH_2)-O-CH_3$	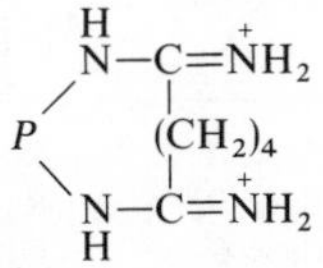	160%	lys-7–lys-37 lys-31–lys-37

[a] For a review of crosslinking and references to individual reagents see F. Wold, "Bifunctional Reagents," in C. H. W. Hirs, (ed.) *Meth. Enzymology*, **11**, 617 (1967).

3.2 CHYMOTRYPSIN (COVALENT E-S COMPLEX)

Chymotrypsin is discussed next to illustrate certain phenomena and experimental procedures: (1) the occurrence of an enzyme in an inactive form, a zymogen, which can be converted to the active form by the action of another enzyme; (2) the formation of a formal, covalent enzyme-substrate intermediate in the catalytic process; and (3) the definition of the active site and chemical mechanism of the catalytic function of an enzyme by kinetic analysis and chemical modification.

Chymotrypsin belongs to a large group of hydrolytic enzymes, known as proteolytic enzymes or proteases, which catalyze the breakdown of proteins. Some of the common proteolytic enzymes are listed in Table 3.5. Since these enzymes provide a gentle and in some cases specific method of affecting hydrolysis of peptide bonds, they are used extensively as hydrolytic agents in protein research; they also have uses in food processing, in the detergent industry, and in medicine. It should be emphasized that most of the proteolytic enzymes hydrolyze native-protein substrates very slowly, whereas denatured (even insoluble) proteins are attacked and degraded much more readily.

TABLE 3.5 COMMON PROTEOLYTIC ENZYMES

Endopeptidases: $\sim\overset{\displaystyle O}{\overset{\|}{C}}-\underset{H}{N}-\underset{H}{\overset{R_1}{C}}-\overset{\displaystyle O}{\overset{\|}{C}}\underset{\uparrow}{-}\underset{H}{N}-\underset{H}{\overset{R_2}{C}}-\overset{\displaystyle O}{\overset{\|}{C}}-\underset{H}{N}\sim$

endopeptidase

Source	*Enzyme*	*Specificity*
Mammalian	Trypsin	R_1 = arg or lys
Mammalian	Chymotrypsin	R_1 = phe or tyr (preferred)
Mammalian	Pepsin	R_2 = phe, tyr, or try
Mammalian	Prolidase	R_1 or R_2 = glycine (preferred)
Plant	Papain	Nonspecific[b]
Plant	Ficin	Nonspecific[b]
Bacterial	Subtilisin[a]	Nonspecific[b]
Bacterial	Nagarse[a]	Nonspecific[b]
Bacterial	Pronase[a]	Nonspecific[b]
Bacterial	Several other proteases	Nonspecific[b]
Exopeptidases: $H_3\overset{+}{N}-\underset{H}{\overset{R_1}{C}}-\overset{\displaystyle O}{\overset{\|}{C}}\overset{\downarrow}{-}\underset{H}{N}\sim\sim\sim\overset{\displaystyle O}{\overset{\|}{C}}\overset{\downarrow}{-}\underset{H}{N}-\underset{H}{\overset{R_2}{C}}-COO^-$ (amino peptidase; carboxypeptidase)		
Mammalian	Carboxypeptidase *A*	R_2 = Any amino acid except arg, lys, and pro
Mammalian	Carboxypeptidase *B*	R_2 = lys or arg
Plant	Carboxypeptidase *C*	Unknown
Mammalian	Iminopeptidase	Liberates proline
Mammalian	Leucine amino peptidase	R_1 = Any amino acid except pro and gly
Mammalian	Aminopeptidase *M*	Nonspecific[b]

[a] Commercial commonly used names.

[b] Most proteolytic enzymes are inactive toward the imide link of proline and attack glycine peptides very slowly.

Activation

Chymotrypsin is synthesized in the pancreas of all higher species as a zymogen, chymotrypsinogen. The zymogen is a single polypeptide chain having a molecular weight of 23,200. The activation to chymotrypsin involves a second pancreatic proteolytic enzyme, trypsin, in a process that is now well understood and is illustrated schematically in Figure 3.8. It is interesting from at least two points of view: (1) Chymotrypsin and chymotrypsinogen are themselves substrates for the active enzyme, but only to a very limited extent; and (2) cleavage of a single peptide bond converts an inactive protein into an active enzyme. The latter feature must be a major point in the discussion of structure-function relationships.

The critical activation step is the tryptic cleavage T_1 between arg-15 (arginine) and ile-16 (isoleucine) giving active π-chymotrypsin. Subsequent cleavages are the result of chymotrypsin attacks to give the final end product, α-chymotrypsin, which is resistant to further attack by either enzyme. If trypsin is left out the C_1, C_2, and C_3 cleavages catalyzed by chymotrypsin give *neochymotrypsinogens* with no catalytic activity; thus it is clear that the tryptic cleavage is responsible for the activation. In producing α-chymotrypsin from chymotrypsinogen then, two dipeptides are released and the single polypeptide chain, the inactive zymogen, is converted to the active structure containing three polypeptide chains held together by S—S bonds. For the better understanding of the mechanism of the activation, we would like to have precise answers to several questions: (1) Why are only four bonds cleaved? (2) How does the cleavage of one peptide bond lead to an active site—by a conformational change or simply by liberating a new functional group (the C terminal of Arg-15 or the N terminal of Ile-16)? Unfortunately, we must make do mostly with speculations instead of answers. We do know (Table 3.6) that trypsin cleaves peptide bonds involving the carboxyl group of the basic amino acids lysine and arginine; it is the most specific of the common proteolytic enzymes. Chymotrypsin is less specific: It shows considerable preference in terms of rates for peptide bonds involving the carboxyl groups of phenylalanine and tyrosine, but it also cleaves leucyl, isoleucyl, glutaminyl, asparaginyl, and valyl peptides at appreciable rates. Considering the fact that proteolytic enzymes do not readily hydrolyze native enzymes, the surprising element in the activation is perhaps not so much that only four bonds are broken, but that the four bonds are attacked at all. In terms of specificity the trypsin cleavage at the carboxyl side of arg-15 reflects the normal specificity of the enzyme as does the chymotrypsin cleavage at tyr-146. The C_1 and C_3 cleavages at leu-13 and asn-148 (asparagine) are, however, less typical. There are six phenylalanine and three other tyrosine residues that would have been much more susceptible to chymotrypsin action if the protein had been denatured. It is reasonable to conclude that the residue sequences 13 through 16 and 146 through 149 are in some way exposed on the surface on the zymogen and therefore are particularly susceptible to proteolytic attack. A similarly high specificity has already been discussed in the cleavage of a single bond in ribonuclease by subtilisin.

We can now consider the second question, regarding how an active site results from cleavage of the arg-15–ile-16 bond. It has been clearly demonstrated that the activation process is accompanied by a change both in the absorption spectrum and in the optical rotation of the enzyme; therefore the activation does cause a change in conformation. In addition it appears that the new α-amino group of the isoleucine may play an essential role in the active site of chymotrypsin; thus the overall activation process may be visualized as involving both the liberation of an active-site component (α-amino group) and the release of covalent restraint to allow an active conformation to form.

(a)

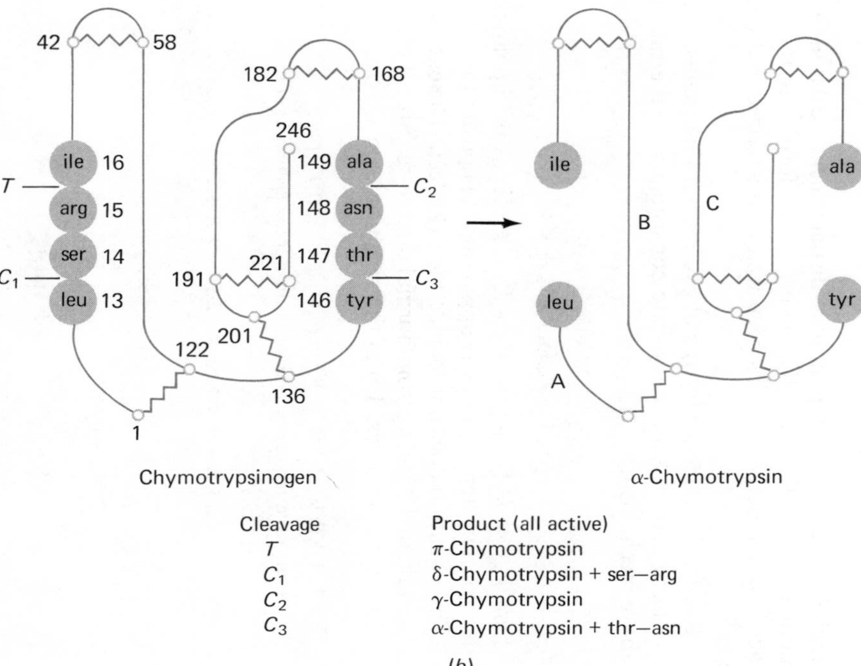

Cleavage	Product (all active)
T	π-Chymotrypsin
C_1	δ-Chymotrypsin + ser—arg
C_2	γ-Chymotrypsin
C_3	α-Chymotrypsin + thr—asn

(*b*)

Figure 3.8 (*a*) The covalent structure of bovine chymotrypsinogen A. (*b*) The activation to different chymotrypsins. The polypeptide chain in (*a*) also indicates the activation steps and the three polypeptide chains (A, B, C) held together by disulfide bonds, which constitute the active α-chymotrypsin after removal of the two dipeptides (in boxes) in the activation process. (Taken from a review by Hartley, B. S., in Goodwin, T. W., J. I. Harris and B. S. Hartley (Eds.), *Structure and Activity of Enzymes*, New York: Academic Press, 1964.)

Active-Site Studies

No enzyme has been probed more extensively or more penetratingly than has chymotrypsin in the many attempts that have been made to elucidate its mechanism of action. As with other proteolytic enzymes, chymotrypsin catalyzes the hydrolysis of both amide and ester bonds. The general reaction for either ester or amide hydrolysis can be written

$$
\begin{array}{ccccccc}
 & & E\text{:::}\,\mathrm{RCOX} & & \mathrm{O} + X & & \\
 & & \updownarrow K_b^1 & & \Vert & & \\
EH + \mathrm{RCOX}_1 & \underset{k_{-1}}{\overset{k_1}{\rightleftharpoons}} & EH\text{:::}\,\mathrm{RCOX} & \underset{k_{-2}}{\overset{k_2}{\rightleftharpoons}} & EHO\mathrm{C{-}R} & \underset{k_{-3}}{\overset{k_3}{\rightleftharpoons}} & EH + \mathrm{R{-}C({=}O){-}O^-} \\
 & & \updownarrow K_a^1 & & \updownarrow K_a^2 \;\; \mathrm{O} & & \\
 & & & & \Vert & & \\
 & & EH_2\text{:::}\,\mathrm{RCOX} & & EH_2\mathrm{OC{-}R} & &
\end{array}
$$

representing a two-step double-displacement reaction where $X = \mathrm{NH{-}R}$ (amide) or O—R (ester). The intermediate formed in the k_2 step, the covalent acyl-enzyme derivative, has been well characterized and must be considered a reality.

Briefly, the evidence for the above reaction mechanism is outlined below.

*Kinetic tests.** Kinetic tests are consistent with the rate law derived for the above mechanism:

$$v_0 = \frac{V_{\max} k_3'/(k_2' + k_3')}{1 + k_m/(S_0)k_3'(k_2' + k_3')}$$

where $V_{\max} = k_2 E_o$, $K_m = (k_2' + k_{-1})/k_1$, k_2' is the apparent rate constant for the acylation step, and k_3' the apparent rate constant for the deacylation step.†

This equation does not include the pH effect. The easiest way to handle that is to evaluate k_2' and k_3' from normal double-reciprocal plots of $1/v_0$ versus $1/S_0$ at different pH values. From the resulting data we can then evaluate the pH-

* See F. J. Kézdy, and M. L. Bender, *Biochem.*, **1**, 1097 (1962).

† This rate equation can be derived in the normal fashion from

$$E + S \underset{k_{-1}}{\overset{k_1}{\rightleftharpoons}} ES \underset{k_{-2}}{\overset{k_2}{\rightleftharpoons}} \underset{P_1}{\overset{}{ES' +}} \overset{k_3}{\longrightarrow} E + P'$$

when

$$v_0 = k_2'[ES]$$

$$E = \frac{K_m[ES]}{S_0} \text{ (steady state)}$$

and

$$E_0 = [E] + [ES] + [ES']$$

where

$$[ES'] = \frac{k_2'}{k_3'}[ES](k_{-3} \sim 0)$$

independent constants k_2 and k_3 as functions of k'_2 and k'_3 as well as pK_a^1, pK_b^1, and pK_a^2:

$$k_2 = k'_2\left(1 + \frac{[H^+]}{K_a^1} + \frac{K_b^1}{[H^+]}\right)$$

and

$$k_3 = k'_3\left(1 + \frac{[H^+]}{K_a^2}\right)$$

The mechanism predicts a bell-shaped pH-dependence curve for the acylation step and a sigmoid pH-dependence curve for the deacylation step (Figure 3.9) and this has been confirmed by the experimental tests (Figure 3.9). This type of analysis with several different substrates has given the following values for the three acid-dissociation constants: $pK_a^1 = 6.6$ to 6.7, $pK_a^2 = 7.0$ to 7.3, and $pK_b^1 = 8.4$ to 8.9; it has also shown that $k_2/k_3 = 200$.

Because of their dependence on the dielectric constant and also in part because of their heat of ionization, the most acidic ionizations (pK_a^1 and pK_a^2) have been proposed to result from the deprotonation of an imidazolyl group in

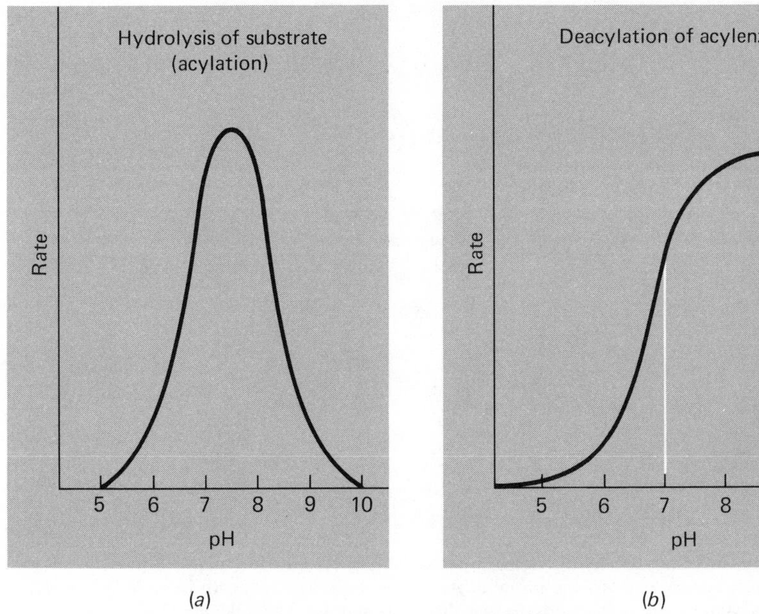

Figure 3.9 The pH-dependence of chymotrypsin catalysis. (*a*) Rate of substrate hydrolysis: the rate-limiting step at high pH is the acylation step. (*b*) The rate of deacylation of acyl-enzyme which can be produced and kept at low pH. (The deacylation step is rate limiting at low pH.)

the protein. The more basic group (pK_b^1) also behaves like a cation-neutral ($BH^+ \rightleftharpoons B + H^+$) ionization in different solvent systems. Moreover there is good correlation between the descending limb of the acylation-pH curve and a plot of optical-rotation change versus pH, indicating that the two processes (activity loss and conformation change) are parallel and involve the same residue. It has been suggested that the α-amino group of the N-terminal isoleucine residue is the critical one. We shall return to this question below. Tentatively then, we propose that the active enzyme requires a deprotonated specific histidine residue for both acylation and deacylation as well as the protonated α-amino group of isoleucine-16 for proper conformation.

The acyl intermediate. The above mechanism suggests that since $k_2 > k_3$ and since the k_2 step also has a lower pH activation than does the k_3 step, it should be possible to stabilize the proposed acyl enzyme intermediate at low pH. This prediction has been tested and found to be correct,* it thus presents one of the key features in the evidence for an acyl enzyme intermediate. Figure 3.10 shows the hydrolysis-rate curve when *p*-nitrophenyl acetate (NPA) was used as substrate for chymotrypsin. One mole of NPA per mole of enzyme was hydrolyzed rapidly, followed by a slower rate of steady-state hydrolysis.

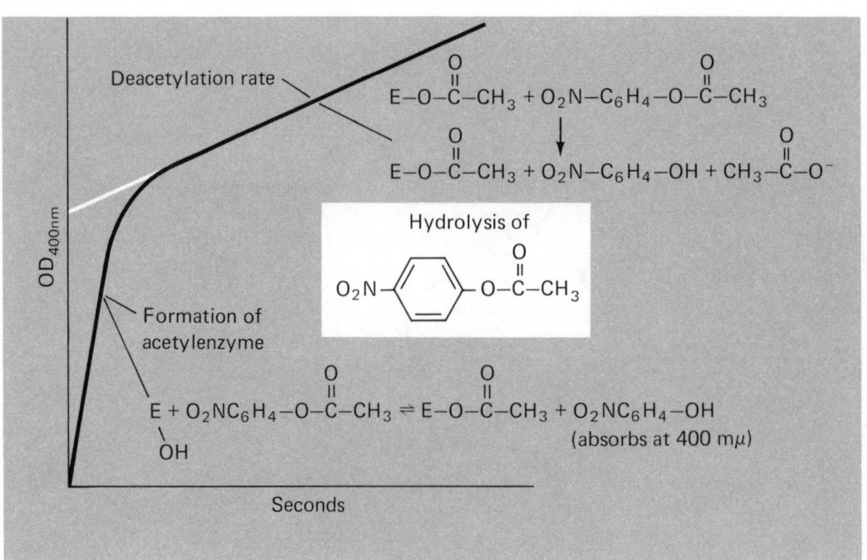

Figure 3.10 The Chymotrypsin catalyzed hydrolysis of p-nitrophenylacetate at low pH. The initial *burst* corresponds to the initial acylation of the enzyme, and the slower steady state rate corresponds to the deacylation rate. The OD at 400 nm is due to the formation of free p-nitrophenol. [From Spencer, T. and J. M. Sturtevant, *J. Am. Chem. Soc.*, **81**, 1874 (1959).]

* G. R. Schonbaum, *et al.*, *J. Biol. Chem.*, **236**, 1930 (1961).

Acidification of the steady-state reaction mixture gave a good yield of a stable monoacetyl chymotrypsin derivative. At neutral pH the acetyl group was rapidly lost, presumably according to the normal deacylation step. At lower pH hydroxylamine released acetohydroxamic acid in a manner characteristic of acetyl esters. Another series of useful substrates in establishing the acyl enzyme intermediate is a group of *trans*-cinnamate esters

$$C_6H_5CH{=}CH{-}\overset{\overset{\displaystyle O}{\|}}{C}{-}OR$$

With these substrates both the acylation and deacylation of the enzyme can be observed directly by spectrophotometry since the cinnamoyl group absorbs strongly at a wavelength of 310 nm.

Chemical Modification

The kinetic evidence, then, has implicated a histidine residue and the N-terminal isoleucine as important for the catalytic ability of the enzyme; in addition there is an indication that the acyl enzyme derivative involves an ester bond. In order to establish the specific residues involved, extensive chemical-modification experiments have been carried out on chymotrypsin. Some of these will be considered below.

The reactive serine. One of the classical examples of the ideal chemical modification of enzymes is the inactivation of a number of ester-hydrolyzing enzymes through the action of diisopropylfluorophosphate (DFP). The loss of enzyme activity is a linear function of amount of DFP reacted and the end point of zero activity corresponds to the incorporation of exactly 1 mole of the diisopropylphosphoryl (DIP) group per mole of enzyme. Chymotrypsin is one of the enzymes that reacts in this way. Using radioactive DFP as reagent the reaction can be written as

$$E{-}OH + \underset{\text{DFP}}{F{-}{}^{32}P({=}O)(OCH(CH_3)_2)_2} \longrightarrow \underset{\text{DIP-enzyme}}{E{-}O{-}{}^{32}P({=}O)(OCH(CH_3)_2)_2}$$

The inactive product is stable over a broad pH range but can be reactivated by treatment with strong nucleophiles (hydroxylamine or hydroxamic acids, for example), which displace the dialkylphosphoryl group. In order to determine the site of reaction in the enzyme, the inactive DIP-enzyme was degraded by enzymatic cleavage and the P-labeled peptide was isolated. Complete sequence analysis revealed that the phosphorylation had occurred at a serine residue, specifically ser-195 (see Figure 3.8).* All the other esterases inhibited

* R. A. Oosterbaan, *et al.*, *Biochim. Biophys. Acta*, **27**, 549, 546 (1958), and *Biochim. Biophys. Acta*, **63**, 204 (1962).

by DFP (other dialkylphosphoryl reagents work as well) are also derivatized at a single specific serine residue, indicating the presence of a uniquely reactive and catalytically important serine residue in all these enzymes.

Since the DIP-chymotrypsin is inactive it is reasonable to propose that the reactive serine is also the site of acylation in the catalytic process. This hypothesis has been tested by submitting the acetylchymotrypsin formed in the low-pH reaction with substrate to the same degradation as above. The isolated acetylpeptide is the same as the DIP-peptide and again ser-195 is the acetylated group. This then is a case of chemical modification that can be interpreted completely unequivocally and the specific role of ser-195 is thus explained.

The reactive histidine and methionine. Only the chemical modification by photooxidation will be discussed. Several light-activated oxidation systems have been used in chemical-modification experiments. The system used with chymotrypsin involved methylene blue (MB); it can be illustrated as

$$\mathrm{MB} + X\mathrm{H_2} \rightleftharpoons \mathrm{MBH_2} + X$$

$$\mathrm{MBH_2} + \tfrac{1}{2}\mathrm{O_2} \xrightarrow{\text{light}} \mathrm{MB} + \mathrm{H_2O}$$

In a protein X can be tryptophan, histidine, methionine, or cysteine, as the most readily oxidized residues.

When chymotrypsin was subjected to photooxidation, activity was rapidly lost but, since both histidine and methionine were destroyed in the process, it was not possible to make a simple direct interpretation of the data and special methods had to be used (Figure 3.11). First it was necessary to distinguish between different types of inactivation. The rate assay could, for example, show an 80-percent loss of activity, but this could be due to either complete inactivation of 80-percent of the total enzyme, or an 80-percent decrease in the catalytic ability of all the enzyme molecules (the active site could still be intact). To distinguish between these two possibilities an all-or-none assay was designed taking advantage of the data in Figure 3.10. According to this assay, if the acylating site is still intact then, at the proper pH, the site can be titrated with the cinnamoyl substrate; thus only the end point of the titration, and not rate, is important. Hence in the example above the first possibility would show an 80-percent activity loss both by the rate assay and by the all-or-none assay, while the second possibility would show an 80-percent loss in initial-rate assay and no loss in the all-or-none assay.

Figure 3.11 Photooxidation of α-Chymotrypsin. (*a*) The rate of oxidation of histidine and methionine. In both cases the observed rate of amino acid disappearance (black curve) reflects the oxidation of two residues, one fast and one slow. Analysis of the rate curves allowed the determination of the individual rate constants. The rate constant for the loss of catalytic activity toward the synthetic substrate N-acetyltyrosine ethyl ester (ATEE) was found to be 0.81 min^{-1} [see (*b*)] in the same experiment. The loss of activity therefore seems to be the result of the oxidation of both one fast histidine ($k = 0.49$ min^{-1}) and one fast methionine ($k = 0.32\ \mathrm{min}^{-1}$). (*b*) Correlation of rate of

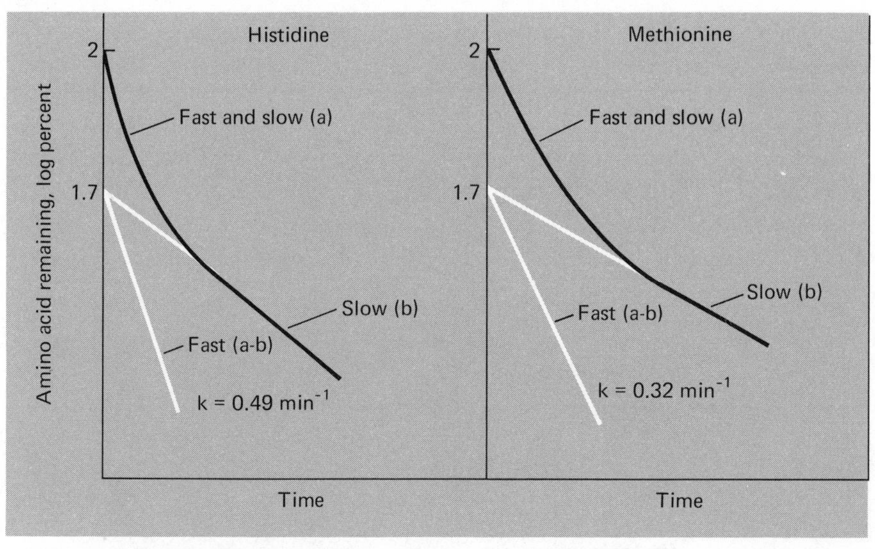

(*a*)

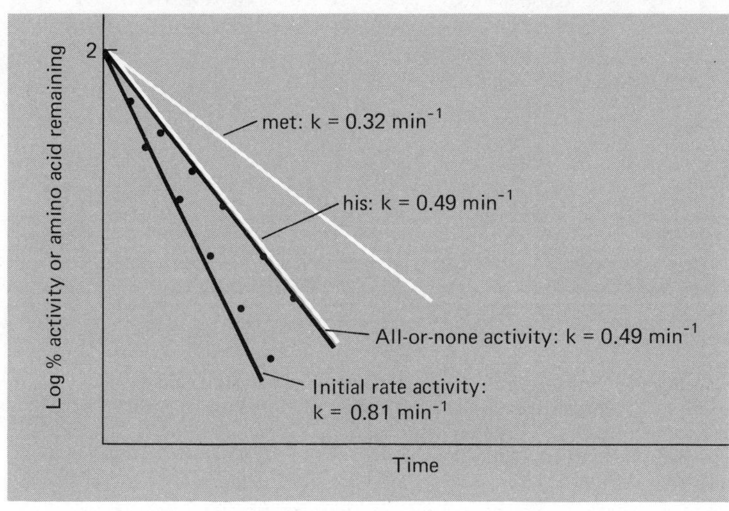

(*b*)

amino acid destruction (colored lines) with loss of initial rate activity and all-or-none activity. Since the all-or-none assay does not distinguish between partially active and fully active derivatives of the enzyme, the conclusion of this analysis must be that oxidation of histidine leads to a completely inactive enzyme (rates of his disappearance and loss of all-or-none activity are identical) while the oxidation of methionine yields a partially inactivated enzyme which is detected as a loss in initial rate, but is not detected in the all-or-none assay. [Taken from Koshland, D. E., *et al.*, *Brookhaven Symp. in Biol.*, **13**, 135 (1960), and **15**, 101 (1962).]

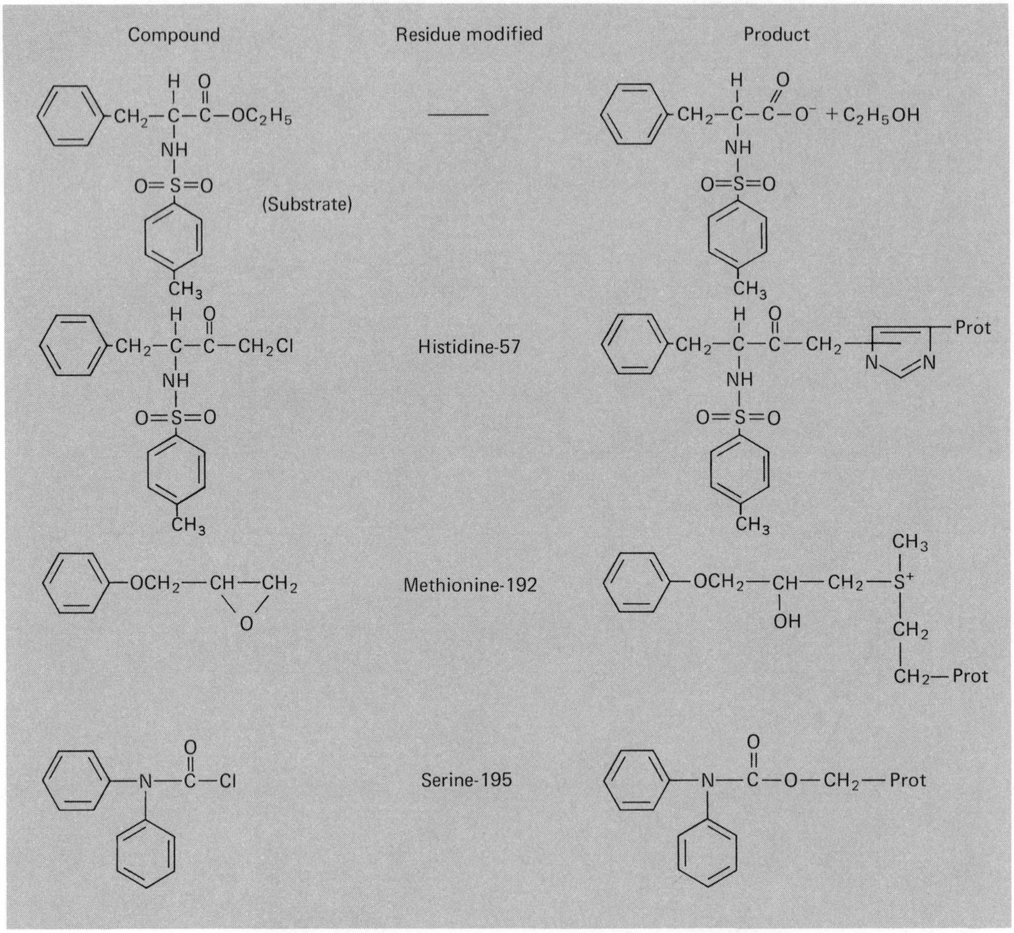

Compound	Residue modified	Product
$C_6H_5CH_2-CH(NH-SO_2-C_6H_4-CH_3)-C(=O)-OC_2H_5$ (Substrate)	—	$C_6H_5CH_2-CH(NH-SO_2-C_6H_4-CH_3)-C(=O)-O^- + C_2H_5OH$
$C_6H_5CH_2-CH(NH-SO_2-C_6H_4-CH_3)-C(=O)-CH_2Cl$	Histidine-57	$C_6H_5CH_2-CH(NH-SO_2-C_6H_4-CH_3)-C(=O)-CH_2-$(imidazole)$-$Prot
$C_6H_5-OCH_2-CH-CH_2$ (epoxide, O bridging CH and CH_2)	Methionine-192	$C_6H_5-OCH_2-CH(OH)-CH_2-S^+(CH_3)-CH_2-CH_2-$Prot
$(C_6H_5)_2N-C(=O)-Cl$	Serine-195	$(C_6H_5)_2N-C(=O)-O-CH_2-$Prot

Figure 3.12 Affinity labeling of α-chymotrypsin. Different compounds (substrate analogs) which inactivate the enzyme specifically in the "active site" are shown together with the residue specificity. [From Hartley, B. S., in T. W. Goodwin, J. I. Harris and B. S. Hartley (Eds.) *Structure and Activity of Enzymes*, New York: Academic Press, 1964, p. 57.]

Using both assays it was possible to analyze the inactivation data and get the results given in Figure 3.11, showing that the rate of destruction of one histidine residue corresponds exactly to the rate of activity loss by the all-or-none assay while the loss of catalytic ability in terms of the initial-rate assay corresponds to the sum of the rates of oxidation of two methionine residues, one fast and one slower. The residues have all been identified in the primary sequence as his-57, met-192 (fast), and met-204 (slow). (These are the only two methionine residues in chymotrypsin.) It thus appears that there is an absolute requirement

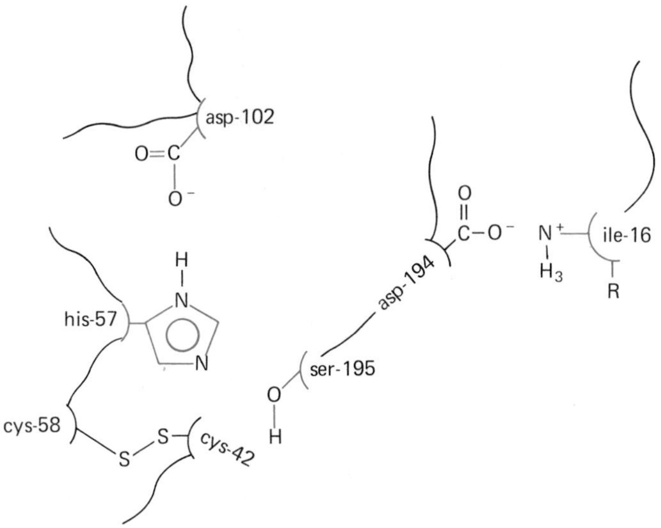

(a)

Figure 3.13 (*a*) A schematic diagram illustrating the approximate conformation of the active site components of α-chymotrypsin. Serine-195, histidine-57, isoleucine-16, which were discussed in the text, are emphasized, as is aspartate-102, which has been implicated from X-ray diffraction data as part of the catalytic site. The role of isoleucine-16 has been proposed to be that of forming an ion pair with aspartate-194, thereby pulling serine-195 into the proper "active" position. (The α-amino group of isoleucine-16 is liberated in the activation step.)

for his-57 in the active site of chymotrypsin. The modification of the methionine residues causes an increase in K_m (and therefore a decrease in initial rate) without affecting the total activity (in terms of available sites) of the enzyme.

The N-terminal isoleucine. In addition to the indirect evidence for the involvement of the N-terminal isoleucine (p. 78) there is only one chemical-modification experiment that implicates this group in the active form of the enzyme. If chymotrypsinogen is exhaustively acetylated so that all the amino groups are blocked, trypsin treatment will still give an active chymotrypsin derivative (having about 80-percent of the activity of normal chymotrypsin). The active derivative has only one free amino group (apparently the activation does not proceed beyond π-chymotrypsin), which must be the α-amino group of isoleucine. The derivative reacts normally with DFP and catalyzes the hydrolysis of N-acetyltyrosine ethyl ester (a common synthetic substrate). If this single remaining amino group is now acetylated, all activity is lost, demonstrating the active enzyme's requirement for the α-amino group of ile-16.

Affinity labeling. From the above experiments we can start to visualize an active site involving ser-195 (and perhaps met-192 and met-204), his-57, and ile-16. This implies that these residues are brought into proper juxtaposition

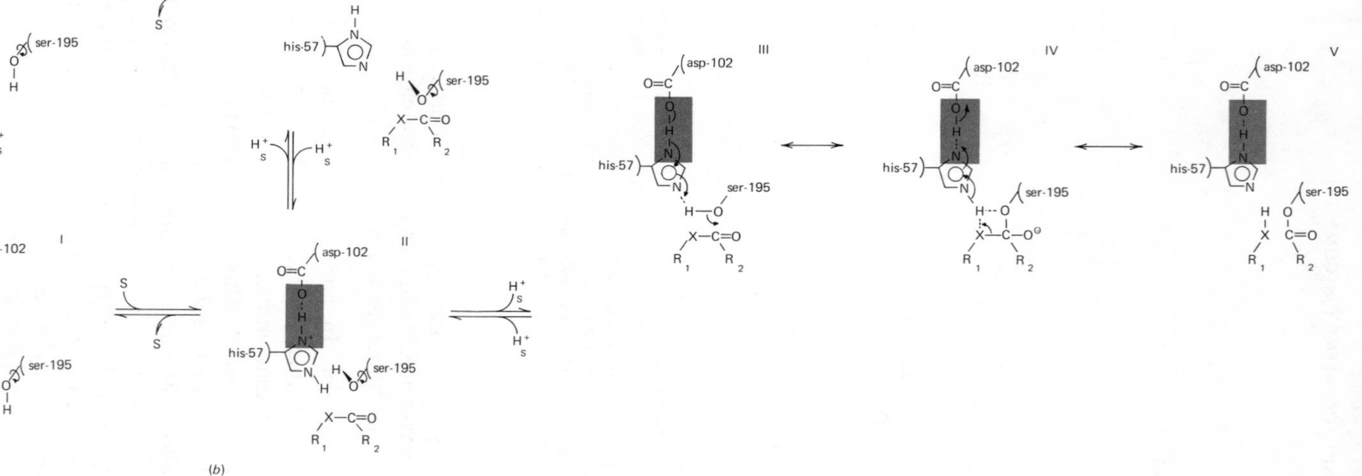

(b)

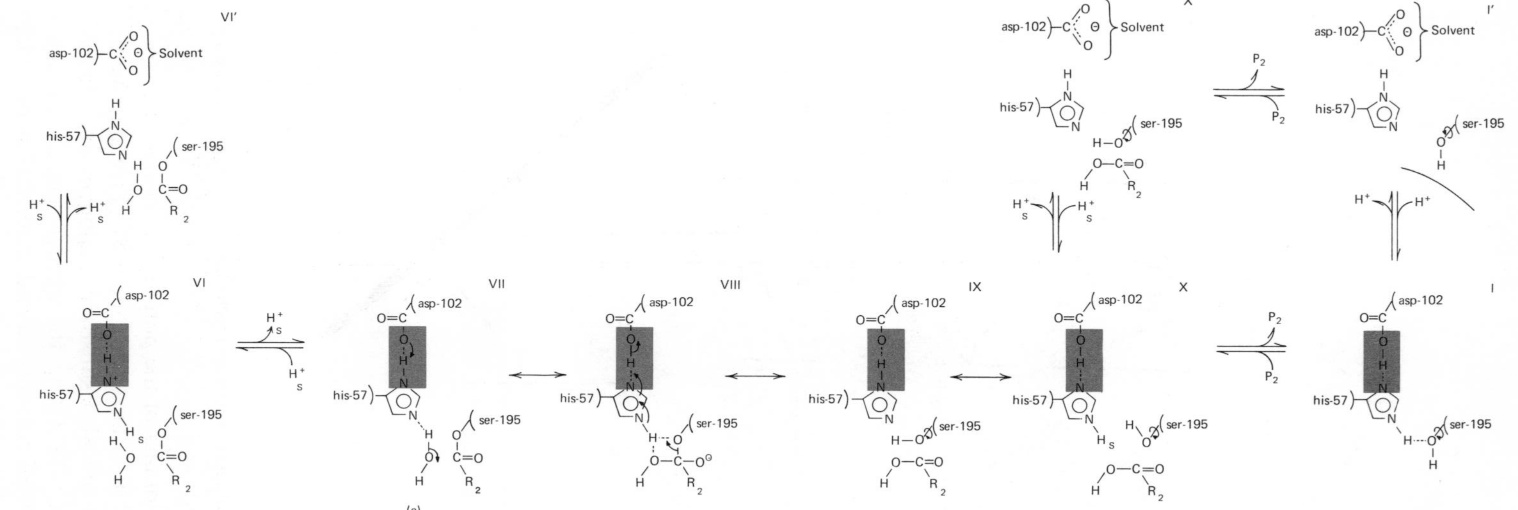

Figure 3.13 (*b*)–(*c*) A representation of a possible mechanism for the α-chymotrypsin catalyzed hydrolysis of an ester or an amide (R_1—X—COR_2, where X is either —O— or —NH—). The horizontal reactions represent the sequence of steps involved in the catalysis; the vertical reactions illustrate several highly probable deprotonated (and inactive) forms of the enzyme (indicated with primed numbers). Based on X-ray crystallography data, it is clear that one side of histidine-57 is "buried" in an hydrophobic region, and this feature is emphasized by the shaded boxes in the figure. (*b*) illustrates the proposed acylation steps, and (*c*) the proposed deacylation steps. V and VI are equivalent, except that the R_1—X—H product in V has been replaced by water in VI. The V′ deprotonated form has also been omitted. This mechanism was contributed by Dr. Leigh Wyborny, whose work is gratefully acknowledged. The structures used in this figure come from the X-ray crystallography data given in Figure 3.14, as well as more recent work: Blow, D. M., J. J. Birktoft, and B. S. Hartley, *Nature*, **221**, 337 (1969); and Freer, S. T., J. Kraut, J. D. Robertus, H. T. Wright, and Ng. H. Xuong, *Biochem.*, **9**, 1997 (1970).

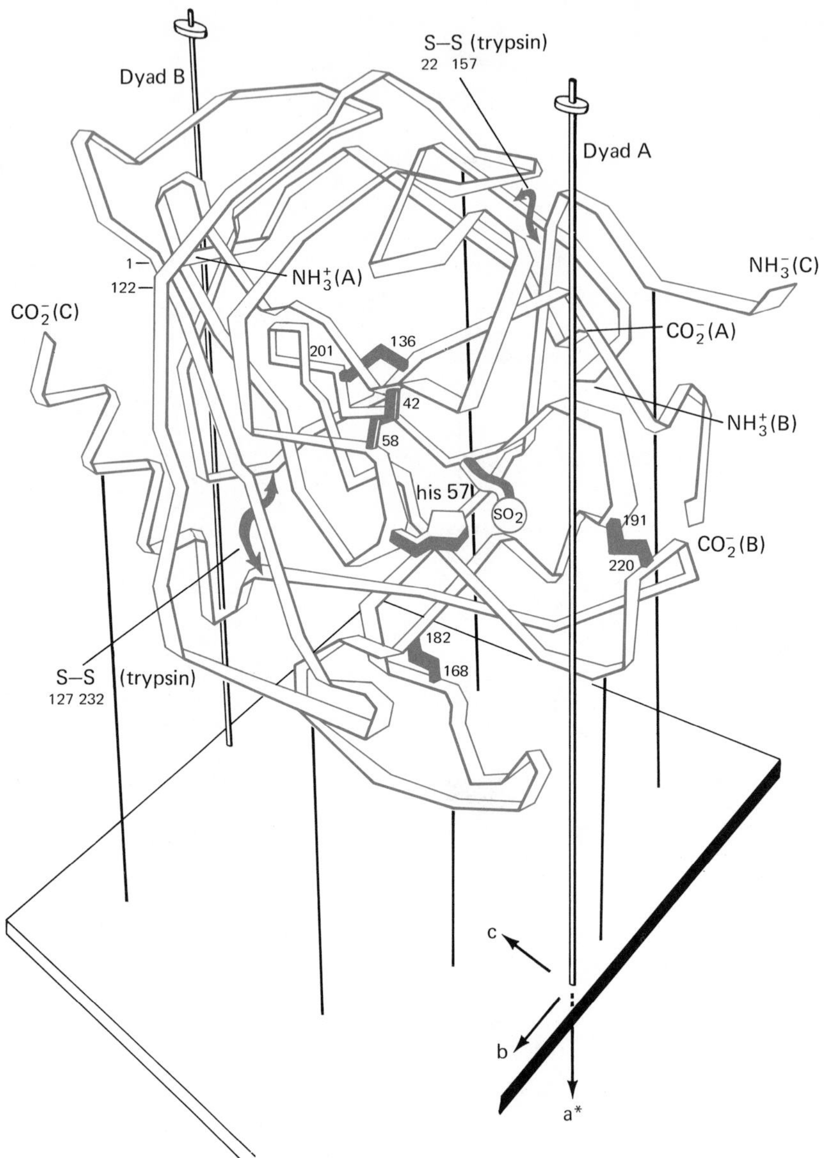

Figure 3.14 The X-ray crystallography model of α-chymotrypsin. [Reprinted with permission from Sigler, P. B., *et al.*, *J. Mol. Biol.*, **35**, 143 (1968).]

through the folding of the polypeptide chains. Unexpected evidence for such a spatial relationship was obtained through chemical modification with reagents that are also substrate analogues and therefore have high affinity for the active site. Using the phenyl ring as the active-site affinity group, different types of reagents caused inactivation by reacting at three different residues, namely his-57, met-192, and ser-195 (Figure 3.12). Assuming that the phenol group of all compounds really seeks out the same site in the enzyme, these three amino acid residues are clearly all within the reagent's reach from its position in the active site. This type of reagent, with affinity for the active site of a specific enzyme, offers many promising possibilities for chemical modifications with clear-cut interpretations.

The Active Site and the Mechanism

The model in Figure 3.13 has been proposed to explain all the data discussed above. More important, it is derived from known enzyme structure based on X-ray crystallography (Figure 3.14); hence the fact that all aspects of the active-site model are consistent with the structural picture is gratifying indeed.

TABLE 3.6 THE SERINE PEPTIDES AND HISTIDINE PEPTIDES OF DIFFERENT PROTEOLYTIC ENZYMES

Enzyme	*Serine Peptides*		*Histidine Peptides*
Chymotrypsin (Bovine)	GVSSCMGDSGGPLVCK	(1)	VTAAHCGVTTSD
Trypsine (Bovine)	NSCQGDSGGPVVCSGK	(2)	VSAAHCOKSGIQ
Elastase (Porcine)	SGCQGDSGGPVVCGQQL	(3)	TAAHCVDRE

Key to the one-letter code:

If only one amino acid begins with a certain letter, that letter is used.

C = cys = cysteine
H = his = histidine
I = ile = isoleucine
M = met = methionine
S = ser = serine
V = val = valine

If more than one amino acid begins with a certain letter, the letter is assigned to the most common amino acid.

A = ala = alanine
G = gly = glycine
L = leu = leucine
P = pro = proline
T = thr = threonine

For other amino acids letters close to the initial are used.

D = asp = aspartic acid
N = asn = asparagine
B = asx = either of above two
E = glu = glutamic acid
Q = gln = glutamine
Z = glx = either of above two
K = lys = lysine

Finally, the last group is assigned phonetically or at random.

F = phe = phenylalanine
R = arg = arginine
O = tyr = tyrosine
W = try = tryptophan

One interesting feature of the established histidine and serine involvement is related to comparative aspects of esterases and peptidases. It has long been known that DFP inhibits a number of enzymes by phosphorylating a serine residue. What is remarkable is that the serine residue appears to be located in very similar peptide sequences in the different enzymes. Moreover, if a histidine residue is responsible for the high reactivity of the serine residue, then one should expect that a similar histidine peptide should also be found in different enzymes. As shown in Table 3.6, this is indeed the case. Thus it may well be that members of a whole family of hydrolytic enzymes have identical catalytic mechanisms and differ only in the parts that determine their substrate specificity.

3.3 LACTATE DEHYDROGENASE (ISOZYMES)

Lactate Dehydrogenase (LDH) is discussed to illustrate some aspects of two-substrate kinetics, multisite enzymes, and a typical case of an enzyme that requires a free sulfhydryl group in its active site. It has also been included here to introduce the phenomenon of *isozymes*, multiple forms of an enzyme produced by a single cell.

General Description

Dehydrogenases catalyze the general reaction

$$SH_2 + X \overset{E}{\rightleftharpoons} S + XH_2$$

where the oxidizing agent, X, can be either a flavin coenzyme, a nicotinamide coenzyme, or in fact Cu + O_2, in the case of cuproprotein dehydrogenases (Table 3.7). Lactate dehydrogenase is one of a large number of enzymes in the group of dehydrogenases that are linked to the coenzyme nicotinamide adenine dinucleotide (NAD). It specifically catalyzes the reversible reaction

$$\underset{\text{L}(+)\text{-Lactate}}{\text{HO—}\overset{\overset{\displaystyle COO^-}{|}}{\underset{\underset{\displaystyle CH_3}{|}}{\text{C}}}\text{—H}} + NAD^+ \rightleftharpoons \underset{\text{Pyruvate}}{\overset{\overset{\displaystyle COO^-}{|}}{\underset{\underset{\displaystyle CH_3}{|}}{\text{C}}}\text{=O}} + NADH + H^+$$

TABLE 3.7 A SURVEY OF DEHYDROGENASES (OXIDOREDUCTASES)

1. Nicotinamide[a]—linked Dehydrogenases: $E_0' \sim -0.32$ volts[b]

		Coenzyme
Alcohol Dehydrogenase	most species	NAD (*A*-specificity)[c]
L-Lactate Dehydrogenase	most species	NAD (*A*-specificity)
L-Glutamate Dehydrogenase	plants	NAD
	yeast	NADP
	liver	both (*B*-specificity)[c]
D-Glucose 6-phosphate Dehydrogenase	yeast and animal sources	NADP (*B*-specificity)

2. Flavoproteins[a] and metalloflavoproteins: $E_0' \sim -0.3$ to 0 volts[b]

		Coenzyme	Electron Acceptor
D-amino acid oxidase	mammalian	FAD	O_2
Glucose oxidase	molds	FAD	O_2
L-amino acid oxidase	mammalian	FMN	O_2
Lipoyl dehydrogenase	mammalian	FAD	NAD
Xanthine oxidase	mammalian	FAD + Fe + Mo	O_2
NADH dehydrogenase	mammalian	FMN + Fe	cytochrome
Succinate dehydrogenase	mammalian and yeast	FAD + Fe	cytochrome

3. Cuproproteins[a]: $E_0' \sim +0.3$ to 0.55 volts[b]

Tyrosinase (phenolase)	plants	1 Cu/34,000
Ascorbate oxidase	plants	6 Cu/160,000
Cytochrome oxidase (Cu-protein component)	mammalian	1 Cu/25,000

[a] The coenzymes or electron carriers are:

2.

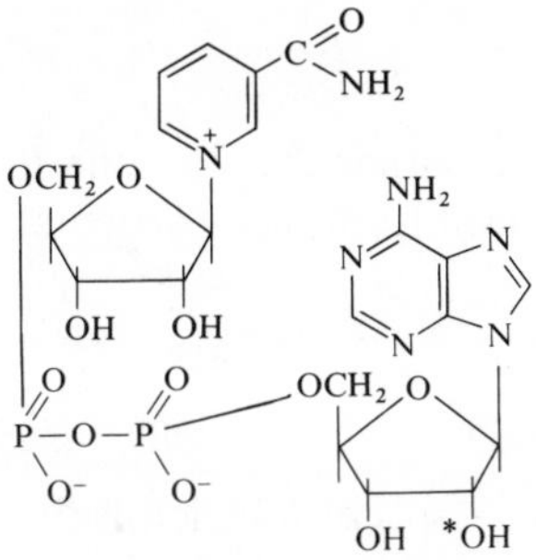

NAD$^+$

(NADP$^+$ has an additional Phosphate in the position marked with *)

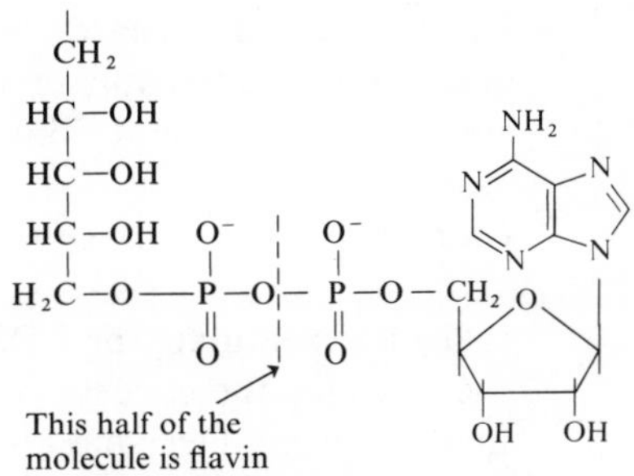

This half of the molecule is flavin mononucleotide (FMN). The total structure is flavin adenine dinucleotide (FAD).

3. Cu^+/Cu^{++}

[b] E_0' is a measure of the reducing power of the coenzyme. The higher negative the value of E_0' is, the stronger reducing agent the compound. From the values given in the table, the normal electron flow should therefore be from substrate to NAD to Flavin to Cu^{2+} to O_2. References: 1. H. Sund, H. Dickman, and K. Wallenfels, *Adv. Enzymol.*, **26**, 115 (1964). 2. G. Palmer, and V. Massey, "Mechanisms of Flavoprotein Catalysts," in T. P. Singer (Ed.), *Biological Oxidations*, Wiley, New York (1969). 3. A. S. Brill, R. B. Martin, and R. J. P. Williams, "Copper in Biological Systems," in B. Pullman (Ed.), *Electronic Aspects of Biochemistry*, Academic Press, New York (1964).]

[c] Refers to the specificity of the hydrogen addition to the nicotinamide ring (see page 90).

This reaction occurs as part of glycolysis in the majority of all living cells, and the NAD-linked LDH's are found throughout the biosphere.

Lactate dehydrogenase has been isolated from a large variety of sources and appears to be quite uniform in most of its properties regardless of the biological origin. It has a molecular weight of 140,000, is specific for L-lactate, and shows a preference for NAD over nicotinamide adenine dinucleotide phosphate (NADP) as the coenzyme.

It is interesting to note one aspect of the specificity of LDH, which in fact applies to all NAD- and NADP-linked dehydrogenases. In reducing the pyridine ring of the coenzyme, all the dehydrogenases carry out stereo-specific hydrogen transfer. If the substrate is deuterated, the deuterium bonds to the 4 position of the ring in only one of the two stereochemically distinct orientations.

H_2N | $C{=}O$ | NH_2 | $C{=}O$ | D | H | N—R | $D^+ + S$ | SD_2 | H— | $\overset{+}{N}$—R | SD_2 | $D^+ + S$ | NH_2 | $C{=}O$ | H | D | N—R

A isomer

B isomer

Different dehydrogenases may produce either the *A* or the *B* isomer but only one or the other is apparently produced from any particular enzyme.* LDH always results in the *A* isomer.

Lactate Dehydrogenase Isozymes

Before examining the LDH isozymes specifically, let us first consider briefly the basis for the concept of isozymes and the nomenclature that is currently in use. Different enzymes that carry out the same catalytic function are referred to as isodynamic. If one compares isodynamic enzymes from different biological sources it becomes evident that the same catalytic activity can be expressed by quite different protein molecules. Thus alcohol dehydrogenases (closely related to LDH) from yeast and from liver differ in molecular weight by a factor of two. In other cases where the molecular weight may be much more similar the isodynamic enzymes show significant differences in other properties such as electrophoretic mobility and amino acid composition (see Section 3.4 on cytochrome c). In some rare instances isodynamic enzymes from different sources may indeed be identical. Isodynamic enzymes from different sources

* B. Vennesland, and F. H. Westheimer, "Hydrogen Transport and Steric Specificity in Reactions Catalyzed by Pyridine Nucleotide Dehydrogenases," in W. D. McElroy and B. Glass (Eds.), *The Mechanism of Enzyme Action*, Johns Hopkins, Baltimore, 1954.

are also referred to as *heteroenzymes.* If a substance with a given enzymatic activity is isolated from a single homogeneous biological source and subjected to, say, high-resolution electrophoresis, the isolated product may also be found to consist of several isodynamic forms as seen in studies over the past decade; this is the basis for the concept of isozymes, or isoenzymes. The definition of isozymes as different forms of the same enzyme arising from a single cell or tissue thus becomes a strictly operational one based on a set of specific experimental parameters. Thus before one can speak of true isozymes one must establish that the starting material is homogeneous and that the multiple forms of the enzyme observed are not artifacts of isolation. It is interesting to note that the whole field of isozyme biochemistry has developed in conjunction with improved analytical techniques. Twenty years ago one of the major criteria of enzyme purity was crystallization and recrystallization to constant specific activity. As chromatographic methods improved it became evident that a crystalline protein was not necessarily pure and later, with the very refined techniques of electrophoresis on starch gel and acrylamide gel and most recently with isoelectric focusing,* crystalline proteins were found to contain several distinct isodynamic forms. As the techniques improve it is likely that more and more isozymes will be found. In virtually all the isozyme systems known today the different forms are distinguished by electrophoretic mobility.

The operational definition of isozymes is obviously not very precise but any attempt at improvement leads to difficulties. In some systems we know, for instance, that two major forms of a particular enzyme in a given cell have distinct origins, one from the mitochondrion, the other from the soluble fraction. Although two forms such as these fit the isozyme definition, it would appear that the concept of isozymes in this case serves no purpose. Knowledge of the subcellular origin makes it possible to refer to the two forms in precise terms such as "mitochondrial" and "soluble" and classification as "isozyme 1" and "isozyme 2" would be meaningless. Perhaps with improved understanding of the significance of isozymes the term will become completely obsolete and be replaced by designations based on biological and biochemical facts. In the meantime the phenomenon of multiple forms of a given enzyme from a single homogeneous biological source is well established and presents an interesting challenge in determining the chemical and biological significance of the different forms.

The LDH isozymes represent one of the best-studied and best-understood groups of isozymes. The phenomenon has been investigated for a large number of animal species and for individual tissues from each species and it is clear that there are at least five isozymes present in different proportions in all LDH-containing tissues of all species studied.

The molecular model for the five isozymes is based on the three experiments given in Figure 3.15: The first simply illustrates the phenomenon; the second provides the evidence for the subunit structure of LDH—the enzyme, having

* See T. Flatmark, and O. Vesterberg, *Acta. Chem. Scand.*, **20**, 1497 (1966), and W. A. Susor, M. Kockman, and W. J. Rutter, *Science*, **165**, 1260 (1969).

a molecular weight of 140,000 is made up of four subunits having a molecular weight of 35,000; and the third provides the test of the model by hybridization of one isolated form to produce all five original forms. Figure 3.15 also shows the isozyme model as the five possible tetramers of two different types of subunits. The model is clearly consistent with the gradual change in electrophoretic mobility from the H_4 to the M_4 form; other properties such as heat stability and catalytic properties also show a continuous and stepwise change between the extremes of M_4 and H_4 in exactly the ratio predicted from the proposed hybrid content of M and H (M_2H_2 is generally intermediate between M_4 and H_4, M_3H is intermediate between M_4 and M_2H_2, and MH_3 is intermediate between M_2H_2 and H_4). The fact that all forms of LDH can bind four moles of coenzyme is consistent with the existence of one binding site per subunit. So far no one has been able to demonstrate activity for the individual subunits but this could be because quite strong denaturing agents (such as guanidinium chloride or detergents) are required to keep the subunits dissociated. Individual subunits, if they indeed ever exist in a native form, could still be active.

Having established the phenomenon and arrived at a molecular model, the next logical step is to ask about the biological significance of the five isozymes. According to natural selection it makes sense to assume that nature is not wasteful and that, if the isozymes are there, they are there for a purpose. However, the answer to this question is yet far from complete, though some of the quantitative data on the proportions and kinetic properties of the LDH isozymes indicate some very interesting possibilities (Table 3.8).

First of all it early became apparent that tissues operating under essentially anaerobic conditions are rich in M subunits and low in the H subunits whereas for aerobic tissues the opposite relationship is true. This, together with the kinetic data given in Table 3.8(*a*), led to the proposal that the two subunit types exist to ensure functional LDH enzymes under widely different physiological conditions. Thus the M_4 isozyme clearly is ideally suited for glycolysis under anaerobic conditions; it can turn over pyruvate at a high rate under conditions of high pyruvate concentrations and is not readily inhibited by normal concentrations of accumulated pyruvate. In this type of tissue (notably skeletal

---

Figure 3.15 The molecular bases of LDH isozymes. (*a*) A tracing of the 5 electrophoretically distinct forms of beef LDH from heart and skeletal muscle observed after electrophoresis at pH 7 on starch gel. (*b*) Molecular weights of the native enzyme (mixture of all isozymes or with a single isolated form) and of the dissociated enzyme showing the tetrameric complex of the native enzyme. (*c*) Tracing of starch gel electrophoretic patterns at pH 7 of LDH form 3 [from (*a*) above] after exposure to pH 4 for increasing length of time. (*d*) The tetrameric model providing the basis for the isozymes, showing the proposed dissociation at low pH, followed by the random reassociation at neutral pH demonstrated in (*c*). The enzyme bands traced in (*a*) and (*c*) were developed with a combination of substrate and coenzyme and dye, so that all protein bands with *active* LDH give a positive reaction. [Data from Anderson, S., and G. Weber, *Arch. Biochem. Biophys.*, **116**, 207 (1066).]

muscle), lactate is the end-product of glycolysis and must be transported to other tissues for further reactions. Tissues that require a continuous supply of energy for sustained mechanical work (such as heart muscle) rely much more on the aerobic oxidation of pyruvate. With the H_4 isozyme operating in the aerobic tissues, very little lactic acid is produced since pyruvate inhibits the *H*-type LDH at low concentrations (Table 3.8) and is thus turned over relatively

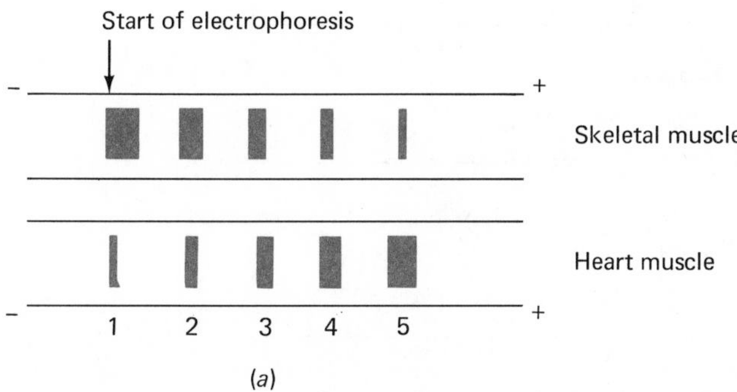

(*a*)

MW of native LDH in pH 7 buffer: 140,000
MW of LDH after treatment with dissociating reagents: 35,000

(*b*)

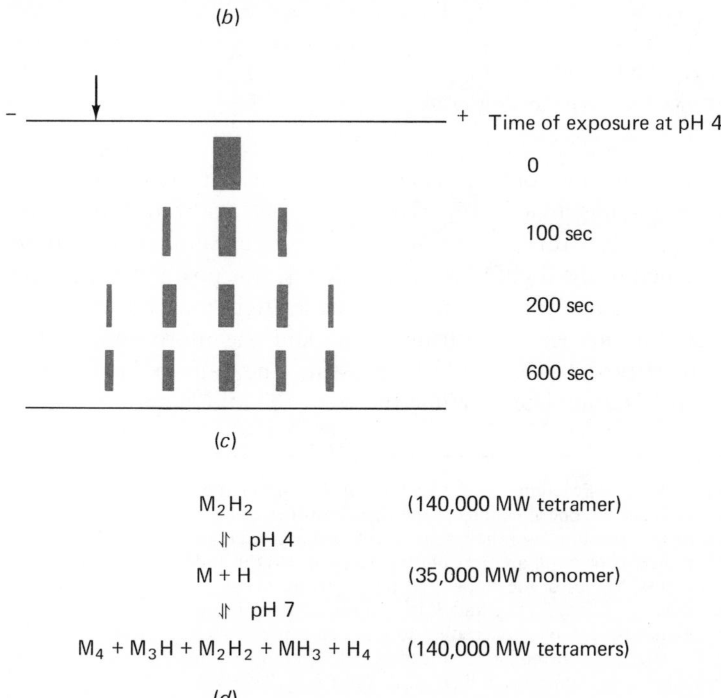

(*c*)

$$M_2H_2 \quad \text{(140,000 MW tetramer)}$$
$$\rightleftharpoons \text{ pH 4}$$
$$M + H \quad \text{(35,000 MW monomer)}$$
$$\rightleftharpoons \text{ pH 7}$$
$$M_4 + M_3H + M_2H_2 + MH_3 + H_4 \quad \text{(140,000 MW tetramers)}$$

(*d*)

TABLE 3.8 DIFFERENT PROPERTIES OF LDH ISOZYMES[a]

(*a*) Kinetic Properties (chicken LDH)

Isozyme	K_m *(Pyruvate)(M)*	*Turnover number for Pyruvate*	*Pyruvate Concentration Needed for 50-percent Inhibition of Isozyme*
H_4	8.9×10^{-5}	45,000	5×10^{-3}
H_2M_2	5.2×10^{-4}	73,200	—
M_4	3.2×10^{-3}	93,400	4×10^{-2}

(*b*) LDH Isozymes in Tissues of Different Ducks, percent *H* Type

	Breast muscle	*Liver*	*Heart*
Pekin duck (domestic)	<5	56	95
Mallard (wild)	35	54	92
Lay san duck[b]	7	54	96
Black duck	45	53	98
Pintail duck	39	50	100

(*c*) Effect of Oxygen Tension on LDH-subunit Synthesis in Cultured Monkey-heart Cells

Gas Phase, 95 Percent	*Subunit Synthesized, Percent*
O_2	42
Air	50
N_2	72

[a] N. O. Kaplan, *Brookhaven Symp. in Biol.*, **17**, 131 (1964).

[b] The Lay san duck is essentially a nonflying close relative of the mallard.

slowly. Therefore the function of this form of the enzyme is probably to oxidize lactate to pyruvate and ensure a supply of pyruvate to be oxidized through the aerobic pathways. To test this hypothesis the relative amount of *H* and *M* types were determined in the flight muscle of ducks. It was postulated that the more active a flyer the duck was, the more aerobic its flight muscles should be to sustain the continuous energy consumption; and the more aerobic the muscles, the more *H*-type LDH should be present. The data in Table 3.8(*b*) support the hypothesis completely. While the *H* to *M* ratio is quite constant

Figure 3.16 The kinetic analysis of the LDH-reaction (see also Figure 2.6). The initial rates are first determined at different concentrations of the two substrates in both the forward and the reverse reaction, and the results are plotted according to (*a*). The slopes and intercepts from (*a*) are next plotted according to (*b*), and the numerical experimental values of the slopes and intercepts in (*b*) can in turn be used to calculate the values for K_m and K (the dissociation constant for the first substrate-enzyme complex) according to (*c*). The experimental values for some of the relevant constants determined at pH 7.8 are given in parenthesis and were taken from Anderson, S. R., *et al.*, *J. Biol. Chem.*, **239**, 2991 (1964).

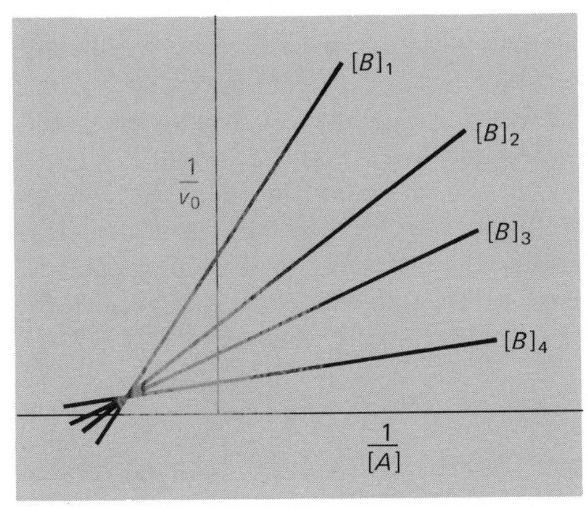

Experiment 1 : A = Lactate, B = NAD
Experiment 2 : A = NAD, B = Lactate
Experiment 3 : A = Pyruvate, B = NADH
Experiment 4 : A = NADH, B = Pyruvate

(*b*) For each experiment, plot the intercepts in *a* vs. $\frac{1}{[B]}$ and slopes in *a* vs. $\frac{1}{[B]}$:

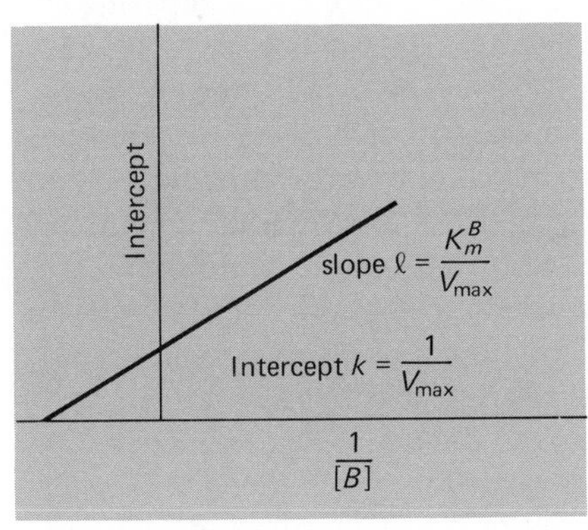

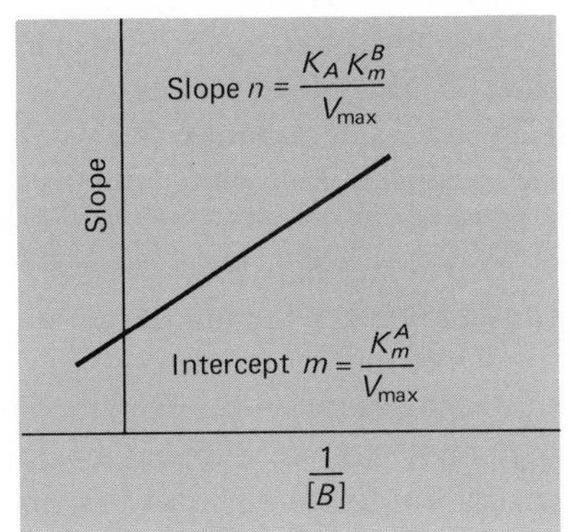

(*c*) From numerical values of slopes and intercepts, calculate the various constants:

	Experiment 1	Experiment 2	Experiment 3	Experiment 4
$\frac{\ell}{k} = K_m^B \longrightarrow$	K_m^{NAD} ($1.1 \times 10^{-4} M$)	$K_m^{lactate}$	K_m^{NADH} ($1.5 \times 10^{-5} M$)	$K_m^{pyruvate}$
$\frac{n}{\ell} = K_A \longrightarrow$	$K_{lactate}$	K_{NAD} ($1.7 \times 10^{-4} M$)	$K_{pyruvate}$	K_{NADH} ($1.3 \times 10^{-5} M$)
$\frac{m}{k} = K_m^A \longrightarrow$	$K_m^{lactate}$ ($3.6 \times 10^{-2} M$)	K_m^{NAD}	$K_m^{pyruvate}$ ($6.6 \times 10^{-4} M$)	K_m^{NADH}

in the heart and liver of the different ducks, the breast muscle ratio varies in exactly the predicted manner.

It is not clear at what stage of development the preference for a given LDH subunit is expressed. It is interesting to note, however, that a given tissue will respond directly to environmental influence and produce different proportions of the two subunit forms [Table 3.8(*c*)].

It must be stressed at this point that there are several cases of different isozymes which seem to have no metabolic significance. Although such negative evidence is difficult to accept, one cannot completely dismiss the possibility that some isozyme systems may result from genetic "accidents."

There are several fascinating aspects of isozyme biochemistry that will be left out of this discussion. The embryology of isozymes and the dramatic changes in isozyme distribution under pathological conditions, on the biological side, and the chemical and biochemical basis for the different forms, on the chemical side, will continue to challenge workers in this area of research.

Kinetic Analysis

The kinetic analysis of LDH is shown in Figure 3.16. It can be demonstrated that the results of this analysis are consistent with the general mechanism for an ordered two-substrate reaction as discussed in Chapter 2. The distinction between ordered and random substrate addition is based on the product-inhibition results. If the whole idea of kinetic analysis is correct and the significance of the kinetic constants is properly defined, the two-substrate systems can be subjected to experimental test by directly determining the dissociation constant for the binary complex of enzyme first substrate and comparing this with the kinetic constant. This has in fact been done for many dehydrogenases and the results for LDH are given in Figure 3.17. The data for the H_4 enzyme fit in the Adair equation (2.3) with a maximum of four binding sites and give a good linear Hill plot indicating four identical sites with binding constant $\mathbf{K} = 2.6 \times 10^6\ M$. The M_4 enzyme, however, gives a completely anomalous binding curve that cannot be accommodated by any simple binding equation. The end point of the binding data indicates again that 4 moles of coenzyme are bound per mole of enzyme; however, they are bound by a very complex type of interaction between the binding sites, perhaps best explained in terms of slow interconversion of different conformations at different stages of saturation. The hybrid forms all show this anomalous behavior, which therefore must be characteristic of the M subunit. The behavior of the M enzyme and its hybrids undoubtedly bears an important relationship to its functional characteristics

---

Figure 3.17 The binding of NADH to lactate dehydrogenase (LDH) and alcohol dehydrogenase (ADH) at pH 7.4. The binding data were obtained from measurements of the fluorescence of free NADH; and of the enzyme-NADH complex. (*a*) Titration curves for NADH binding to the two beef LDH isozymes M_4 and H_4 showing 4 binding sites per 140,000 daltons of enzyme. (*b*) Hill plot of the data in (*a*). The Hill coefficient for H_4-LDH is 1.2, indicating that the

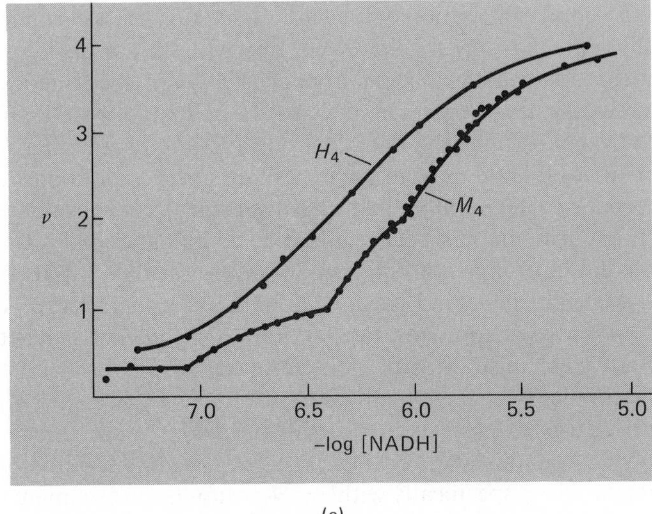

(a)

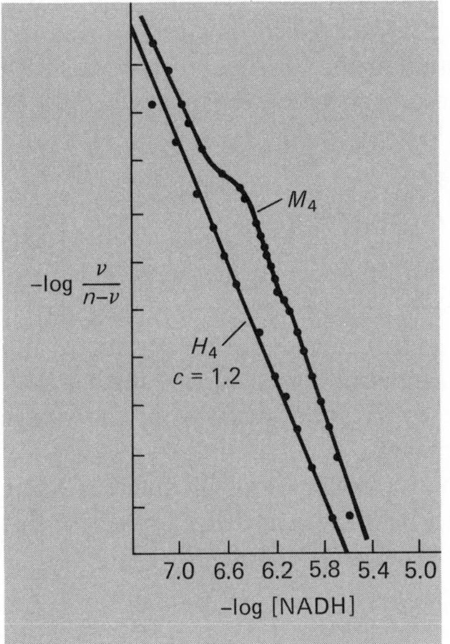

(b)

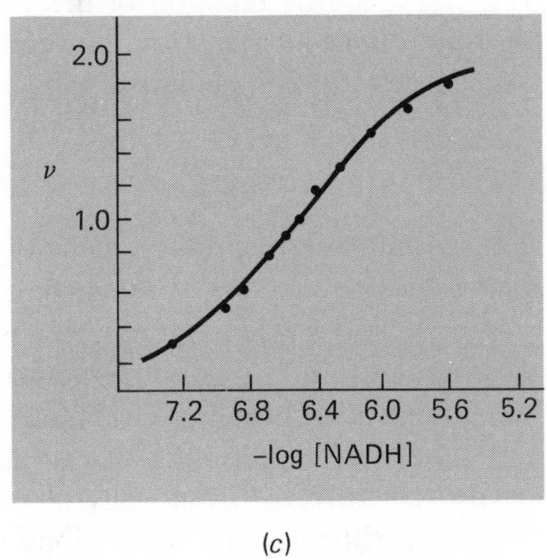

(c)

4 sites are very nearly independent ($c \cong 1$) with an intrinsic binding constant $\mathbf{K} = 2.6 \pm 0.4 \times 10^6\ M$. The complex titration behavior of the M_4-LDH is not amenable to a similar simple interpretation. (*c*) Titration curve for NADH binding to ADH showing 2 binding sites per 80,000 daltons with a binding constant $\mathbf{K} = 3.4 \times 10^6\ M$ in excellent agreement with a K_A (NADH) of $3.1 \times 10^{-7}\ M$ ($\mathbf{K} = 1/K$). The data are taken from Anderson, S. R. and G. Weber, *Biochem.*, **4**, 1948 (1965).

but there has been no good explanation offered as to what this relationship may be. It is important to note that even for the H_4 enzyme, with the normal binding curve, the agreement between the dissociation constant and the kinetically determined K_m for coenzyme is very poor. This has been found for LDH from several sources and demonstrates the danger in always simply equating K_m, a complex function of several rate constants, with a simple dissociation constant. However, the discrepancy will undoubtedly be resolved when proper evaluation of all individual rate constants has been made by rapid-kinetic analysis. To illustrate that this complication is not universal, however, the data for alcohol dehydrogenase have been included in Figure 3.17. In this case there is excellent agreement between the dissociation constant and the K_m, indicating that the ordered two-substrate mechanism with all its assumptions adequately describes the reaction for the simpler two-site dehydrogenase.

Inhibitors have been used extensively in the study of LDH. Perhaps the most significant conclusions from this work are based on very strong inhibition by a number of reagents that react specifically with —SH groups (such as *p*-mercuribenzoate, bromoacetate, and so forth; see Barker in this series) and thus strongly implicate an —SH group in the catalytic action of this enzyme.* All models attempting to formulate a molecular mechanism of LDH action include a free —SH group in the active site of the enzyme, involved in thioacetal formation with the carbonyl function of the substrate. In the absence of more detailed information about structure, however, these proposed mechanisms should be considered only as speculative.

3.4 CYTOCHROME c (HEMEPROTEIN AND EVOLUTION)

A large number of proteins depend on the presence of the heme group for their biological activity. Cytochrome c is discussed briefly as a representative of this group of proteins; it has been isolated from a wide variety of sources and the comparative study of its structure and function has advanced further than that for any other protein. The importance of this type of study to our knowledge of biochemical evolution is also briefly reviewed.

The structure of the porphyrin prosthetic group found in different heme proteins is shown in Table 3.9 along with a list of some other members of this group of molecules. Hemoglobin will be discussed in Chapter 4.

General Description

Cytochrome c is a constituent of the electron-transport chain in the mitochondrion (see Chapters 9 and 10 and also Larner in this series):

$$\begin{array}{ccccccccc} & 2H^+ & & & & & & & \\ \text{Flavin} & \nwarrow\!\!\uparrow\!\!\nearrow & 2\text{ cyt b Fe}^{2+} & \swarrow & 2\text{ cyt c}_1\text{ Fe}^{3+} & \nwarrow\nearrow & 2\text{ cyt c Fe}^{2+} & \swarrow & 2\text{ cyt a (a}_3\text{) Fe}^{3+} & \nwarrow\nearrow & O^{=} \\ & \times & & \times & & \times & & \times & & \times & \\ \text{Flavin H}_2 & \swarrow\searrow & 2\text{ cyt b Fe}^{3+} & \searrow & 2\text{ cyt c}_1\text{ Fe}^{2+} & \swarrow\searrow & 2\text{ cyt c Fe}^{3+} & \swarrow\searrow & 2\text{ cyt a (a}_3\text{) Fe}^{2+} & \swarrow\searrow & \frac{1}{2}O_2 \end{array}$$

* G. Pfleiderer, *et al.*, *Biochem. Z.*, **329**, 104 (1957).

TABLE 3.9(a) BROAD STRUCTURAL AND FUNCTIONAL CLASSIFICATION OF HEME PROTEINS

The tetrapyrrol nucleus: Porphyrin with Fe^{2+} forms heme, with Fe^{3+} forms hemin. Four of the six coordinations of iron are bound to the four pyrrol N's, the fifth is generally bound to a protein, histidine-N, and the sixth to histidine, H_2O, O_2, or so forth.

$V = -CH{=}CH_2$
$M = CH_3$
$E = -CH(-S\text{-Protein})-CH_3$
$F = -C{=}O$ (with H)
$X = -CHOH-CH_2(CH_2-CH_2-CH(CH_3)-CH_2)_3H$

	Porphyrin Structure 2	4	8	Porphyrin Link to Protein	Molecular Weight Protein	Function
Ferrous iron (Fe^{2+})						
Hemoglobin	*V*	*V*	*M*	noncovalent	65,000	O_2 transport
Myoglobin	*V*	*V*	*M*	noncovalent	17,000	O_2 transport
Ferric iron (Fe^{3+})						
Catalases	*V*	*V*	*M*	noncovalent	220,000–250,000	$2H_2O_2 \longrightarrow 2H_2O + \frac{1}{2}O_2$
Peroxidases	*V*	*V*	*M*	noncovalent and covalent	~50,000	$H_2O_2 + AH_2 \longrightarrow 2H_2O + A$
$Fe^{2+} \rightleftharpoons Fe^{3+}$						
Cytochromes a	*X*	*V*	*M*	noncovalent	~70,000	Electron transport and redox
Cytochromes b	*V*	*V*	*M*	noncovalent	30,000–170,000	Electron transport and redox
Cytochromes c	*E*(*V*)	*E*(*V*)	*M*	covalent	13,000–35,000	Electron transport and redox

Other important porphyrin derivatives:
Mg-containing porphyrin derivatives: Chlorophylls
Co-containing porphyrin derivatives: Cobamide (vitamin B_{12})

Whether cytochrome c actually qualifies as an enzyme has been debated since it is fairly stably lodged in the lipoprotein structure of the mitochondrion (see Chapter 10) and has as its substrates other cytochromes of the electron-transport chain. This is a rather fine distinction to make; here we shall consider cytochrome c an enzyme in good standing. For our discussion it is sufficient to define cytochrome c in terms of its spectrum and redox potential [Table 3.9(*b*)]. The activity assay is based on high-resolution spectral analysis: The characteristic reduced spectrum appears in the presence of the proper reducing agents

TABLE 3.9(b) SOME PROPERTIES OF CYTOCHROMES (MAMMALIAN SOURCE)

Cytochrome	*Reduced Form Absorption Maxima*[a] α	β	γ	E'_0*(Volts)*	*Molecular Weight*	*Reaction with* O_2
b	563	530	430	+0.07	30,000	–
c_1	554	523	418	+0.22	37,000	–
c	550	521	416	+0.25	13,000	–
a	605	517	414	+0.29	70,000 (× 4) } 400,000[b]	–
a_3	600	–	445		70,000 (× 2) }	+

[a] All the heme proteins are characterized by their absorption spectrum. Most of them show a typical three-band spectrum, the three absorption bands being referred to as the α, β, and γ (or Soret) bands. The three bands may be found within a certain wavelength region depending on the structure of the individual proteins. Thus the α band is generally found in the range of 550 to 565 nm with a relative intensity of 1, the β band in the range of 520 to 535 nm with a relative intensity of 0.5 (in some hemeproteins the β-band is missing), and the most intense γ-band in the range of 400 to 420 nm with a relative intensity of 3. The classification of cytochromes into a, b, or c types is based entirely on their spectral properties and has no direct implications regarding function or structure.

[b] The complex of cytochrome a and cytochrome a_3 is often referred to as cytochrome oxidase. The smallest a subunit containing one heme group has a molecular weight of about 70,000 and the functional (in vivo) cytochrome oxidase appears to be a hexamer.

(cyt b ⟶ cyt c_1 in vivo while ascorbic acid ⟶ cyt c in vitro) and disappears in the presence of the proper oxidizing agent (cytochromes a and a_3, also referred to as cytochrome oxidase and O_2).

Structure

Cytochrome c is structurally unique among the cytochromes in that the iron porphyrin prosthetic group is covalently linked to the protein. The linkage can be considered to be the result of the addition of two —SH groups to the two vinyl groups of protoheme (one of the most common porphyrin derivatives):

The molecular weight of cytochrome c as reported from a large variety of sources falls in the range of 13,000 to 14,000.

Comparative Aspects

Cytochrome c is readily purified from mitochondria and thus it has been obtained from a large number of species. With this large group of low-molecular-weight heteroenzymes available, it becomes possible to ask a very important structure-function question in a unique way. Suppose the biological function expressed by redox potential and spectral properties is strictly dependent on the presence of a unique amino acid sequence; then this sequence should be common to all heteroenzymes of cytochrome c. Other regions of the molecule not directly involved in the functional site might vary among different species. If this supposition is correct, can the active site of cytochrome c be pinpointed by comparing the primary structure of a large number of heteroenzymes?

The primary sequences of more than 20 types of cytochromes c have been determined and some of the main features are given in Figure 3.18. One feature of the comparison with direct functional implications is the constancy of the heme-binding sites cyS-14 (half-cystine), cyS-17, and his-18. Residues 19, 20, and 21 are also common to most species with some conservative* substitutions. Perhaps even more significant, residue 13 is lysine in most species and whenever a substitution occurs it is arginine, another conservative substitution. Thus it appears that the part of cytochrome c involved in the binding of the heme group is quite constant and perhaps can be considered as the active site.

Perhaps more remarkable is the absolute constancy found in the sequence 70 through 80. If our general proposition of the significance of a constant sequence is correct, we must conclude that this sequence is an essential part of the functional makeup of the enzyme. One likely role for this sequence has been suggested to be interaction with cytochrome oxidase (the next component on the oxidizing side of the electron-transport chain). All the variations of cytochrome c with the invariant 70-through-80 sequence interact with and are oxidized by cytochrome oxidase. One bacterial cytochrome c has been found that is very similar to yeast cytochrome c except for replacements in the 70-through-80 peptide; the bacterial cytochrome c, however, does not react with cytochrome oxidase. Other single residues throughout the cytochrome c structure are also constant (try-59 (tryptophan) arg-38, arg-91, his-26, and a surprisingly large number of glycine residues, for example). No obvious functional role can be assigned to these residues at present. It appears, however, that the general approach illustrated here—the direct correlation of constant structural features with the characteristic common functional properties—may well represent an excellent method for elucidating the active-site components

* The terms "conservative" and "radical" substitutions have been introduced as a very useful and reasonable but perhaps also presumptive way of distinguishing between similar and different amino acids. Thus replacing leucine with valine or isoleucine, arginine with lysine, glutamate with aspartate, and so forth, would all be considered conservative substitutions since the chemical characteristics of the side chains in each case would be kept very similar. On the other hand, replacing glutamate with lysine or leucine with serine would constitute a radical substitution because of the chemical differences between the side chains.

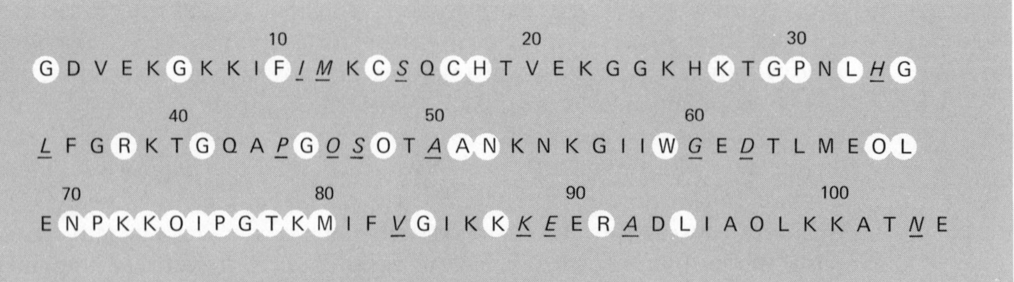

Figure 3.18 The sequence of the 104 amino acids in human cytochrome c is given using the one-letter abbreviations (see Table 3.6 for the key to the one-letter code). Comparison of the sequence of cytochrome c from human, Rhesus monkey, horse, donkey, pig, cow, sheep, dog, gray whale, rabbit, kangaroo. The residues in italic lettering in the human sequence above (16 residues) are variable in the above 11 species. The remaining 88 residues are common to all 11 species. Extending the comparison further to include 14 more species: chicken, turkey, penguin, pekin duck, snapping turtle, rattlesnake, bullfrog, tuna fish, screw-worm fly, silkworm moth, wheat, *Neurospora crassa*, *Candida krusei*, and baker's yeast, there remain only 35 residues which are common to all 25 species. These are shaded in the structure above. (Based on work by Margoliash and others, quoted in Dayhoff, M. O. and R. V. Eck, *Atlas of Protein Sequence and Structure* 1967–68, National Biomedical Research Foundation, Silver Spring, Md., 1968.)

of enzymes. The forthcoming results of the analysis by X-ray crystallography of the cytochrome c structure will hopefully provide some further evidence bearing on the role of the yet unexplained constant features of the primary structure.

Evolutionary Implications

Data of the kind presented in Figure 3.18 have become very important in suggesting a quantitative biochemical method for estimating evolutionary relationship between species. It is indeed remarkable that such a large number of residues are constant in the cytochrome c sequence. This finding suggests very strongly that all the cytochrome c molecules have evolved from a single ancestral form and that cytochrome c has "survived" as a functional and structural entity for some 2 billion yr. If the complete structures from each individual species are carefully examined, it becomes immediately apparent that there is a direct correlation between taxonomic kinship and the number of amino acid substitutions: The more closely related the species, the more similar the amino acid sequence is. Based on this relationship, illustrated in Table 3.10, we shall assume that the number of different residues in the cytochrome c of two different species is a *direct measure* of the time elapsed since the divergence of the evolutionary lines leading to the two species. If we can

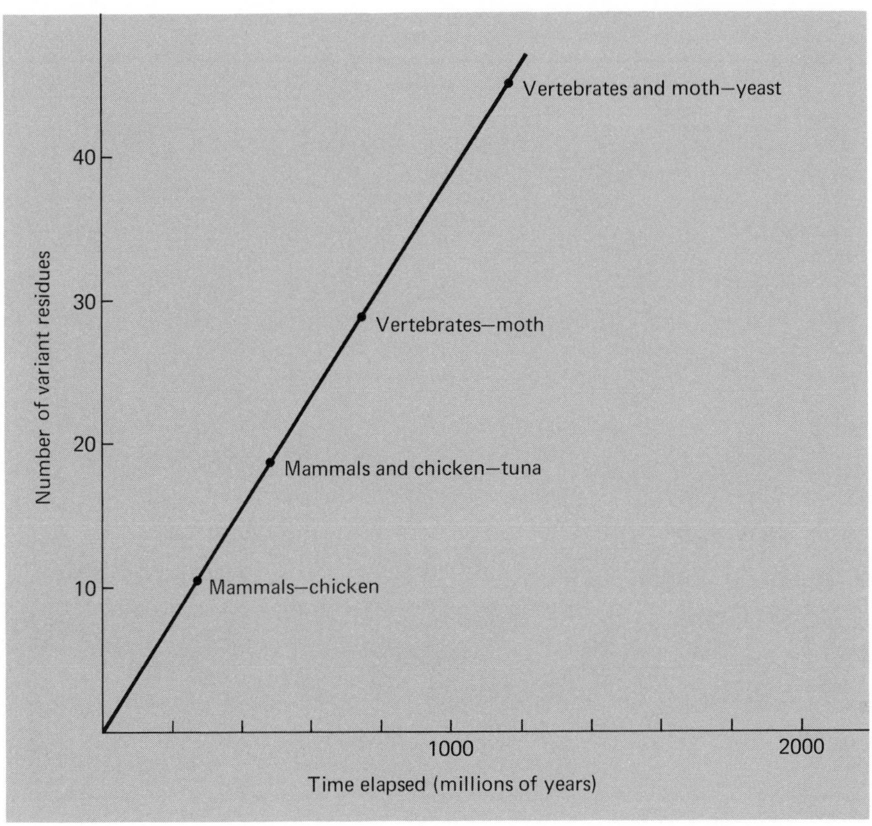

Figure 3.19 A possible method for estimating evolutionary age. From the amino acid sequence of cytochrome c from many different organisms, the average number of residue differences between different phylogenetic lines is readily obtained. Based on the assumption that each amino acid replacement represents an evolutionary period of 26.2 million years (see text), the number of different amino acids between two lines gives a direct estimate of the time elapsed since the two lines diverged. [Redrawn from the data in Margoliash, E. and A. Schejter, *Adv. Prot. Chem.*, **21**, 113 (1966).] This type of analysis is clearly fraught with hazards. In the strictest sense the time scale applies only to this one protein, cytochrome c, and the total amount of information that has gone into the analysis, impressive though it may be, is comparatively small and inadequate.

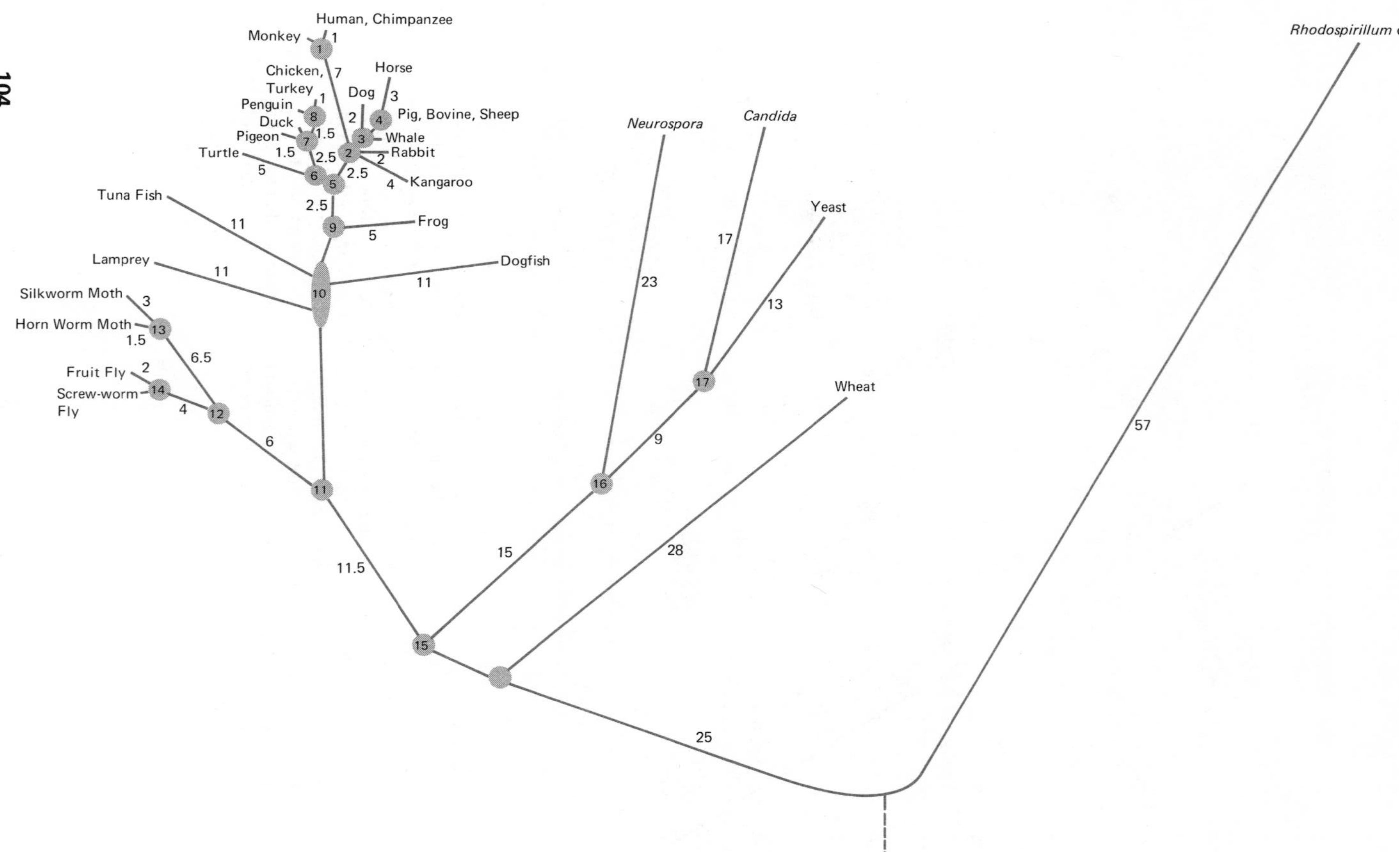

Figure 3.20 The phylogenetic tree based on cytochrome c analysis. The basis for the construction of the tree is the same as in Figure 3.19; the length of each line is proportional to elapsed time. [Reprinted from *Atlas of Protein Sequence and Structure* 1969, M. O. Dayhoff (Ed.), National Biomedical Research Foundation, Silver Spring, Md., 1969, with permission.]

TABLE 3.10 EVOLUTION OF CYTOCHROME c

Comparison	*Average Number of Variant Residues*
Group *A*,[a] Chicken	10.7
Group *B*,[b] Tuna	18.7
Group *C*,[c] Moth	28.7
Group *D*,[d] Yeast	45.2

[a] Mammals, (Rabbit, pig, sheep, cow, dog, horse, kangaroo, monkey, and man)

[b] Group A and chicken

[c] Group B and tuna

[d] Group C and moth

next estimate by some other means the time elapsed since one such divergence took place we should then be able to calculate a universal *unit evolutionary period*, the average period required for one amino acid replacement to occur. According to the data in Table 3.10 the average residue difference between mammals and chickens is 10.7. The time elapsed since the divergence of avian and mammalian lines is estimated to be 280 million yr and the unit evolutionary period is thus 26.2 million yr. Using this as the standard unit, the data on amino acid differences can be plotted as a straight line relationship with elapsed time (Figure 3.19), indicating that 490 million yr have elapsed since the divergence of the line leading to tuna from that leading to mammals and birds, 750 million yr since the divergence of invertebrate and vertebrate lines, and 1,200 million yr since the divergence of the lines leading to yeast and to higher species. A more detailed analysis from species to species allows the construction of complete phylogenetic trees such as the one shown in Figure 3.20. Some of the hazards inherent in this approach are given in the legend to Figure 3.19. In spite of the many uncertainties, this is a promising and exciting information bonus obtained from the comparative work on cytochrome c.

Before leaving this brief excursion into evolutionary biochemistry, other methods of studying relationships between species should be mentioned in order to be considered in context. Studies of similarities in DNA's (see Table 6.2 and Figure 6.2) and in specificities of amino acyl–tRNA synthetase (see Chapter 7) both give essentially the same type of information as the cytochrome c work.

3.5 ASPARTATE TRANSCARBAMYLASE (ALLOSTERIC ENZYMES)

Aspartate transcarbamylase is one of the best-described *allosteric* enzymes (for definition of allosteric regulation see Chapter 8). It offers an excellent illustration of feedback regulation and is one of the first enzymes in which the presence of a regulatory site chemically and physically completely distinct from the substrate site was established. This section will deal with the experimental basis for the aspartate transcarbamylase allosteric model. The discussion is based entirely on work with the enzyme from *Escherichia coli.*

General Description

Aspartate transcarbamylase (ATCase) catalyzes the transfer of the carbamyl group from carbamyl phosphate to aspartate:

$$H_2N{-}\overset{O}{\overset{\|}{C}}{-}O{-}P(=O)(O^-)_2 \; + \; {}^-OOC{-}CH_2{-}CH(NH_3^+){-}COO^- \xrightarrow{\text{ATCase}} H_2N{-}\overset{O}{\overset{\|}{C}}{-}\underset{H}{N}{-}CH(COO^-){-}CH_2{-}COO^- + HPO_4^{2-} + H^+$$

(inhibited by CTP)

It is the first of a sequence of steps leading to the biosynthesis of pyrimidine nucleotides; the final end product of this sequence is cytidine triphosphate (CTP). Genetic regulation of the concentration of ATCase was found in *E. coli* (see Chapter 8). In addition the activity of the isolated enzyme can be regulated by different effectors; significantly, the most powerful of the inhibitory effectors was CTP, the end product of the biosynthetic pathway that ATCase initiates.*

Kinetic Analysis

The plot of the catalytic behavior of ATCase given in Figure 3.21 shows the sigmoidal curve characteristic of allosteric systems. One way to explain the demonstrated behavior mechanistically is to propose an "autocatalytic" activation by substrate according to the substrate-activation model illustrated in Chapter 2.†

The reactions

$$E + S \overset{K_S}{\rightleftharpoons} SE^*$$

$$SE^* + S \underset{k_{-1}}{\overset{k_1}{\rightleftharpoons}} SE^*S \xrightarrow{k_2} SE^* + P$$

gives the rate equation (2.12), where E^* = active form of enzyme and E = inactive form of enzyme

$$v_0 = \frac{V_{max}}{1 + (K_m/[S])(1 + K_S/[S])} = \frac{V_{max}}{1 + K_m/[S] + K_mK_S/[S]^2}$$

* J. C. Gerhart, and A. B. Pardee, *Cold Spring Harbor Symp.*, **28**, 491 (1963).

† For simplicity the two-substrate ATCase reaction is treated as a one-substrate system. In practice this is essentially how the experiments are done. One of the substrates (carbamyl phosphate) is added in excess and the effect of the concentration of the second substrate (aspartate) on the rate is evaluated.

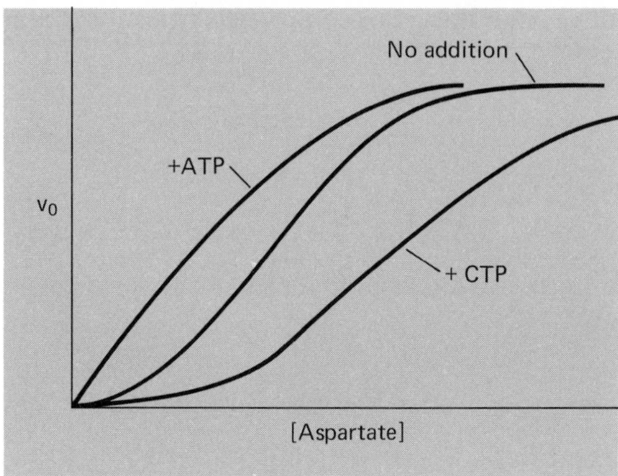

(*a*) Native enzyme

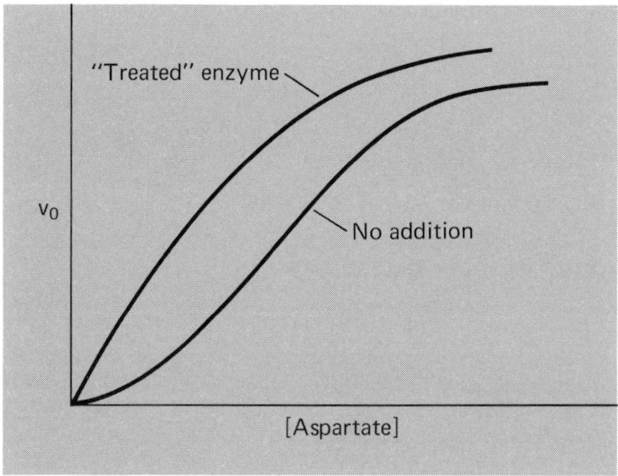

(*b*) Effect of different enzyme treatments

Figure 3.21 The kinetic behavior of aspartate transcarbamylase (ATCase). The curves are drawn in a slightly exaggerated manner to illustrate the characteristic features of the different effects. The rates were measured as a function of the aspartate concentration at constant, high levels of carbamyl phosphate, the second substrate. (*a*) The effect of the negative effector, CTP, and the positive effector, ATP, on the rate of the reaction. Note that V_{max} is constant, while there is a pronounced influence of the effectors on the apparent K_m. (*b*) A comparison of the kinetic characteristics of the reaction of the native enzyme and "treated" enzyme. Several treatments (see text) will give the observed effect. In this particular case the treatment was simply addition of 10^{-8} *M* Hg^{++}. [Taken from the data of Gerhart, J. C., *Brookhaven Symposium in Biol.*, **17**, 222 (1964).]

In the presence of CTP the sigmoidal shape is even more pronounced but it must be noted that at high substrate concentrations the same rate is reached in the presence and absence of CTP and, by definition, this is the criterion for a competitive inhibition. The competitive type of inhibition by CTP, represented by I, can be described by the mechanism

$$E + I \overset{K_I}{\rightleftharpoons} IE^+ \text{ (inactive)}$$

$$E + S \overset{K_S}{\rightleftharpoons} SE^* \text{ (active)}$$

$$SE^* + S \underset{k_{-1}}{\overset{k_1}{\rightleftharpoons}} SE^*S \xrightarrow{k_2} SE^* + P$$

This represents two allosteric forms of the same enzyme (designated E^+ and E^*), and

$$v_0 = \frac{V_{max}}{1 + (K_m/[S])[1 + (K_S/[S])(1 + [I]/K_I)]}$$

Interestingly it was also found that ATP acts as an activating effector in this system; thus when ATP is added the sigmoidal curve is altered to give the normal Michaelis–Menten hyperbolic rate curve. Again the mechanism can be expressed most simply as

$$E + A \overset{K_A}{\rightleftharpoons} AE^*$$

$$AE^* + S \underset{k_{-1}}{\overset{k_1}{\rightleftharpoons}} AE^*S \xrightarrow{k_2} AE^* + P$$

where A = activator = ATP implying that the same conformation, E^*, is formed in the activation with either S or A but not necessarily that the site is the same for S and A. Then,

$$v_0 = \frac{V_{max}}{1 + (K_m/[S])(1 + K_A/[A])}$$

Figure 3.22 Molecular properties of ATCase, as observed in the ultracentrifuge. (*a*) The sedimentation pattern of untreated and mercury treated enzyme (10^{-8} M Hg^{++}) in the analytical centrifuge. (*b*) Sedimentation behavior of the mercury treated enzyme in a sucrose density gradient in the preparative ultracentrifuge. In this experiment it is possible to collect the content of the centrifuge tube by punching a small hole in the bottom and allow the content to slowly drip out. The two components can thus be separated. (*c*) Demonstration of CTP binding to component R. The profiles represent an idealized scan of absorbance (298 nm) across the ultracentrifuge cell. The protein is transparent, and only CTP concentration (the 5-Br-derivative was actually used) is measured. The tracing at the left shows the distribution of CTP across the cell in the control sample (no protein added). Addition of C-protein gives an identical pattern. When R-protein is added, however, part of the free CTP (60 percent in the illustration) is removed and sediments with the protein. From the amount of CTP bound by different amounts of protein, a titration curve can be constructed and the number of sites can be determined. [The data is from Gerhart, J. C., and H. K. Schachman, *Biochem.*, **4**, 1054 (1965).]

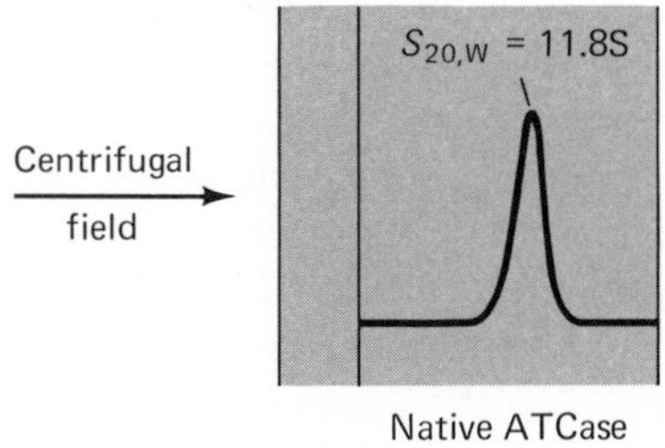

Native ATCase

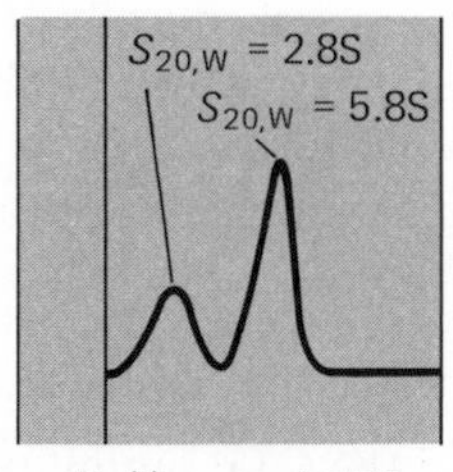

Hg^{++}-treated ATCase

(*a*)

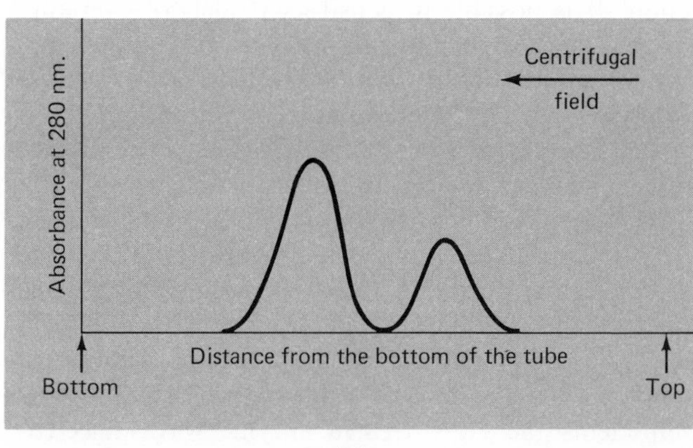

(*b*)

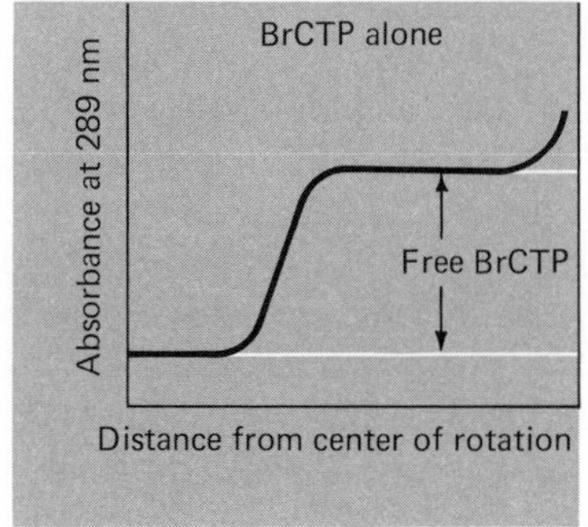

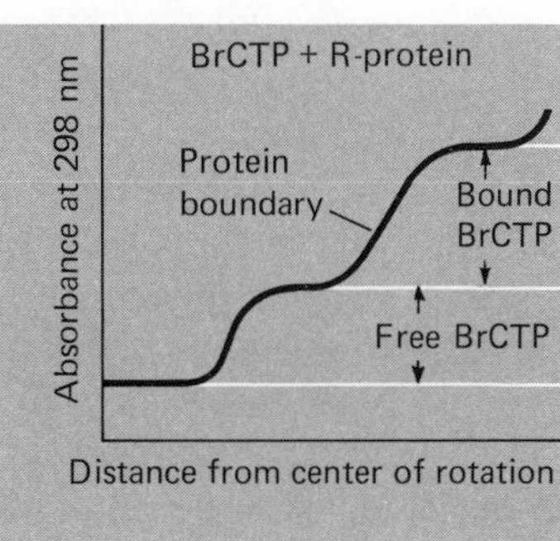

(*c*)

The Allosteric Model

The involvement of more than one binding site and more than one conformational form of the enzyme is implicit in all these mechanisms but the fact that the kinetic data are in agreement with the above mechanism does not constitute proof that the mechanisms are correct. A demonstration that E^* and E^+ are indeed different conformations of the same enzyme, and that the CTP site and the substrate site are chemically distinct would, however, provide direct evidence for the proposed regulatory mechanism. Both of these points have now been unequivocally established. First of all a number of treatments that would convert the sigmoidal rate curve to the normal hyperbolic curve were established (Figure 3.21). For example, by heating the enzyme at 60°C for 4 min, by treating it with urea solutions or simply by assaying in the presence of Hg^{2+} or *p*-mercuribenzoate, it was possible to eliminate the substrate activation, the ATP activation, and the CTP inhibition.

These results indicate that the regulatory sites are indeed distinct from the catalytic site in that physical and chemical modification apparently destroyed one without destroying the other. It is not, however, clear what "distinct" means in this connection. It is possible to construct models of overlapping sites, for example, where modification of only one bit of structure unique to the inhibitor could destroy the inhibitor site without affecting catalysis. Such a model does not absolutely require any conformational modification.

The ultimate proof for the allosteric mechanism was obtained from the physical characterization of the enzyme before (when inhibited by CTP and showing a sigmoidal rate curve) and after (without CTP inhibition, showing a hyperbolic rate curve) treatment (Figure 3.22). Before treatment the enzyme in the ultracentrifuge had a molecular weight of 310,000 and sedimented as a single symmetrical band with a sedimentation coefficient, $S_{20,w}$, of 11.8 S (Svedberg units). After treatment the sedimentation-velocity patterns showed two distinct peaks, one with $S_{20,w} = 5.8$ S and one with $S_{20,w} = 2.8$ S. In order to separate the two components the Hg^{2+}-treated enzyme was subjected to sucrose density-gradient sedimentation, giving the pattern shown in Figure 3.22(*b*). The separated fractions could now be studied individually and certain characteristics were established. Component *C* (the catalytic protein) makes up 68 percent of the original enzyme, has an $S_{20,w}$ of 5.8 S, and has a molecular weight of 100,000. It shows full transcarbamylase activity with a normal hyperbolic rate curve and is not affected by CTP or ATP. Direct-binding experiments with succinate (an inactive analogue of aspartate) in the presence of an excess of carbamyl phosphate, showed 2 moles of succinate bound per 100,000 g (grams) of component *C*. Component *R* (the regulatory protein) accounts for the remaining 32 percent of enzyme mass, has an $S_{20,w}$ of 2.8 S, and has a molecular weight of 34,000. It shows no transcarbamylase activity but the experiments (again in the ultracentrifuge) shown in Figure 3.22(*c*) demonstrate that it binds 1 mole of CTP per 34,000 g of protein.

From these observations one can construct a model of the native enzyme.

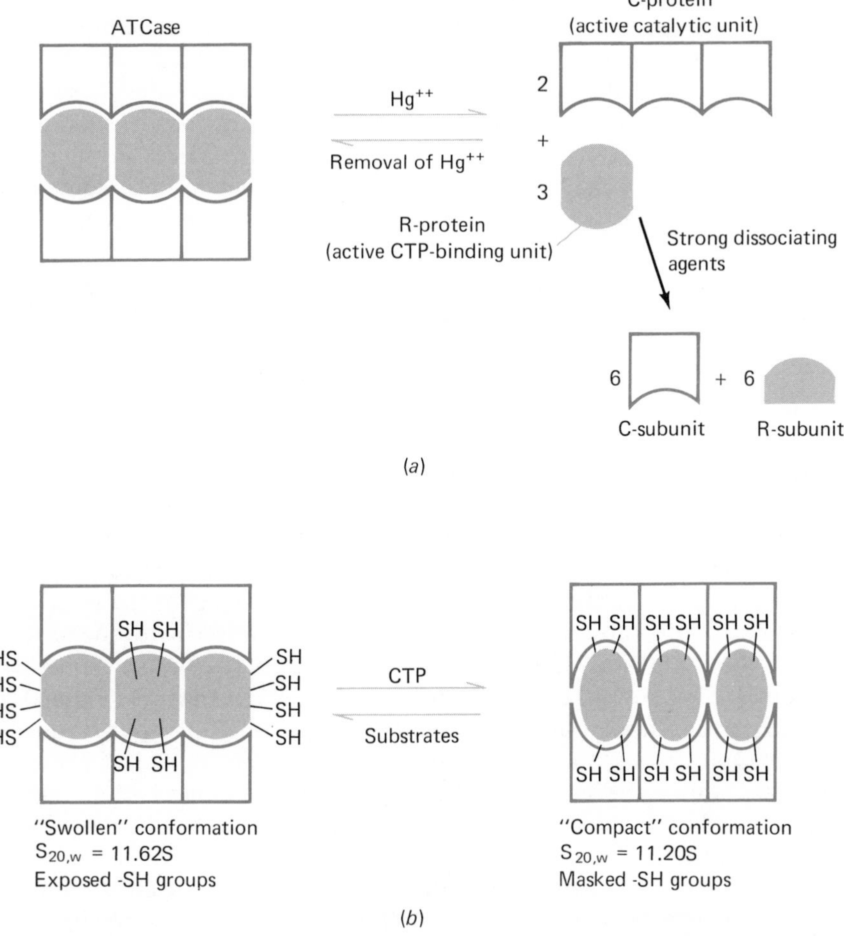

Figure 3.23 A model of ATCase. Based on recent data pertaining to the subunit structure of C-protein and R-protein, including amino acid sequences, and the finding of a three-fold and two-fold symmetry axis in the crystal structure, it is now clear that the ATCase molecule is made up of six 33,000 MW subunits of C-protein and six 17,000 subunits of R-protein. The stable, active (catalytic) C-protein thus is a C_3 trimer, and the stable, active (CTP-binding) R-protein an R_2 dimer, and the intact ATCase molecule has the molecular composition $(R_2)_3 - (C_3)_2$. (*a*) An attempt to illustrate in two dimensions the architectural make-up of ATCase. (*b*) An illustration of the two distinct conformational (allosteric) forms of ATCase.

In order to account for the molecular weight of 310,000 we must use two *C* and three *R* components in some kind of array to present the active molecule in such a way that the two kinds of sites interact (see model in Figure 3.23). In some manner Hg^{2+} or *p*-mercuribenzoate disrupts the forces of interaction between the two kinds of subunits, resulting in dissociation into individual subunits.

As a final piece of evidence for this model the mercury was removed from each of the isolated subunits and recombination experiments were attempted. The Hg^{2+}-free component C still showed an $S_{20,w}$ of 5.8 S, and the Hg^{2+}-free component R showed an $S_{20,w}$ of 2.8 S; hence the isolated subunits show no tendency to recombine with themselves to give higher-molecular-weight aggregates. When Hg^{2+}-free C and R components were mixed in the proper ratio, however, a component with $S_{20,w} = 11.8$ S resulted and this component showed all the properties of the native enzyme (a sigmoidal rate curve, CTP inhibition, and ATP activation).

The second part of the evidence for the allosteric model is the establishment of two distinct conformations of ATCase (E^* and E^+ in the above mechanism). The experimental approach used in obtaining such evidence was based on chemical reactivity and sedimentation velocity (see Table 2.2) and the results summarized in Figure 3.23 clearly demonstrate that two distinct forms of ATCase exist. Aspartate transcarbamylase contains 32 cysteine residues (24 in the regulatory subunits and 8 in the catalytic subunits). The free —SH groups of these cysteine residues react readily with *p*-mercuribenzoate and the reaction leads to dissociation. When the rate of reaction of *p*-mercuribenzoate with ATCase was determined in the presence of carbamyl phosphate and succinate a sixfold increase in the rate was observed over that found for the reaction in the absence of the substrates. Carbamyl phosphate or succinate alone had no effect. When CTP was added the rate enhancement by the substrates was eliminated. Similarly, when the sedimentation coefficient was determined for free enzyme and for enzyme in the presence of carbamyl phosphate and succinate, the presence of substrates was found to cause a decrease of 3.6 percent in the sedimentation coefficient. Since a single homogeneous form of the enzyme was observed, no dissociation can be involved; the decrease in sedimentation velocity thus must be due to an increase in the frictional coefficient of the enzyme or, in other words, to a "swelling" of the enzyme molecule. Again, the addition of CTP opposed the substrate effect. Both of these observations demonstrate the existence of two different and distinct conformations of ATCase (Figure 3.23).

It should be restated that the above results were obtained with the *E. coli* ATCase. The same enzyme from mammalian sources shows none of these regulatory features.

This system illustrates several unique properties of proteins. Subunit structure as the basis for regulatory mechanism is one of these. It is certainly a significant concept that an organism having a low-molecular-weight catalytic protein has evolved, in addition, a regulatory subunit that specifically recognizes the catalytic unit and forms a stable aggregate; thus the enzyme can be turned on and off as the end products of its catalytic function decrease or increase. Several cases of feedback-regulated enzymes are now known (see Table 3.12) but few are as well characterized as aspartate transcarbamylase.

Considering all the features of the ATCase model it is clear that the protein subunits making up the complex enzyme have several types of sites. Thus in addition to substrate-catalysis sites there are substrate-activation sites, CTP-inhibition sites, ATP-activation sites (*probably* not the same as the substrate-activation site), and finally there must be a number of specific *subunit-recognition-and-binding sites.** These last-mentioned sites determine both the specific stoichiometric composition of the functional complex and the long-range interactions between the various binding sites that lead to regulation. With all these functional properties resulting from unique amino acid sequences in specific conformational forms, the old idea that the major part of the enzyme molecule was an unnecessary appendage to the important active-site structure can certainly be laid to rest for good. With regard to a better-defined and more detailed model for both the subunit interactions and the different binding sites of ATCase, it is noteworthy that the complete primary structure of the regulatory subunit has recently been completed† and that of the catalytic subunit is nearly completed. Work is also in progress on the X-ray crystallography of the intact enzyme; thus ATCase may continue to provide important new information for structure-function relations.

3.6 PYRUVATE DEHYDROGENASE COMPLEX (MULTIENZYME COMPLEXES)

The pyruvate dehydrogenase (decarboxylase) complex is discussed next as an example of an isolated enzyme complex. Understanding the operation of this type of complex, made up of several individual enzymes to form a complicated functional unit, may well represent the first step toward understanding how enzymes really function in vivo.

Practically all our knowledge of structure-function relations of proteins and in particular, of enzymes, is derived from studies of isolated, pure proteins. This stage of biochemical development was and is an inevitable and important one. However, as already mentioned in Chapter 1 the interpretation of this type of information only applies to "the pure enzyme in dilute aqueous solution;" the extrapolation from there to the intact living cell is a long and hazardous one. Just as inevitably as fractionation and definition of the isolated units of function had to be a step in the process of biochemical development, the current attempts at putting the proteins back into their native environments to study their action in the living cell are logical and inevitable consequences of the information obtained from the isolated systems. Any inadequacy of current knowledge is not connected with the study of isolated molecules for the purpose of defining their individual functional properties but rather with the attempts to understand the biochemistry of the intact cell in terms of the information

* The 70 to 80 sequence in cytochrome c must, according to the proposal in Section 3.4, represent such a protein-protein–interaction site, determining the proper interaction with its functional partner, cytochrome oxidase. Other examples of specific protein interactions are the trypsin-trypsin-inhibitor complexes, and the type of enzyme complexes discussed in the next section.

† K. Weber, *Nature*, **218**, 1116 (1968).

obtained from the individual units. The biochemical operation of a cell requires the integration of many reactions and the main pitfall inherent in the current approach is the assumption that the properties of any series of reactions is a simple sum of the properties of the individual reactions.

Several enzyme complexes—functional and structural units built up from several different enzymes—have been studied in recent years and it now is possible to see the first outlines of the very involved operational units that must make up the functional apparatus of the cell. On the one extreme of complexity are the thoroughly studied respiratory and photosynthetic electron-transport chains, tightly associated lipoprotein complexes bound in the structure of the mitochondrial or chloroplast membranes (see Chapter 10). Next on the scale of complexity are the "soluble" multienzymic complexes, readily extractable functional units such as mammalian and yeast fatty acid synthetase (see Table 3.12), and the α-keto acid–dehydrogenase complexes, one of which will be considered here.

General Description

The oxidative decarboxylation of pyruvate to yield acetyl-S-CoA can be described by the following steps:

$$CH_3{-}\overset{\overset{\large O}{\|}}{C}{-}COO^- \xrightarrow{\text{(E}_1\text{)}-TPP} CH_3{-}\underset{\underset{\large OH}{|}}{\overset{\overset{\large H}{|}}{C}}{-}TPP{-}\text{(E}_1\text{)} + CO_2$$

$$\text{(E}_2\text{)}{-}\text{lipoate}{-}S_2$$

$$CH_3{-}\overset{\overset{\large O}{\|}}{C}{-}S{-}\text{lipoate}{-}SH{-}\text{(E}_2\text{)}$$

$$\text{CoA}{-}SH \longrightarrow HS{-}\text{lipoate}{-}SH{-}\text{(E}_2\text{)} \qquad \text{(E}_3\text{)}{-}FADH_2 \qquad \text{(E}_3\text{)}{-}FAD \qquad NAD^+ \rightarrow NADH + H^+$$

$$CH_3{-}\overset{\overset{\large O}{\|}}{C}{-}S{-}\text{CoA}$$

The complex that catalyzes the total reaction involves three enzymes: pyruvate decarboxylase (E_1), with thiamin pyrophosphate (TPP) as coenzyme; lipoyl reductase–transacetylase (E_2) with two coenzymes, lipoic acid* and CoA

* Lipoic acid,

$$\underset{CH_2CH_2CH{-}(CH_2)_4{-}COO}{S\text{———}S}$$

is covalently bound to the enzyme through an amide linkage with a lysine ε-amino group in the protein and should therefore be classified as a prosthetic group. Coenzyme A is used up in the series of steps indicated and should therefore be viewed in the strictest sense as a substrate rather than a coenzyme. Similarly for E_3, FAD is a prosthetic group and NAD is a substrate.

(coenzyme A); and dihydrolipoyl dehydrogenase (E_3), with two coenzymes, FAD and NAD (Table 3.11). A complex of these enzymes has been isolated from different sources; the *E. coli* complex has been studied most extensively.*

Application of some of the conventional isolation procedures (such as protamine treatment or isoelectric precipitation) to extracts of *E. coli* allows the intact pyruvate dehydrogenace multienzyme complex to be obtained in pure form. (Other conventional methods such as ammonium sulfate fractionation will cause dissociation; flavoprotein, E_3, is readily released from the complex.) The molecular weight of the isolated, fully active complex is approximately 4.8×10^6 and when it is added to a mixture of pyruvate, NAD^+, and CoA—SH it catalyzes the overall reaction

$$CH_3COCOO^- + NAD^+ + CoA{-}SH \longrightarrow CH_3{-}\overset{O}{\overset{\|}{C}}{-}SCoA + NADH + H^+ + CO_2$$

Dissociation and Reassociation

The flow sheet in Table 3.11 indicates how the intact complex can be dissociated and gives some properties of the component enzymes.

Based on the molecular weights in Table 3.11 the native complex should consist of 12 molecules E_1 (183,000 × 12 = 2,200,000), 6 molecules E_3 (112,000 × 6 = 670,000) and 24 E_2 subunits (24 × 70,000 = 1,600,000).

* See reference in legend to Figure 3.24 and L. J. Reed, and D. J. Cox, *Ann. Revs. Biochem.*, **35**, 57 (1966).

TABLE 3.11 DISSOCIATION OF THE PYRUVATE DEHYDROGENASE COMPLEX

(*a*) Some properties of the component enzymes.

$[E_1 - E_2 - E_3]$ Complex

Chromatography in 4 *M* urea → $E_3 + [E_1 - E_2]$ (MW = 112,000); $[E_1 - E_2]$ —pH 9.5→ $E_1 + E_2$

pH 9.5 → $[E_2 - E_3] + E_1$ (MW = 183,000); $[E_2 - E_3]$ —4 *M* urea→ $E_3 + E_2$ (MW = 1.6×10^6) $\underset{pH\,7}{\overset{pH\,2.6}{\rightleftharpoons}}$ E_2 (inactive) (MW = 70,000)

(*b*) Relative reaction rates of individual enzymes:

Reaction	*Relative rate*
1. $CH_3{-}COCOO^- + CoA\text{-}SH + NAD^+ \xrightarrow{E_1, E_2, E_3} CH_3{-}CO\text{-}S\text{-}CoA + NADH + H^+ + CO_2$	100
2. $CH_3COCOO^- + lipS_2(free) \xrightarrow{E_1}$ acetyl-S-lipSH(free) + CO_2	1
3. $CH_3CO\text{-}S\text{-}CoA + lip(SH)_2(free) \xrightarrow{E_2} CH_3CO\text{-}S\text{-}lipSH + CoA\text{-}SH$	70
4. $LipS_2 + NADH + H^+ \xrightarrow{E_3} Lip(SH)_2 + NAD^+$	700

If the three individual enzymes are mixed together in any proportion, a high-molecular-weight complex forms spontaneously, resembling the native complex in composition, activity, and morphology. It is interesting to note that E_1 and E_3 do not aggregate when mixed but that either one will form aggregates with E_2. This is in complete accord with the observations on the dissociation and must be accounted for in the model of the complex.

The study of dehydrogenaseless *E. coli* mutants (see Chapter 7) has been useful in developing the model.* These mutants fall into three groups: (1) those lacking E_1; (2) those lacking E_2, and (3) those lacking E_1, E_2, and E_3. Although E_2–E_3 complexes can be isolated from mutants of type 1, only free E_1 and E_3 (no complexes) can be obtained from mutants of type 2, just as predicted by the dissociation-reconstitution data.

It is also of interest that in the wild type the structural genes of E_1 and E_2 are closely linked (close neighbors on the chromosome), E_1 is synthesized first, and in spite of the unequal quantities of E_1, E_2, and E_3 required for intact complex it appears that the rate of synthesis is regulated so that the components are produced in exactly the right proportions for complex formation.

The fact that the three enzymes form complexes according to a fixed stoichiometry has been used as an argument both for and against the proposition that the isolated complex is the native functional unit. Certainly the ease with which the dissociated system is reconstituted clearly demonstrates that even if the enzymes existed free and separate in the cell they could well aggregate during the isolation. The evidence that the isolated complex is indeed the native complex is indirect and rests entirely on the above observations that the three enzymes are synthesized in the right proportion for complex formation and that quantitatively very little of any of the three activities are found free of the complex structure. If one accepts the proposition that life is efficient and avoids waste then the argument that the enzymes are synthesized in fixed proportions for a purpose (complex formation) is not a bad one.

The data presented above strongly suggest that the pyruvate dehydrogenase complex is not an artifact of random aggregation but rather a highly specific association of enzymes in a fixed stoichiometry. The forces of interaction are noncovalent and if one accepts the proposition that the interaction is specific, then the amino acid sequences of the individual enzymes must contain the information that goes into the protein-protein recognition and interaction.

What is the architecture of the pyruvate dehydrogenase complex? The answer to this question has been suggested by the application of the rapidly developing techniques of electron microscopy and specific specimen shading and staining. The results of studies involving these techniques are given in Figure 3.24 and led to the current model. It is of interest that the model suggests the central coupling role of E_2 relative to E_1 and E_3.

At the beginning of this section it was suggested that the weakness of in vitro and in vivo extrapolations must be primarily in the attempt to understand how a chain of events can be related to each individual component step as studied

* U. Henning, *et al.*, *Cold Spring Harbor Symp. Quant. Biol.*, **31**, 227 (1966).

separately. It was proposed that the whole is not necessarily the simple sum of the components and it was implied that functional complexes as might exist in vivo would be superior to the corresponding mixture of individual enzymes. The pyruvate dehydrogenase complex should provide an excellent test system for this hypothesis since the three component enzymes can be studied separately as well as in the complex. Unfortunately only limited relevant information is

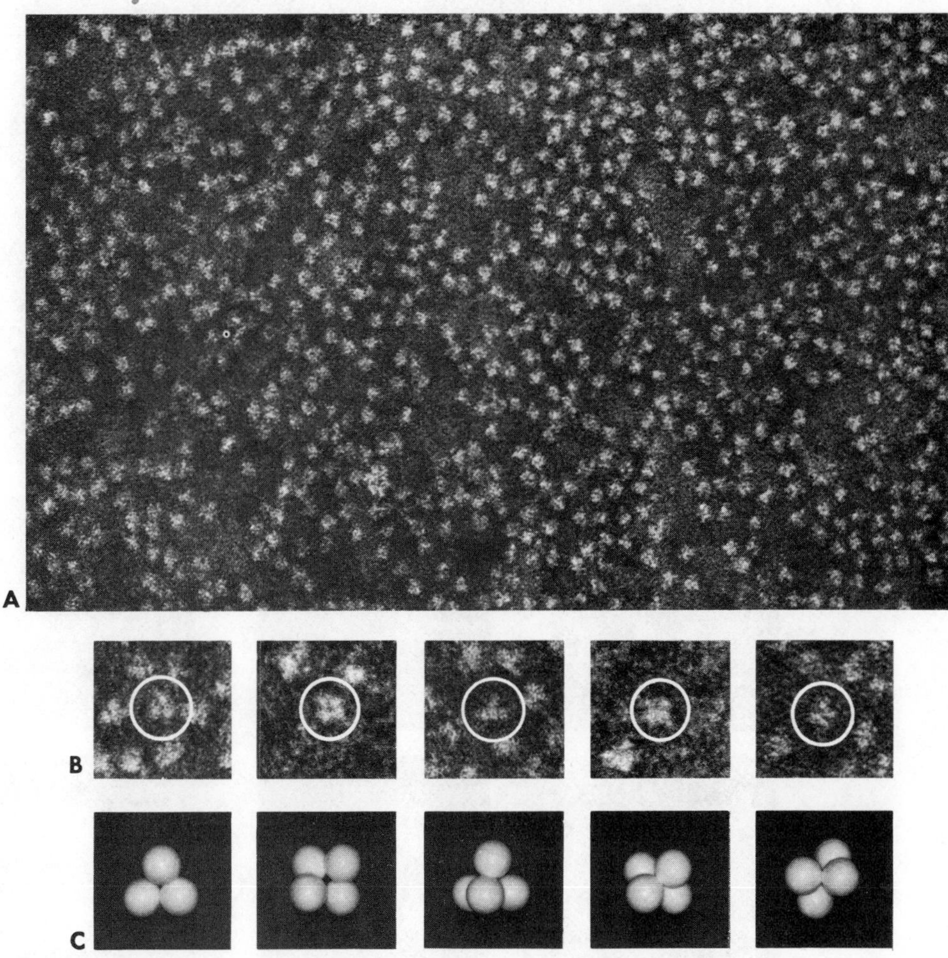

Figure 3.24 Electron micrographs of individual components, subcomplexes, and the complete (*E. coli*) pyruvate dehydrogenase complex. Reproduced with permission from Reed, L. J., and R. M. Oliver, *Brookhaven Symp. in Biol.*, **21**, 397 (1969). Frame 1: The tetrameric aggregate of the pyruvate decarboxylase component (E_1, also referred to as "pyruvate dehydrogenase"). A. Electron micrograph ($\times$ 400,000). B. Selective images from A, showing different orientations ($\times$ 700,000). C. Interpretative models corresponding to the images in B.

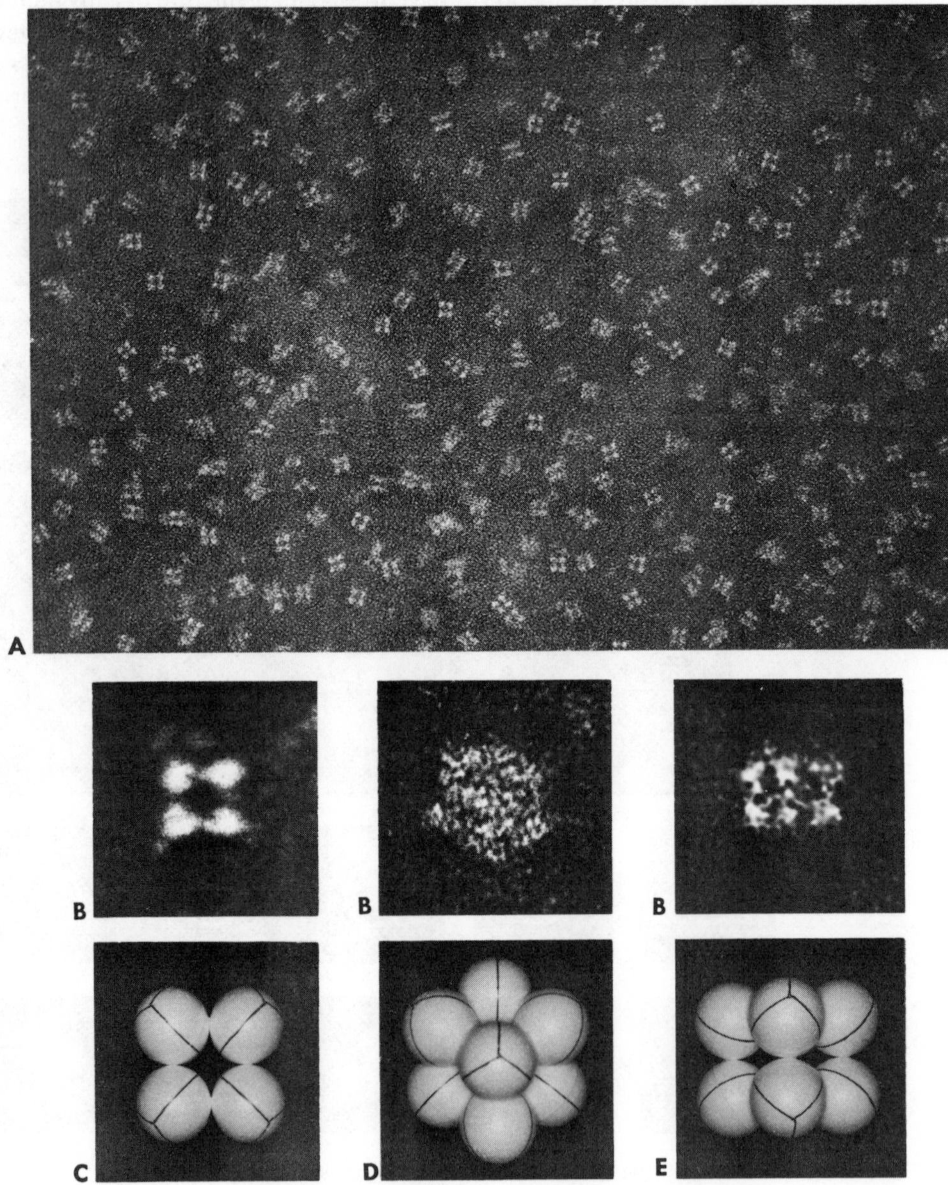

Figure 3.24 Frame 2: The "octameric" aggregate of the dihydrolipoyl transacetylase component (E_2). A. Electron micrograph (×200,000). B. Selected predominant images from A (×870,000). C, D and E. Interpretative models corresponding to the images in B. Each sphere is made up of 3 identical subunits (indicated by black lines) and the aggregate is interpreted to consist of 8 spheres (24 subunits) at the vertices of a cube.

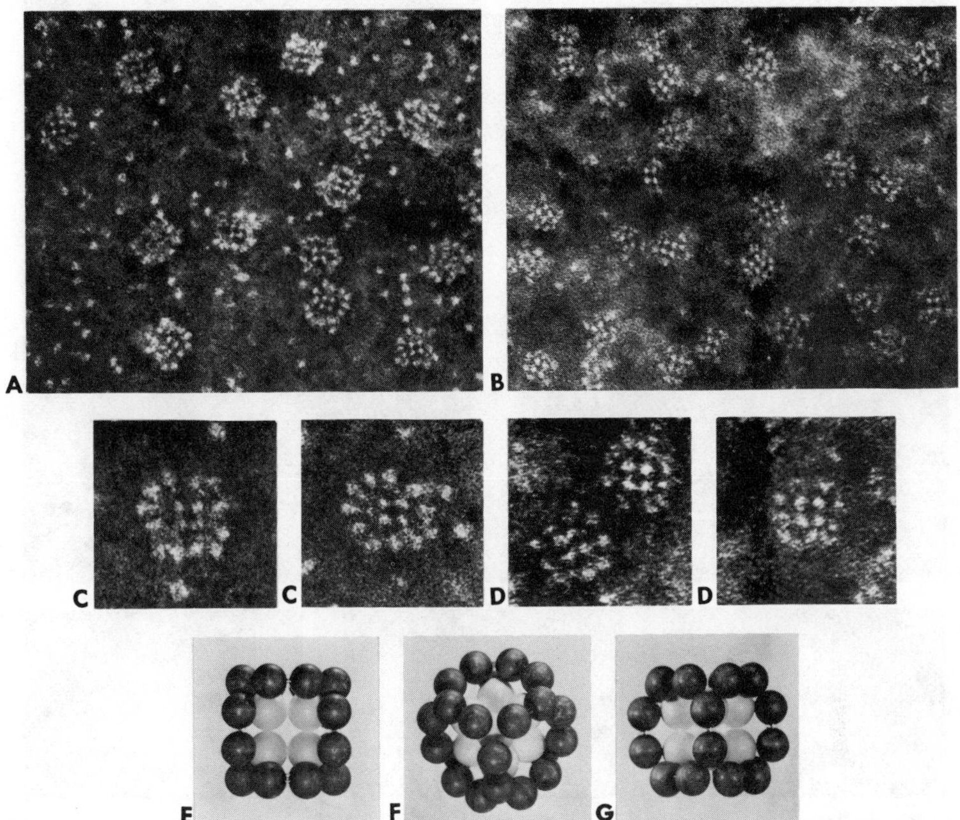

Figure 3.24 Frame 3: A. Electron micrographs of the transacetylase–decarboxylase subcomplex ($E_1 - E_2$) ($\times$200,000). B. The transacetylase–dihydrolipoyl dehydrogenase subcomplex ($E_2 - E_3$) ($\times$200,000). C and D. Selected images of the two subcomplexes ($\times$400,000). E, F, and G. Interpretative model of the transacetylase–decarboxylase subcomplex, representing the decarboxylase as 24 black spheres (24 $\times$ 90,000 monomer units or twelve 182,000 dimers) orderly arranged on the transacetylase cube from Frame 2.

available on this point. The model reactions given in Table 3.11 with relative overall rates indicate the status of this aspect of the problem.

The main drawback with this comparison is that free lipoic acid (or lipoamide) was used as a substrate in reactions (2), (3) and (4) in Table 3.11 whereas in (1) it is the covalent lipoyl moiety of E_2 that is functional. The results suggest that the rate-limiting step may be the decarboxylation and reductive acetylation of lipoic acid (transfer from α-hydroxyethyl thiamin pyrophosphate to lipoic

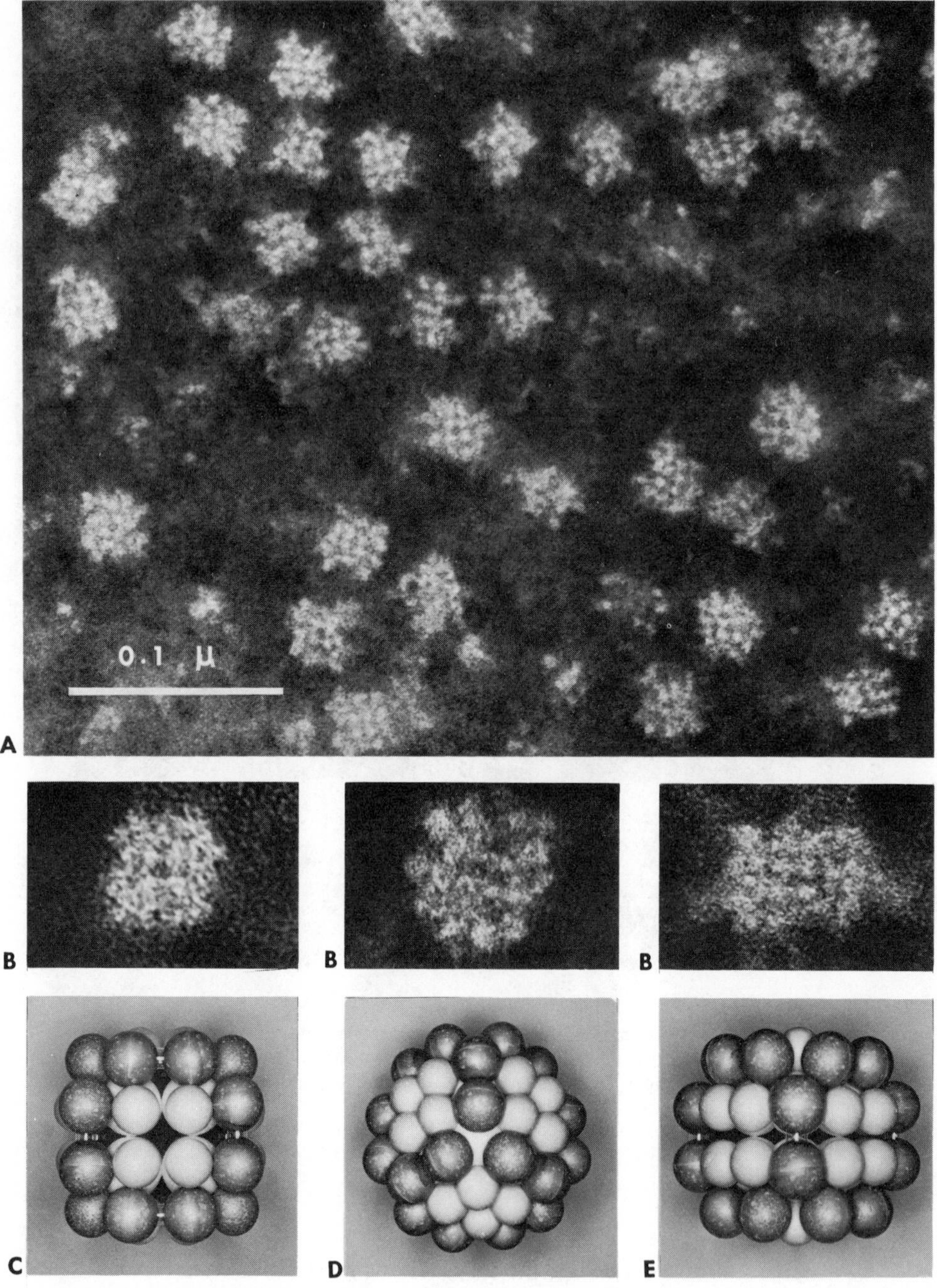

Figure 3.24 Frame 4: The pyruvate dehydrogenase complex. A. Electron micrograph (× 300,000). B. Selected images (× 600,000). C, D, and E. Interpretative model of the complex. The three views of the transacetylase cube (Frames 2 and 3) again represent the core, the 24 black spheres the 24 monomers of the decarboxylase (as in Frame 3), and the 6 aggregates of 4 small white spheres occupying the remaining space represent the six 112,000 MW dihydrolipoyl dehydrogenase components (twenty-four 30,000 MW subunits).

acid). This would intuitively seem right. If the only acceptor available were the E_2-bound lipoyl groups then a molecular complex of E_1 and E_2 would be much more efficient than a system in which interaction between free E_1 and free E_2 would be controlled by diffusion. The subsequent reoxidation of E_2-lip(SH)$_2$ would also be more efficient in an E_2-E_3 complex for the same reason. The functional complex could therefore be viewed as

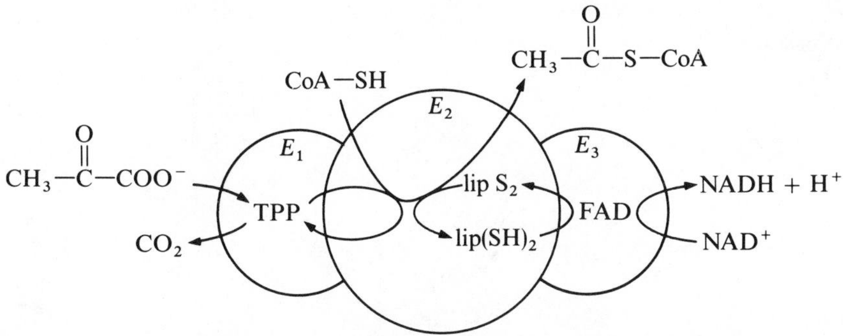

in good general agreement with the observed architecture.

It should be mentioned that pyruvate dehydrogenase complexes from other sources are very similar to the one described here. Also, the α-ketoglutarate dehydrogenase complexes are essentially the same except that specificity of E_1 and E_2 for the acyl group and the stoichiometry of enzymes making up the complexes are different.

3.7 CONCLUSIONS

The "case histories" discussed in this chapter were selected to illustrate some relevant features of our present knowledge of enzymology and protein structure and function. A very large number of other enzymes could have illustrated the same points equally well and the selection for this chapter obviously is quite arbitrary. For anyone interested in a broader and more complete documentation of studies on the structure and function of enzymes, the list in Table 3.12 should provide some starting points for further study.

The discussion of the six enzymes in this chapter perhaps can be summarized as follows:

1. Enzymes can always be described in terms of at least one functional parameter, catalysis; the structural components immediately involved in catalysis—the active site—can in principle be characterized. Based on model reactions with simpler components this information can be translated into mechanistic models, which are extremely difficult to prove.

2. In addition to the catalytic site, which constitutes a rather small part of the total protein molecule, we must consider a number of other sites or

TABLE 3.12 SOME ENZYME SYSTEMS OF INTEREST FOR FURTHER STUDIES

1. Extensively Studied Relatively Simple Enzymes (crystal structure of free enzyme as well as enzyme-substrate analog complexes, extensive biochemical and chemical characterization)

   Egg white lysozyme[a]
   Carboxypeptidase A[b]
   Micrococcal nuclease[c]
   Ferredoxin[d]

2. Isozymes: multiple forms of enzymes

   Creatin kinase[e]
   Hexokinase[f]

3. Allosteric Systems: regulated enzymes

   Glutamine synthetase[g]
   CTP synthetase[h]
   Fructose 1,6-diphosphatase[i]

4. Enzyme Complexes

   Fatty acid synthesis[j]
   Trypotophan synthetase[k]

[a] L. N. Johnson, et al., *Brookhaven Symposia in Biol.*, **21**, 120 (1968), and R. L. Hill, et al., *Ibid.*, **21**, 139 (1968).

[b] W. N. Lipscomb, et al., *Ibid.* **21**, 24 (1968), H. Neurath, et al., *Ibid.* **21**, 1 (1968), and B. L. Vallee, and J. F. Riordan, *Ibid.* **21**, 91 (1968).

[c] P. Cuatrecasas, et al., *Ibid.*, **21** 772 (1968).

[d] H. Matsubara, et al., *Ibid.*, **21**, 207 (1968).

[e] D. M. Dawson, et al., *Ann. N.Y. Acad. Sci.*, **151**, Art. 1, 616 (1968).

[f] J. Gazieth, et al., *Ibid.*, 307 (1968), R. T. Schimke, and L. Grossbard, *Ibid.*, 332 (1968).

[g] E. R. Stadtman, et al., *Brookhaven Symposium in Biology*, **21**, 378 (1968).

[h] A. Levitzki, and D. E. Koshland, Jr., *Proc. Nat. Acad. Sci.*, **62**, 1121 (1969).

[i] K. Taketa, and B. M. Pogell, *J. Biol. Chem.*, **240**, 651 (1965).

[j] F. Lynen, *Biochem. J.*, **102**, 381 (1967).

[k] D. A. Wilson, and I. P. Crawford, *J. Biol. Chem.*, **240**, 4801 (1965).

functional components. These include regulatory sites, as well as the sites involved in subunit-subunit interactions, those involved in interaction between molecules in multienzyme complexes, and probably also those involved in interactions with other environmental components present in the native milieu in vivo.

3. The types of forces involved in the subunit or protein interactions also play a role in stabilizing the folded, three-dimensional structure of the individual polypeptide chain. Recent developments in methodology have made it possible to determine three-dimensional structures; this may eventually make it possible to evaluate all the interaction parameters and thus permit the calculation of three-dimensional structure from knowledge of only the covalent structure.

4. Living systems apparently require more than one type of a given enzyme, as several cases of isozymes are now known. These multiple forms of enzymes may reflect a compartmentalization within the cell to such a degree that different "compartments" of a single cell may be as different as two unrelated cells.

5. In spite of tremendous technological advances leading to an equally impressive progress in our understanding of protein structure and function relationships, we are yet far away from the ultimate goal of understanding all the fine nuances of structures that are the foundation of biological function. In fact we are just beginning to understand the complexity of properties that make up the total function of a protein. Perhaps the one fundamental obstacle to our having a clearer picture right now is the incompatibility of our methodology with the very nature of functional proteins. If our present model of proteins as flexible, highly dynamic systems is correct, there must be a very definitely limited return from methods that are primarily static in nature.

Our method is essentially like trying to appraise a moving object by taking single snapshots—sometimes even with slow film and too long an exposure time, so the single picture is blurred. Although X-ray crystallography has undoubtedly provided the greatest impetus to understanding of structure and function in recent years by locating the position of virtually every atom in a biologically active protein crystal, this technique also most dramatically exemplifies the problems we are faced with at present. In order to locate an atom by X-ray diffraction, it has to stay still long enough to be observed; if it moves the picture immediately blurs. When substrate is added to the solid crystal the catalytic-site components go to work; thus the most relevant components are those that move the most and hence are most difficult to observe. Using inert substrate analogues allows observation of a static complex of enzyme and substrate analogue that can be and has been described very precisely. With dynamic protein systems, however, one is faced with some kind of macromolecular uncertainty principle: The less active components are well resolved but the resolution decreases as activity increases. The utility of our present methods thus decreases as we approach the active and most dynamic site of a working enzyme.

REFERENCES

General (multivolume) reference sources:

Boyer, P. D., H. Lardy, and K. Myrbäck (Eds.), *The Enzymes, 2nd Edition* (8 volumes), Academic Press, New York, 1959–1963.

Nord, F. F. (Ed.), *Advances in Enzymology*, Interscience Publishers, Inc., New York, 1941–present.

Experimental Techniques:

Colowick, S. P., and N. O. Kaplan (Eds.), *Methods in Enzymology*, Academic Press, New York, 1955–present.

A Comprehensive Text:

Dixon, M., and E. C. Webb, *Enzymes, 2nd Edition*, Academic Press, New York, 1964.

A Classic:

Haldane, J. B. S., *Enzymes*, Longmans, Green and Co., London, 1930. Reprinted by the M.I.T. Press, Cambridge, Mass., 1965.

Mechanism of enzyme action: general reaction mechanisms:

Bruice, T. C., and S. Benkovic, *Bioorganic Mechanisms*, Vols. I and II, W. A. Benjamin, Inc., New York, 1966.

Ingraham, L. L., *Biochemical Mechanisms*, John Wiley and Sons, Inc., New York, 1962.

Jencks, W. P., *Catalysis in Chemistry and Enzymology*, McGraw-Hill, New York, 1969.

Special Topics:

Calvin, M., *Chemical Evolution*, Oxford University Press, New York, 1969.

Ginsburg, A., and E. R. Stadtman, "Multienzyme Systems," *Ann. Rev. Biochem.*, **39**, 429 (1970).

Jukes, T. H., *Molecules and Evolution*, Columbia University Press, New York, 1966.

Vesell, E. S. (Ed.), "Multiple Molecular Forms of Enzymes," *Ann. N.Y. Acad. Sci.*, **151**, Art. 1 (1968).

Wilkinson, T. H., *Isoenzymes*, J. B. Lippincott Co., Philadelphia, 1966.

FOUR | NONCATALYTIC PROTEINS

In this chapter a number of proteins with varied specialized functions are discussed and an attempt is made to show how they are studied and how structure-function models have been developed.

The examples chosen are the immunoglobulins with highly specific recognition-and-binding sites; hemoglobin, a carrier protein; insulin, a regulatory protein; and actomyosin, a contractile protein complex. It should perhaps be emphasized at once that the specific proteins discussed in this chapter are not ubiquitous but are rather specialized proteins found only in higher organisms.

4.1 IMMUNOGLOBULINS

The serum* of higher animals contains proteins that exhibit the ability to selectively bind foreign substances. These proteins are the antibodies; they are generally associated with the γ-globulin fraction of serum (Figure 4.1) though some antibody activity may also be found in the β-globulin fraction. In the ultracentrifuge the γ-globulin fraction is found to be heterogeneous, 80 to 90 percent sedimenting with a sedimentation coefficient, S_{20w}, of 7 S (corresponding to a molecular weight of 150,000) and the rest-sedimenting with an S_{20w} of 19 S (corresponding to a molecular weight of 10^6). By more refined

* Blood minus cells is plasma; plasma minus fibrinogen is serum.

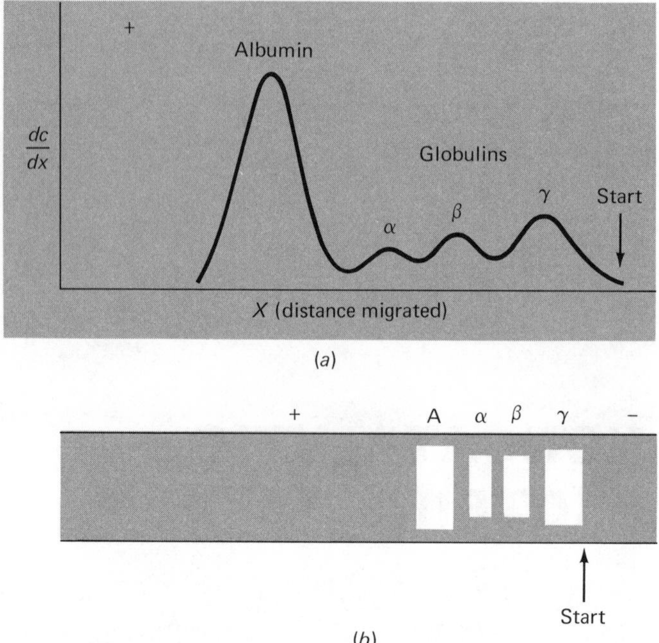

Figure 4.1 Electrophoretic patterns of serum proteins. Serum is an approximately 8 percent protein solution. Upon (*a*) boundary electrophoresis in veronal buffer at pH 8.5 or (*b*) zone electrophoresis on starch or acrylamide gel, four major electrophoretically distinct components are observed: α-, β-, and γ-globulins and albumin. Each globulin component, although electrophoretically homogeneous, represents a very heterogeneous group of proteins. The boundaries are recorded as refractive index change in (*a*), and the protein bands in (*b*) are visualized by staining.

techniques it has also been possible to distinguish a minor component with a high carbohydrate content within the 7-S fraction. All the immunoglobulins are glycoproteins. A new set of nomenclature has been set up according to which the major 7-S fraction will be referred to as immunoglobulin G (IgG or γG), the minor 7-S fraction, immunoglobulin A (IgA or γA), and the 19-S, component as immunoglobulin M (for macroglobulin) (IgM or γM).

The antibodies are synthesized in the organism in response to the presence of foreign substances (antigens). In order to elicit antibody synthesis, the antigen as a general rule must be of high molecular weight (proteins and polysaccharides being the best ones) and must be a foreign substance to the organism into which it is injected. Within this minimum requirement, however, it is possible to make virtually any substance an antigenic agent. Thus antibodies to nucleic acids have recently been made by injecting into an animal an insoluble complex of thermally denatured DNA and methylated bovine serum albumin. Antibodies to small and well-defined molecules can also be made by fixing the small molecule (in immunochemical terms, the *hapten*) to a protein carrier of high molecular weight prior to injection.

When the antiserum from an animal immunized with a certain antigen is exposed in vitro to that antigen, a portion of the γ-globulin fraction in this antiserum will specifically interact with the antigen. The interaction may manifest itself in different ways. For example, if the antigen is a protein a precipitate is formed (precipitin reaction); if the antigen is an intact cell (erythrocytes or bacterial cells) clumping or agglutination occurs; and if the antigen is a simple, monovalent hapten a soluble antibody-hapten complex is formed. The biological significance of this reaction must be the precipitation and immobilization of foreign, often toxic, agents. The insoluble material is subsequently phagocytized by lymphocytes and thus destroyed.

A fairly typical experiment designed to obtain specific antibodies is illustrated below. The antibodies are specific toward the hapten, benzene arsonic acid.

1. *Preparation of antigen-hapten*: The diazonium salt of the hapten, benzene arsonic acid (BA), is coupled to a protein carrier ovalbumin (OA) a protein with a molecular weight of 43,000, from egg white; the free tyrosine residues in the protein are the acceptor for the diazo coupling:

ovalbumin OA (–OH, –OH) + $BF_4^- \; N_2^+$–C₆H₄–$As(=O)(O^-)_2$ ⟶ ovalbumin (–OH, –N=N–C₆H₄–$AsO_3^=$; –OH, –N=N–C₆H₄–$AsO_3^=$)

p-(arsonic acid)-benzene diazonium fluoroborate

BA-diazonium salt

OABA

2. *Immunization*: The OABA is injected into rabbits in the form of an emulsion, allowing slow and steady assimilation of the antigen by the animal; after three injections over a period of several weeks, blood is collected and separated from cells and fibrinogen to give the desired serum. As controls, two separate groups of rabbits are also injected on the same schedule, one with unmodified OA and another with free BA, and the serum is collected in the same manner. Serum collected from the rabbits prior to the first injection is also included as a control.

3. *Precipitin tests*: Each of the four antisera are now tested against the three antigens, OA, BA, and OABA. Typical precipitin curves and qualitative tests are given in Figure 4.2 and the results are summarized and interpreted in Table 4.1 and Figure 4.2(c).

4. *Purification of anti-BA antibodies*: The anti-BA antibodies are clearly only one among many kinds of antibodies in the antiserum from rabbits immunized with OABA. To purify this one type of antibody, use is made of a so-called *immune adsorbent*. This is prepared by coupling BA to an inert solid

(for example, a polyphenolic resin can be treated with the BA-diazonium salt in a reaction analogous to the OABA coupling above). When the OABA antiserum is passed through a column of the BA-resin the antibodies toward BA are bound by interaction with the fixed BA molecules while all the other antibodies pass through the column. When all the other proteins have been washed off with dilute buffer, the anti-BA antibodies are obtained by elution with salt solutions strong enough to dissociate the antibody-BA complex. The resulting purified antibody can next be demonstrated to be homogeneous with respect to its reaction with BA.

We would now like to obtain the answer to the usual questions regarding this protein: How many binding sites does it have and how specific are the sites for the hapten? What is its structure and which structural features make up the binding sites? Obtaining the answers should in theory be straightforward but there is one very significant complication. Even if the isolated antibodies are homogeneous with respect to antigen (or hapten) specificity, they are highly

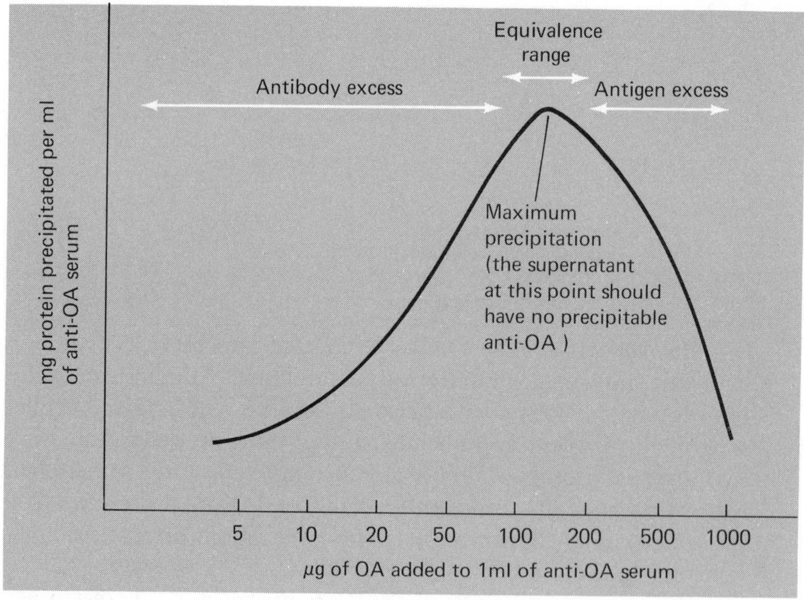

Quantitative precipitin test

Figure 4.2(*a*) The precipitin curve obtained when increasing amounts of antigen (OA) are added to the antibody solution. The precipitate is allowed to form for several hours before it is collected and washed with dilute salt solution. The equivalence point represents a measure of the potency of a given antiserum (its titre) in terms of its capacity to precipitate antigen. The precipitin test can be scaled down to microquantities by the use of a radiactive label in either the antigen or antibody (^{125}I is commonly used).

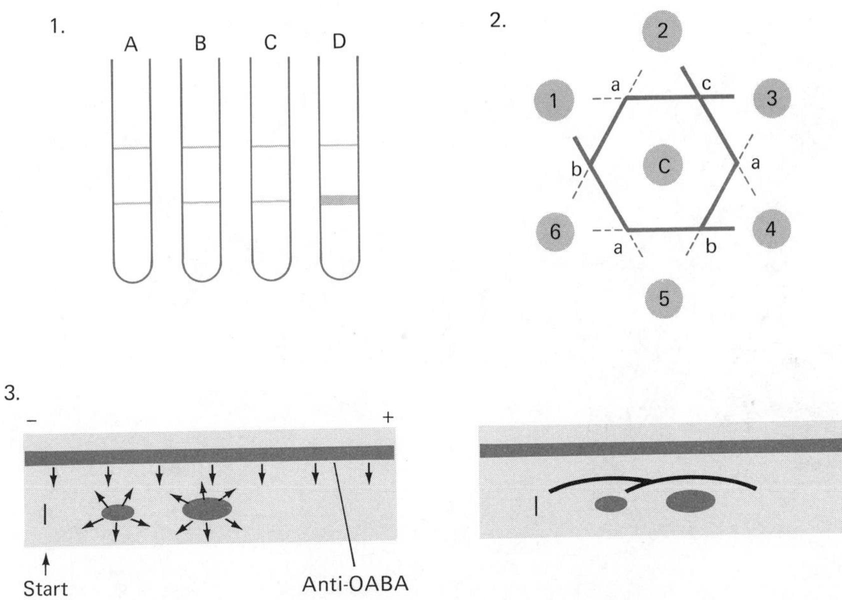

Qualitative and semiquantitative precipitin tests

Figure 4.2(*b*) Qualitative and semiquantitative microtests. (1) The ring test is a rapid test for antibody–antigen interaction. In the example illustrated, tubes A and B contain normal serum and C and D anti-OA serum. Buffer solution is carefully layered on top of the serum in tubes A and C and a solution of OA in buffer in tubes B and D in such a way that there is no mixing at the interphase between the two solution layers. After a few minutes a band of precipitate on or near the interphase is indicative of a positive test. In the example above only tube D gives a positive test. (2) Gel diffusion technique. In this test the antiserum (anti-OABA) is located in the center well (C) cut in a thin slab of poly-acrylamide gel. The antigen solutions (OA, OABA and BSABA, a complex of hapten with bovine serum albumin) are placed in wells concentrically arranged around the antibody (1, 2: OA; 3, 4: BSABA; 5, 6: OABA). The gel is incubated at elevated temperature to allow diffusion to take place. Where antibody and antigen meet (at equivalence concentrations), precipitin bands form. This procedure provides a convenient test for identity of antigens. Thus, the dotted lines represent precipitin bands which do not show up because they are dissolved in the antigen excess region, and a continuous band without "spurs" therefore demonstrates antigen identity (regions marked a). Partial antigen identity, by the same criterion, is indicated by a single spur (regions marked b), and nonidentical antigens by a double spur (region marked c). (3) Immuno-electrophoresis. This test combines gel electrophoresis and gel diffusion. The example above illustrates the separation of a mixture of OA and OABA by electrophoresis (the more negative OABA moves further towards the anode). After electrophoresis, the long well on the top of the gel slab is cut and filled with antiserum (anti-OABA in this case) and the electrophoretic species are visualized by the formation of precipitin bands. Again a test for partial identity is observed.

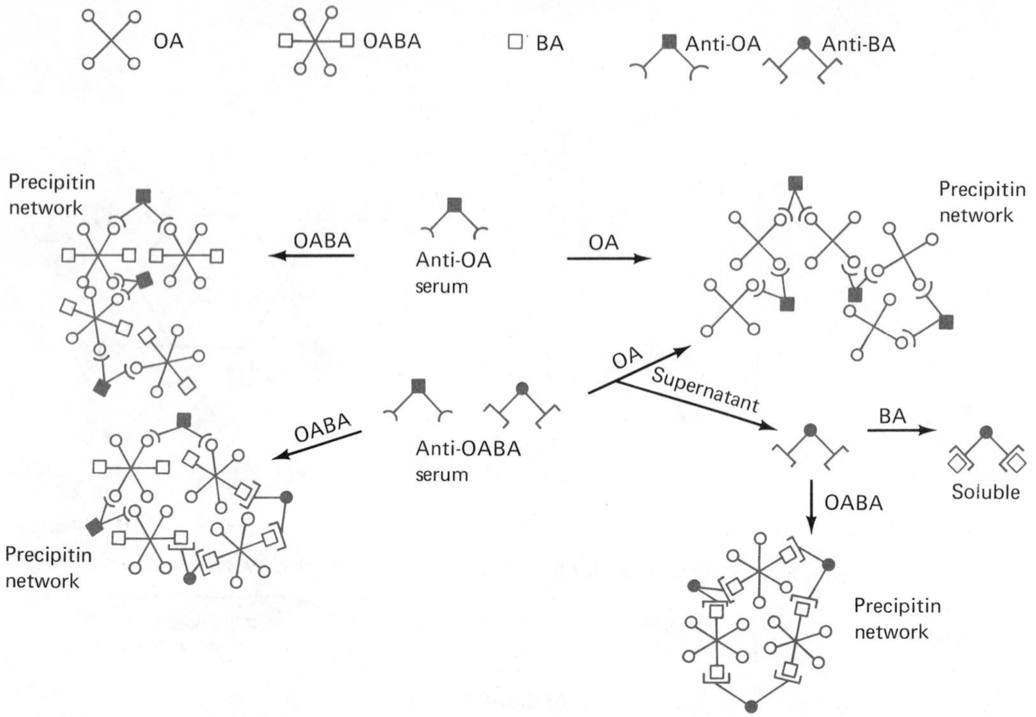

Figure 4.2(*c*) A model for the antibody–antigen interaction. The antigen is illustrated as polyvalent (having many antigenic sites), and the antibody is divalent.

heterogeneous with respect to other criteria. The scheme in Figure 4.3 shows some of the known types of antibodies. When it also appears unquestionable that all these differences are completely separate from the antibody-combining site so that any specificity can be found associated with any or all of the different types, the system becomes complicated indeed: It is like isolating an enzyme to constant specific activity and finding it to contain several hundred isozymes.

There are two reasons why tremendous progress nevertheless has been made in the structural analysis of antibodies. One of these is the fact that the difference between different types apparently is rather minor and the other is based upon the fact that mice and man, in certain pathological cases, produce large quantities of a single type of immunoglobulin. This condition is known as myelomatosis and is a cancer of the immunoglobulin-producing cells. It appears that a single cell starts dividing rapidly and producing large quantities of a single immunoglobulin called myeloma protein. (It can be inferred from this observation that the different types of immunoglobulins arise from different cells and that each cell makes only one type of immunoglobulin.) In addition to finding this high level of a single immunoglobulin in the serum of patients with this condition, some of these patients also excrete large quantities of another protein (called Bence–Jones protein) in the urine, and this protein is now known to be

TABLE 4.1 THE REACTION OF DIFFERENT ANTISERA WITH DIFFERENT ANTIGENS

Antiserum	*Precipitin Reaction With*		
	OA	*BA (Binding but No Precipitin)*	*OABA*
1 Normal Serum	–	–	–
2 Anti-OA	+	–	+
3 Anti-BA	–	–	–
4 Anti-OABA	+	– (+)	+
Supernatant after removing			
5 Precipitate of OA and anti-OA	–	–	–
6 Precipitate of OABA and anti-OA	–	–	–
7 Precipitate of OA and anti-OABA	–	– (+)	+
8 Precipitate of OABA and anti-OABA	–	–	–

The results show:

That normal serum contains no protein that will precipitate any of the antigens; injection with free hapten, benzene arsenic acid (BA), does not elicit antibodies.

That injection of either ovalbumin (OA) or ovalbumin-hapten complex (OABA) causes formation of antibodies capable of binding and precipitating both free OA and OABA. BA is not precipitated but can be shown to bind to the anti-OABA antibodies.

If the supernatant left after removing the precipitates formed at the equivalence point in the positive cases is retested with the same antigens, it is seen that OA precipitated all the anti-OA antibodies in both the anti-OA serum (5 and 6) and in the anti-OABA serum (7, 8). In the latter case, however, a fraction was left that no longer recognized OA, but still reacted with OABA and with BA.

the light chain (see below) of γG. Beginning with the myeloma proteins and the Bence–Jones proteins and subsequently continuing with normal γ-globulins, a large volume of information is now available on the structure of antibodies. Some of this information is summarized in Figure 4.4 and recent studies culminated in the successful elucidation of the complete primary structure of one myeloma protein.* However, the myeloma proteins have no known antibody activity; hence the structural work on these proteins so far has no direct interpretation in terms of function. Good progress is also being made toward determining the structure of specific antibodies in spite of the complexity of the molecules involved; based on the model in Figure 4.4, some of the answers to the questions stated above are now available.

Thus equilibrium-dialysis experiments with purified, specific antihapten antibodies have demonstrated that there are two antibody-combining sites on each molecule and the dissociation constant for the antibody-hapten is of the order of 10^{-9} M. Further data on the dimeric nature of antibodies have also been obtained. By extremely careful reduction of only two S—S groups (only one in rabbit γG), it is possible to reduce the molecular weight of γG from 150,000 to 75,000 at low pH; this corresponds to half of the molecule. By direct binding studies (equilibrium dialysis or gel filtration) it can be shown that the binding sites are still intact, one for each half-molecule, and that no precipitate can form with these monovalent derivatives. Digestion with papain also gives

* See reference in legend to Figure 4.4.

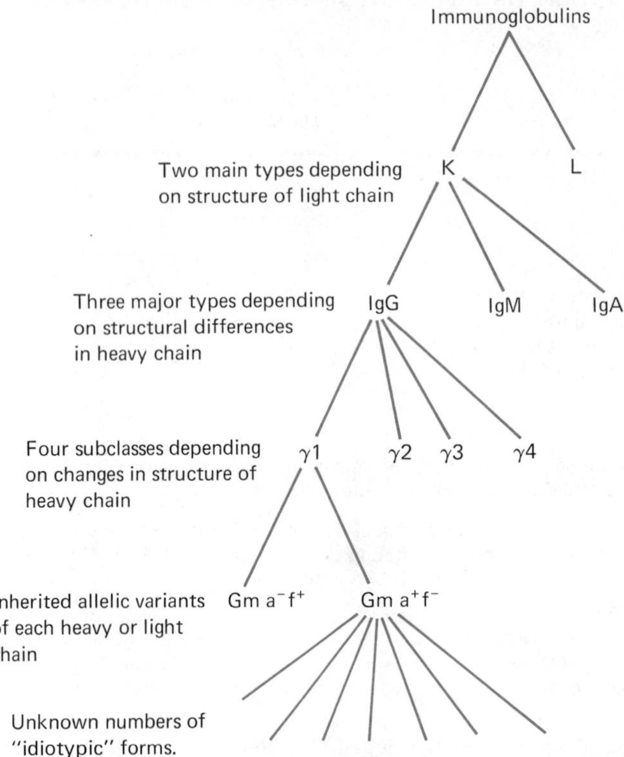

Figure 4.3 Structural variations among immunoglobulins. [Reproduced from Porter, R. R., *Sci. Amer.*, **217**, 81 (1967) with permission of author and publisher.]

Figure 4.4 Some structural properties of human γG antibody. (*a*) The basic structural make-up of γG, showing the two heavy chains and the two light chains (in solid color) that make up the complete molecule and the distribution and arrangement of the disulfide bonds. Each light chain is linked to a heavy chain through one disulfide bond involving the carboxyl terminal cystein residue in the light chain. The heavy chains are held together by two disulfide bonds only a few residues apart. Rabbit γG has the same basic structure, but one of the inter-heavy chain disulfide bonds (in parenthesis above) is missing in rabbit γG. The linear and periodic arrangements of disulfide bonds in both heavy and light chains are noteworthy. The first 100–110 residues from the amino-terminal end of the light and heavy chains vary among antibodies of different specificity, while the remainder of amino acid sequence is remarkably constant for all antibodies. It seems reasonable to propose that the antigen binding site is associated with the variable regions and that the general physiological properties of γG are expressed by the constant regions. (*b*) γG fragments produced by reduction and by enzyme attack. Fab refers to "fraction-antibody binding" and Fc to "fraction-crystalline" (the C-terminal halves of the heavy chain aggregate and crystallize quite readily). This figure is redrawn from Porter, R. R., *Sci. Amer.*, **217**, 81 (1967) and Edelman, G. M., *et al.*, *Proc. Nat. Acad. Sci.*, **63**, 78 (1969). The complete primary sequence of one human γG molecule (myeloma protein Eu) is reported in the latter reference.

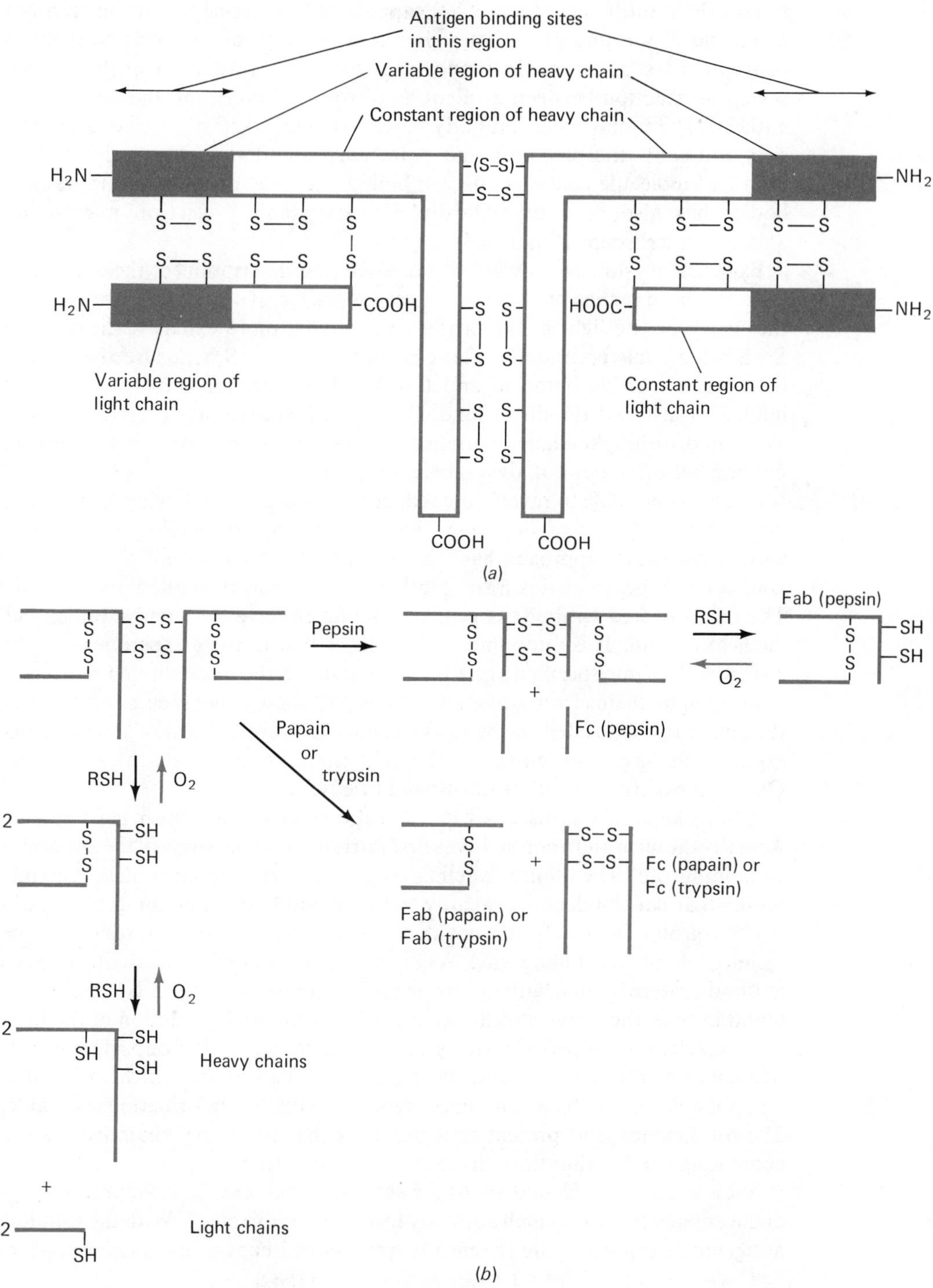
Antigen binding sites
in this region
Variable region of heavy chain
Constant region of heavy chain
H_2N
H_2N
(S-S)
S-S
S—S
COOH
HOOC
NH_2
NH_2
Variable region of
light chain
Constant region of
light chain
COOH
COOH
(a)
Pepsin
Fab (pepsin)
RSH
SH
SH
O_2
Fc (pepsin)
Papain
or
trypsin
RSH
O_2
2
Fc (papain) or
Fc (trypsin)
Fab (papain) or
Fab (trypsin)
Heavy chains
Light chains
(b)

monovalent antibodies (Figure 4.4) capable of binding hapten or antigen but incapable of forming precipitates. In the work with monovalent antibody a convenient test is to use the monovalent piece as an inhibitor of the normal precipitin reaction between divalent analogous antibody and the appropriate antigen. The monovalent antibody blocks an equivalent number of antigenic sites and shifts the precipitin end point correspondingly. This suggests that each half-molecule contains one combining site. The divalent nature of antibodies has also been dramatically demonstrated by electron microscopy and light microscopy (Figure 4.5).

Experiments directed toward a more detailed description of the active site have also been carried out but are not yet conclusive. To determine first whether the heavy or the light chain carries the activity, more extensive cleavage of S—S bridges has been used. After cleavage of five S—S bonds free heavy and light chains can be obtained and tested individually for activity. Using the inhibition test and the direct-binding assay, no activity has ever been found associated with light chains; with heavy chains, however, low levels of antigen binding has often but not always been observed.

Since it is possible to reverse the reductive cleavages by allowing reoxidation of a mixture of reduced heavy and light chains and to thereby obtain active antibody, another approach has been to mix light chains from one antibody (anti-X) with heavy chains from another (anti-Y) and then allow reoxidation. The reconstituted antibody is generally found to have the same specificity as the heavy chain. It is undoubtedly significant that in these experiments, light chains of the same specificity give higher activity in the reconstituted molecules than do light chains from other antibodies. All these experiments suggest that the active site is located in the heavy chain. However, chemical modification experiments have also implicated the light chain as containing the active site. One such experiment will be mentioned briefly.

Affinity labeling was discussed in connection with chymotrypsin in Chapter 3. Actually the term and approach was first introduced in the study of the antibody-combining site. The affinity-labeling reagent contains an antigenic (haptenic) group that can bind noncovalently with the antibody-combining site and a reactive group that can form a covalent bond with an amino acid residue in the vicinity of the combining site. When mouse anti-DNP (anti-dinitrophenyl) antibodies were treated with dinitrophenyl diazonium salt, the DNP group was found in both the heavy and the light chain.* Careful degradation of the light chain resulted in a dipeptide, asp-tyr, labeled with the DNP group. This dipeptide could be the same as residues 85 and 86 in Bence–Jones protein but it is not possible to establish this unequivocally with the information available. The total evidence at present thus indicates that the heavy chain carries the combining site but that the light chain is also involved.

A few words must be said about the antigen. What are the structural features of the antigen towards which antibody formation is directed? With the simplest antigenic determinant, the chemically well-defined hapten, the answer appears

* L. Wofsy, H. Metzger, and S. J. Singer, *Biochem.* **1**, 1031 (1962).

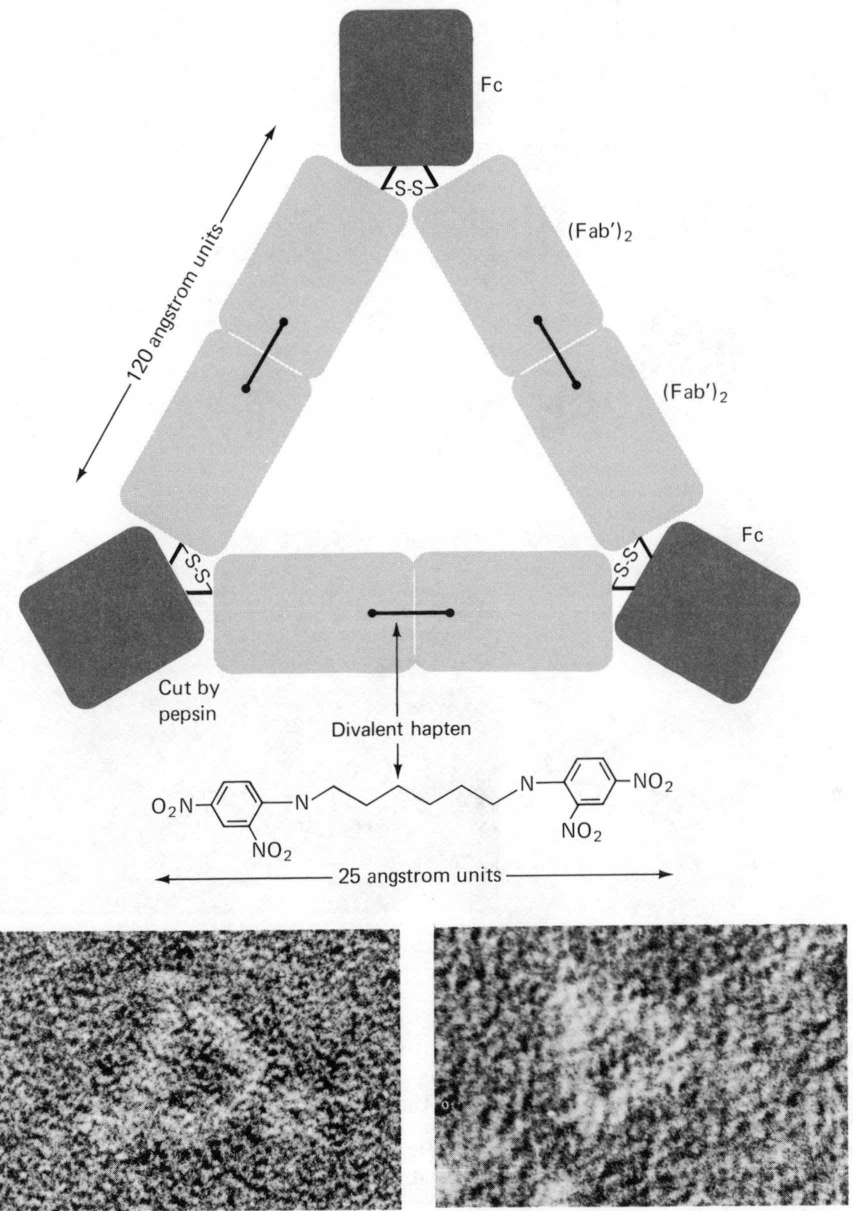

Figure 4.5(*a*) A structural model of divalent antibody based on the electron micrographs of the complex formed between anti-dinitrophenol antibodies and the bifunctional hapten derivative N, N′-(bis dinitrophenyl)-hexamethylene diamine. The electron micrographs (obtained and generously supplied by Drs. Michael Green and Robin Valentine) of the complex with both intact antibody (*left*) and with Fab_2 pepsin (*right*) are shown and suggest that the Fc fragments are located at the corners of the triangle observed with the intact antibody. See the interpretive model. [Reprinted from Porter, R. R., *Sci. Amer.*, **217**, 81 (1967) with permission of the author and the publishers.]

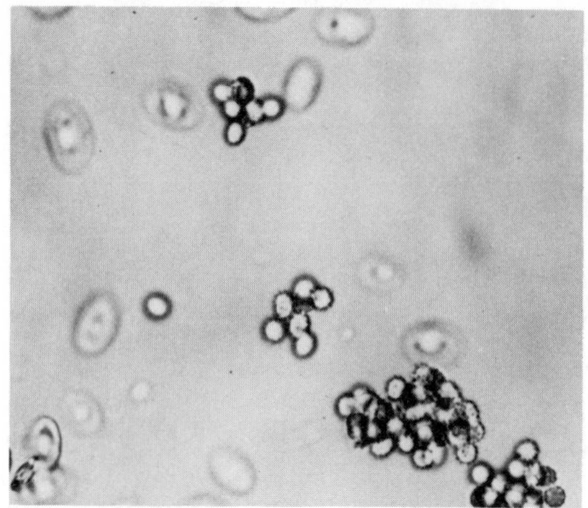

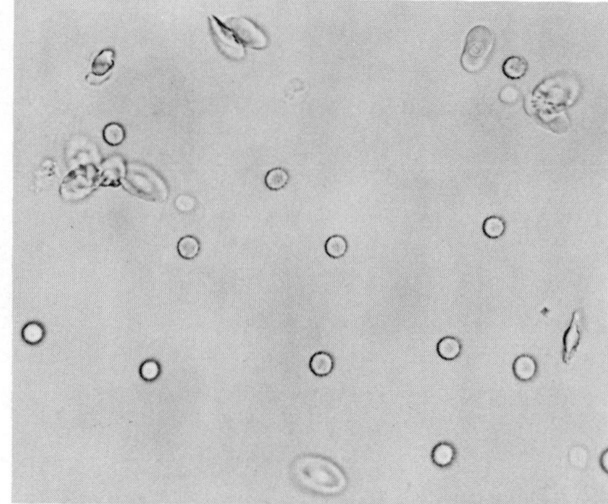

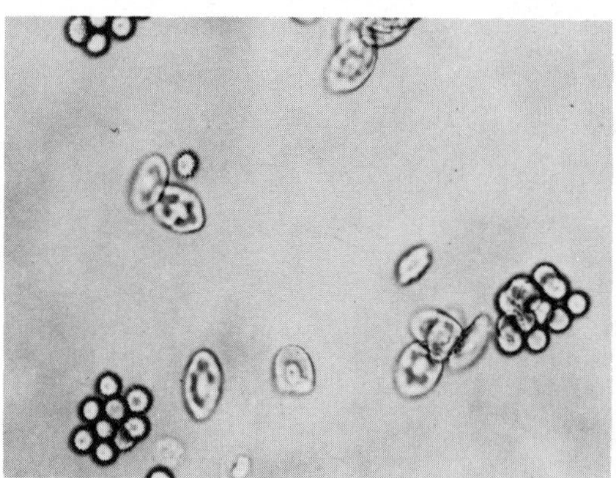

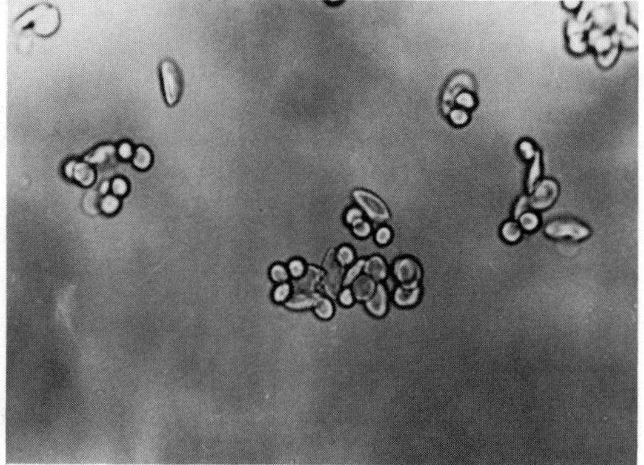

Figure 4.5(*b*) A visual demonstration of the divalent nature of antibodies, and of the reversible reduction–oxidation of the two halves of an antibody molecule. Anti-ovalbumin and anti-bovine γ-globulin were prepared and treated with pepsin to give the respective Fab_2 fractions. These were next reduced to individual Fab-monomeric units and allowed to oxidize either separately to give back the original Fab_2 divalent fractions, or together to give a mixture of hybrids and homogeneous Fab_2 fractions. The hybrids should contain one antibody site for each of the two antigens. To visualize the test for the presence of hybrids, the ovalbumin was coated onto human erythrocytes (small, spherical, and enucleated cells) and the bovine γ-globulin onto duck erythrocytes (large, oval, and nucleated cells). The pictures show the result of the test. The reoxidized anti-ovalbumin Fab_2 molecules clump human erythrocytes (I), and the reoxidized anti-γ-globulin Fab_2 molecules clump the duck erythrocytes (II). A mixture of the two Fab_2 molecules give separate clumps of the two types of cells (III), but the hybrids form clumps where both types of cells are present (IV). This type of experiment should in principle work equally well if whole γG molecules had been hybridized instead of the Fab fractions. [Reprinted with permission from Fudenberg, *et al.*, *J. Exp. Med.*, **119**, 151 (1964).]

simple. Similarly, polysaccharide antigenic structures apparently can be defined quite precisely. As long as the hapten is associated with the high-molecular-weight molecules, it elicits specific-antibody production and the antibodies recognize and combine specifically with the free hapten. If the antigen is a protein, however, the answer is extremely complex—almost paradoxical. The antigenic site cannot be simply an amino acid residue or a short sequence of residues, or the observed specificity would be difficult to explain. In the experiment with ovalbumin, for example, antiovalbumin is a mixture of antibodies directed against different determinants in ovalbumin; however, all these antibodies are specific for ovalbumin and will not mistakenly react with other proteins. It is therefore generally assumed that each antigenic component must contain enough structural components to make it unique and, since unique sequences probably would have to be sizeable, it is further assumed that conformational parameters are involved as well. The paradox here is that the injected antigen is generally quite deliberately and thoroughly denatured prior to injection yet the native antigen will subsequently be recognized. Many attempts have been made to define the constitution of an antigen site. It is clear that certain residues such as tyrosine or charged groups, make better antigenic sites than others. If synthetic polymers are used, a certain minimum chain length is required (attached to the protein) before antibodies are produced and, upon digestion of protein antigen of high molecular weight, there is a definite lower limit to the size of polypeptides that are still recognized by the antibody.* Clearly the structure-function relationship of protein antigen is itself a very fascinating but poorly understood problem.

Antibodies represent very powerful tools in the study of the structure and function of macromolecules. As is described in other sections of this chapter, immunochemical cross-reactivity represents a quick and quantitative comparison of related proteins. Furthermore, with the development of microtechniques and refined purification procedures it is essentially possible to prepare a series of γG molecules, each with specific binding sites for each of a large number of hapten molecules. Once it is possible to chemically define the antibody-combining sites, it should become feasible to correlate many different hapten structures to the specific binding-site structures. With rapid progress toward understanding the complete primary structure of γG and concommitant development of immunochemical methodology, the complete understanding of how the antibody works may not be too far away.

The fascinating problem of antibody biosynthesis is introduced in Chapter 8.

4.2 HEMOGLOBIN (TRANSPORT OR CARRIER PROTEIN)

Hemoglobin is a typical *carrier protein*, a protein that transports ligands from one site to another. This large class of functional proteins includes membrane-bound carrier proteins responsible for transportation of compounds into and

* M. Sela, *Adv. Immunol.*, **5**, 29 (1966).

out of cells (see Chapters 9 and 10). Although the particular example chosen here is found only in multicellular organisms, the class of carrier proteins is represented in all living cells. The basic functional properties of such proteins is a readily reversible binding and release of ligand. It will become apparent that this situation is most effectively achieved for proteins with sigmoidal binding curves and in ligand concentrations around the midpoint of the saturation curve. Hemoglobin illustrates this type of binding very well and its suitability for its oxygen-transport role in higher animals is readily rationalized on the basis of its particular binding characteristics. Hemoglobin is one of the best-characterized proteins in terms of both structure and function. The primary amino acid sequence has been determined, the subunit structure is firmly established, and the three-dimensional structure has been determined by X-ray crystallography. The ligand binding is well understood and can be correlated to protein structure. Because of the extensive knowledge of the primary structure of both normal and abnormal hemoglobins, important findings on hemoglobin genetics and evolution have emerged. All these features of hemoglobin will be discussed. Because of the very similar properties of myoglobin, data will be included for this protein whenever relevant.

General Description

Hemoglobin and myoglobin are *hemeproteins* (see Table 3.9), in which the iron of the protoporphyrin is in the reduced divalent form. Both hemeproteins can bind O_2 reversibly and the iron remains in the divalent form. Oxidizing agents will convert the ferroheme to ferriheme and the resulting *methemoglobin* no longer has the ability to bind O_2. The heme is bound to the protein by noncovalent forces and can be removed by relatively mild measures such as treatment with acid-acetone solution. At neutral pH the protein (globin) and the heme readily recombine to give back normal hemoglobin:

$$\text{globin} + \text{ferroprotoporphyrin} \rightleftharpoons \text{hemoglobin}$$

The oxygenation of hemoglobin is proposed to involve a direct interaction between O_2 and the sixth coordination of Fe^{2+} (the first four coordinations are involved in binding to the four pyrrole nitrogens, the fifth to a protein-imidazole group; see Table 3.9). Representing nonoxygenated hemoglobin as Hb and oxygenated hemoglobin as HbO_2, the reaction

$$\text{Hb} + O_2 \rightleftharpoons \text{HbO}_2$$

is characterized by the fact that both the reactants, Hb (represented as Fe^{2+}—N_5 to illustrate iron with its five coordinated nitrogens) and O_2, have magnetic moments while HbO_2 is diamagnetic (lacking a magnetic moment). Since molecules possessing unpaired electrons are magnetic (O_2 and Fe^{2+}—N_5), the loss of magnetic moment in HbO_2 represents strong evidence for the direct interaction of iron and O_2 with a pairing of the electrons.

Hemoglobin is located in the mature erythrocyte in blood while myoglobin is found in the muscle cells.

Structure

Hemoglobin is a protein with a molecular weight of 65,000 that contains four subunits, each with a molecular weight of about 16,000. Each subunit contains one heme group. There are three major types of normal Hb monomers (subunits), α, β, and γ, and the amino acid sequences of these three as well as that of myoglobin are given in Table 4.2. Normal adult hemoglobin (HbA) is made up of two α and two β subunits and is given the designation $\alpha_2\beta_2$; normal fetal hemoglobin (HbF) is made up of two α and two γ subunits and is given the designation $\alpha_2\gamma_2$. These are the two major normal human hemoglobins.

In addition to these, some 120 abnormal human hemoglobins have been characterized (see Table 4.3 for a partial list). Most of these differ from the normal by the replacement of a single amino acid residue. The different hemoglobins are detected almost exclusively by electrophoretic mobility, which only detects change in charge; hence it is highly probable that a large number of other abnormal hemoglobins exist that have not been detected. A number of the abnormal hemoglobins have retained normal functional properties while others have impaired function. Sickle-cell hemoglobin HbS is one of the classical abnormal hemoglobins, in which the single replacement of glutamic acid with valine in position 6 of the β chain causes a dramatic change in the solubility of the hemoglobins. Thus deoxygenated HbS will actually "crystallize" in the red cell and cause a change in the red cell's shape from the normal biconcave disk to a sickle or crescent shape. This modified red cell is fragile and tends to break. The number of functional red cells is thus reduced, leading to a condition known as sickle-cell anemia. The homozygous state of this disease is generally fatal but individuals heterozygous for this trait get along quite well with their (usually) quite mild, chronic anemia. Nevertheless it is surprising that this genetic mutation should have survived the natural selection process. The reason is undoubtedly the interesting fact that individuals with the sickle-cell trait are resistant to malaria (the malaria parasite lives in the erythrocytes). In regions where this disease is endemic, the sickle cell individuals have a higher life expectancy and higher fertility rate than the normal individuals and thus the sickle cell trait is perpetuated. The genetics of the hemoglobins have been the subject of extensive studies.* Using the same arguments as those set forth for cytochrome c in the previous chapter, a phylogenetic tree of globins has been constructed based on the number of amino acid replacements in the different types of globins from different species. On the assumption of a single ancestral globin molecule, it is proposed that myoglobin diverged first, followed by lamprey hemoglobin. Later the α and β chains diverged by gene doubling; still later the γ chain diverged.

* C. Baglioni in J. H. Taylor (Ed.), *Molecular Genetics*, part I, Academic Press, New York (1963).

TABLE 4.2 THE PRIMARY SEQUENCE OF HUMAN HEMOGLOBIN α, β, γ CHAINS AND MYOGLOBIN[a]

	1	2	3	4	5	6	7	8	9	10	11	12	13	14	15	16	17	18	19	20	21	22	23	24	25
α	V	—	L	S	P	A	D	K	T	D	V	K	A	A	W	G	K	V	G	A	H	A	G	E	O-
β	V	H	L	T	P	E	E	K	S	A	V	T	A	L	W	G	K	V	D	—	—	V	D	E	V-
γ	G	H	F	T	E	E	D	K	A	T	I	T	S	L	W	G	K	V	D	—	—	V	E	D	A-
M			V	(E	E	G	V	V	L	L	L	S	H	W	W)	A	K	V	E	A	D	V	A	G	(H-

	26	27	28	29	30	31	32	33	34	35	36	37	38	39	40	41	42	43	44	45	46	47	48	49	50
α	G	A	E	A	L	E	R	M	F	L	S	F	P	T	T	K	T	O	F	P	H	F	—	D	L-
β	G	G	E	A	L	G	R	L	L	V	V	O	P	W	T	E	R	F	F	E	S	F	G	D	L-
γ	G	G	E	T	L	G	R	L	L	V	V	O	P	W	T	E	R	F	F	D	S	F	G	D	L-
M	G	E)	D	I	L	I	R	L	F	K	S	H	P	E	T	L	E	K	F	D	R	F	K	H	L-

	51	52	53	54	55	56	57	58	59	60	61	62	63	64	65	66	67	68	69	70	71	72	73	74	75
α	S	H	—	—	—	—	—	G	S	A	E	V	K	G	H	G	K	K	V	A	D	A	L	T	D-
β	S	T	P	D	A	V	M	G	D	P	K	V	K	A	H	G	K	K	V	L	G	A	F	S	D-
γ	S	S	A	S	A	I	M	G	D	P	K	V	K	A	H	G	K	K	V	L	T	S	L	G	D-
M	K	T	E	A	E	M	K	A	S	E	D	L	K	K	H	G	V	T	V	L	T	A	L	G	A

	76	77	78	79	80	81	82	83	84	85	86	87	88	89	90	91	92	93	94	95	96	97	98	99	100
α	A	V	A	H	V	D	D	M	P	D	A	L	S	A	L	S	D	L	H	A	H	K	L	R	V-
β	G	L	A	H	L	D	D	L	K	G	T	F	A	T	L	S	E	L	H	C	D	K	L	H	V-
γ	A	I	K	H	L	D	D	L	K	G	T	F	A	E	L	S	E	L	H	C	D	K	L	H	V-
M	I	L	K	K	K	G	H	H	E	A	E	L	K	P	L	A	E	S	H	A	T	K	H	K	I-

	101	02	03	04	05	06	07	08	09	10	11	12	13	14	15	16	17	18	19	20	21	22	23	24	25
α	D	P	V	D	F	K	L	L	S	H	C	L	L	V	T	L	A	A	H	L	P	A	E	F	T
β	D	P	E	D	F	R	L	L	G	D	V	L	V	C	V	L	A	H	H	F	G	K	E	F	T
γ	D	P	E	D	F	K	L	L	G	D	V	L	V	T	V	L	A	I	H	F	G	K	E	F	T
M	P	I	K	O	L	E	F	I	S	E	A	I	I	H	V	L	H	S	R	H	P	G	D	F	G

	126	27	28	29	30	31	32	33	34	35	36	37	38	39	40	41	42	43	44	45	46	47	48	49	50	51	52	53	54
α	P	A	V	H	A	S	L	D	K	F	L	A	S	V	S	T	V	L	T	S	K	O	R						
β	P	P	V	E	A	A	O	E	K	V	V	A	G	V	A	D	A	L	A	H	K	O	H						
γ	P	E	V	E	A	S	W	E	K	M	V	T	G	V	A	S	A	L	S	S	R	O	H						
M	A	D	A	E	G	A	M	D	K	A	L	E	L	F	R	K	D	I	A	A	K	O	K	E	L	G	O	(E	G)

[a] No distinction is made between glutamic acid and glutamine (both represented by E) and between aspartic acid and asparagine (both represented by D). (See Table 3.6 for the key to one-letter symbols.) The sequences are aligned to give the best match of homologous sequences. Sequences in parentheses are uncertain. The sequences are the result of work from many laboratories, quoted in M. O. Dayhoff, *Atlas of Protein Sequence and Structure 1969*. National Biomedical Research Foundation, Silver Springs, Md., 1969.

The subunit structure of the hemoglobin tetramer has been studied extensively; it has a molecular weight of 65,000 and dissociation into both dimers and monomers has been demonstrated with several dissociating agents (urea, guanidinium salts, acid, and alkali). As a general rule it has been possible to reassociate the subunits upon removal of the dissociating agent. One interesting aspect of the dissociation phenomenon is the question about the structure of the intermediate dimer: Does the tetramer dissociate symmetrically to yield two $\alpha\beta$ dimers or is the process asymmetrical to give α_2 and β_2 dimers? No one has ever observed the expected heterogeneity of a mixture of α_2 and β_2 dimers; the dimer population appears to be homogeneous, suggesting that only $\alpha\beta$ dimers are produced.

Hemoglobin and myoglobin were the first proteins for which the complete three-dimensional structure was elucidated through analysis by X-ray diffraction.* The structural models are shown in Barker's book in this series, and represent one of the most significant advances in modern structural analysis.

Function

The two most important characteristic features of the binding of O_2 to hemoglobin, the sigmoidal binding curve and the Bohr effect, are illustrated in Figure 4.6. The sigmoidal curve is symptomatic of cooperative binding and the slope in a Hill plot gives a value for *c*, the cooperativity, of +2.8. The Bohr effect demonstrates a significant difference in the dissociation of at least one ionizable group in Hb and HbO_2; H^+ is released from Hb—H when O_2 is bound (HbO_2 is a stronger acid than Hb). By analyses very similar to those used in the enzyme characterizations in Chapter 3, the pK_a values involved have been found to be $pK_a = 8.2$ (for Hb) and $pK_a = 6.95$ (for HbO_2) at 20°C at an ionic strength of 0.2. This pK_a value is characteristic of histidine but that group has not been definitely identified.

TABLE 4.3 SOME ABNORMAL HUMAN HEMOGLOBINS

Hb Designation	*Position*[a]	*Normal*	*Mutant*
1. α-chain variations			
J Toronto	5 (6)	ala	asp
I	16 (17)	lys	glu
G Honolulu	30 (31)	glu	gln
M Boston	58 (65)	his	tyr
G Bristol	68 (75)	asn	lys
M Shibata	87 (94)	his	tyr
O Indonesia	116 (123)	glu	lys
2. β-chain variations			
S	6 (6)	glu	val
G	7 (7)	glu	gly
J Baltimore	16 (16)	gly	asp
G Saskatoon	22 (24)	glu	lys
Geneva	28 (30)	leu	pro
Hikari	61 (63)	lys	asn
M Chicago	63 (65)	his	tyr
Agenogi	90 (92)	glu	lys
Kansas	102 (104)	asn	thr
D Portugal	121 (123)	glu	gln
3. Abnormal chain aggregations			
HbH = β_4			
Hb Bart's = γ_4			

[a] Position number refers to actual residue position. The numbers in parentheses refer to the position in the alignment in Table 4.2. Selected from the more complete list compiled in M. O. Dayhoff, *Atlas of Protein Sequence and Structure 1969*, National Biomedical Foundation, Silver Springs, Md., 1969.

* J. C. Kendrew, *et al.*, *Nature*, **185**, 422 (1960), and M. F. Perutz, *J. Mol. Biol.*, **13**, 646 (1965).

Chemical modification of and comparative studies on different hemoglobins (and myoglobin) have again been important in the attempt to explain and correlate the cooperativity and the Bohr effect in terms of structural models. Table 4.4 summarizes some of the available information. Even though cooperative binding and the Bohr effect do not always go hand in hand, it seems clear

TABLE 4.4 COMPARISON OF O_2-BINDING CHARACTERISTICS OF DIFFERENT HEMOGLOBINS AND HEMOGLOBIN DERIVATIVES[a]

Protein	c^b	*Bohr Effect*
HbA ($\alpha_2\beta_2$)	2.8	+
HbF ($\alpha_2\gamma_2$)	2.8	+
HbH (β_4)	1	−
Artificial (γ_4)	1	−
Myoglobin	1	−
Hb monomers	1	−
Hb dimers ($\alpha\beta$)	2.6	+
Hb-CPA[c]	~1	low
Hb-CPB[c]	2.8	low
Hb-CP(A + B)[c]	~1	−

[a] A. Rossi-Fanelli, E. Antonini, and A. Caputo, *Adv. Prot. Chem.*, **19**, 73 (1964).

[b] c is the empirical cooperativity index of the Hill equation. $c = 1$ means no interaction (hyperbolic binding curve), $c > 1$ means positive interaction (cooperative binding).

[c] Carboxypeptidase treatment: Treatment with CP*A*, carboxypeptidase *A*, removes histidine and tyrosine from β chains; treatment with CP*B*, carboxypeptidase *B*, removes arginine from α chains (see Table 3.5); CP (*A* + *B*) represents both enzymes.

α chain: thr-ser-lys-tyr-arg
β chain: ala-his-lys-tyr-his

Figure 4.6 Some characteristic properties of the reaction of human hemoglobin and myoglobin with oxygen. (*a*) Oxygen saturation curves for (1) myoglobin at pH 6.5 and 7.5, (2) hemoglobin at pH 7.4, and (3) hemoglobin at pH 6.0. The endpoints of the titrations correspond to a binding of 1 mole of O_2 per mole of myoglobin and 4 moles of O_2 per mole of hemoglobin. (*b*) Hill plot of the data in (*a*). Good linear relationships are obtained for hemoglobin in these plots in the range between 10 and 90 percent saturation. Outside this range the slope of the plots tends toward 1. The Hill coefficient, c (see Figures 2.3 and 3.17), is 1 for myoglobin (a single site) and +2.8 for hemoglobin (interacting sites). (*c*) The Bohr effect: the change in oxygen affinity of hemoglobin with a change in pH [also illustrated by difference between curves 2 and 3 in (*a*)]. The oxygen affinity of myoglobin is essentially independent of pH. Affinity is expressed as $P_{1/2}$, the O_2 pressure at which 50 percent saturation is achieved. This corresponds to the intrinsic dissociation constant K or to the inverse of the binding constant **K**. Such a pH dependence shows a competition between oxygen binding and the binding of a proton, and the data in (*c*) can also give the apparent proton affinities (pK_a values) for oxygenated and deoxygenated hemoglobin. The data in this figure are compiled from several sources [see Rossi Fanelli, *et al.*, *Adv. Prot. Chem.*, **19**, 73 (1964)]. It specifically refers to human myoglobin and hemoglobin, but with minor quantitative differences, the picture has general validity for all species.

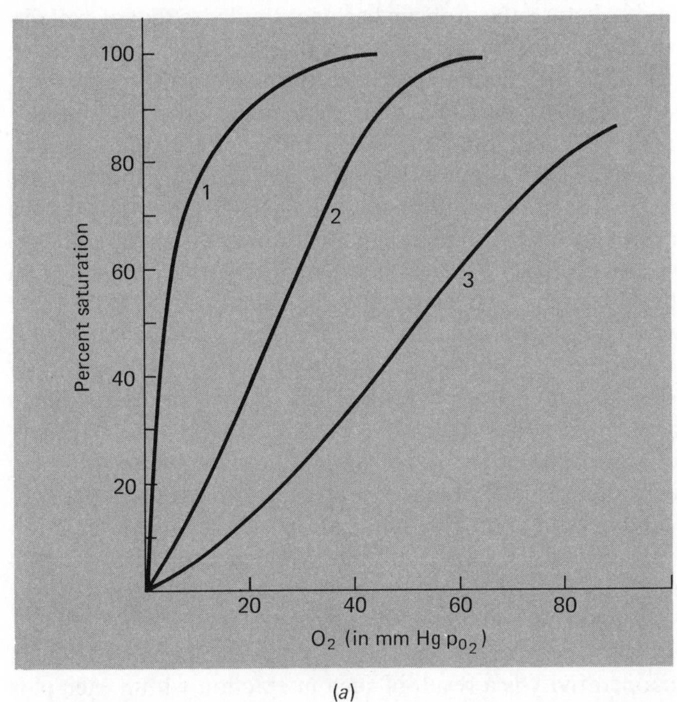
100
80
60
40
20
Percent saturation
1
2
3
20
40
60
80
O_2 (in mm Hg p_{O_2})

(a)

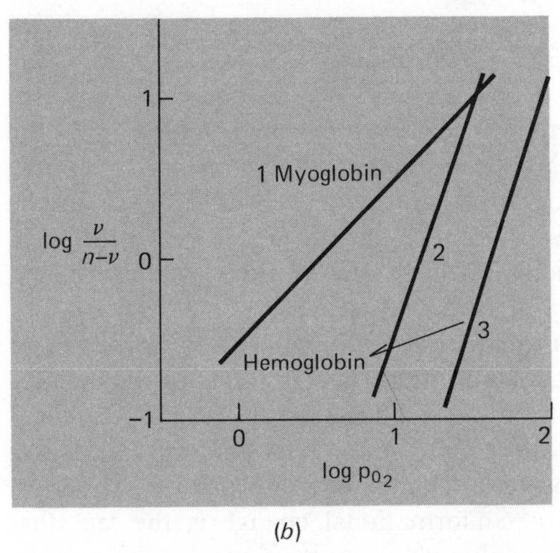
1
0
−1
log $\frac{\nu}{n-\nu}$
1 Myoglobin
2
3
Hemoglobin
0
1
2
log p_{O_2}

(b)

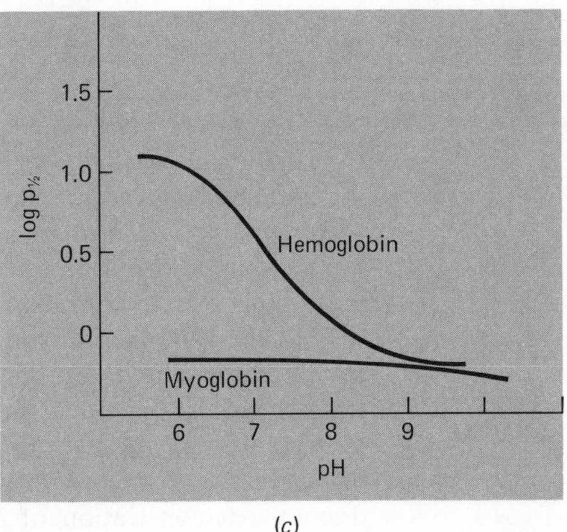
1.5
1.0
0.5
0
log $p_{½}$
Hemoglobin
Myoglobin
6
7
8
9
pH

(c)

that they are somehow related. In the attempt to understand the effects first on the basis of subunit structure, the problem becomes similar to that of explaining the antibody-combining site on the basis of either the heavy or the light-chains in immunoglobins and analogous to the problem of the anomalous binding properties of the M_4-type LDH. It is clear that neither the subunit monomer, exemplified by myoglobin, nor the β_4 and γ_4 hemoglobins, which do not contain α chains, shows cooperativity and Bohr effect. On the other hand the carboxypeptidase modification of α chains (CP*B* treatment; see Table 4.4) has little effect on the two experimental parameters whereas carboxypeptidase treatment of β chains (CPA treatment; see Table 4.4), removing two amino acids from the C-terminal end, essentially abolishes both the Bohr effect and the cooperativity. (It is noteworthy that one of the amino acids removed is histidine and it is possible that this is the group whose pH changes upon oxygenation. If so, the lack of a Bohr effect in HbH (β_4) is still unexplained.) The most reasonable explanation of these observations is that the cooperativity and the Bohr effect are results of the *combination* of α chains with β- (or γ-) chains and that the characteristic O_2-binding properties reflect the proper subunit combination and interaction. The implication is that the interaction between α and β chains is the important one ("interaction" here means functional interaction, and does not necessarily mean that the $\alpha\beta$ dimer physically has the strongest interaction as compared to the α_2 or β_2 dimer). This is in agreement with the strong cooperativity observed for the $\alpha\beta$ dimers (Table 4.4). If the cooperativity is a result of such interaction within each pair of $\alpha\beta$ dimers in the tetrameric HbA molecule we can further predict that in the stepwise oxygenation of HbA (represented as Hb_4 to emphasize the tetrameric structure with four oxygen binding sites),

$$Hb_4 \xrightarrow{1} Hb_4O_2 \xrightarrow{2} Hb_4O_4 \xrightarrow{3} Hb_4O_6 \xrightarrow{4} Hb_4O_8$$

steps 1 and 2 represent the oxygenation of one $\alpha\beta$ pair and steps 3 and 4 the oxygenation of the second pair. With the postulated interaction (cooperativity), there should be no accumulation of Hb_4O_2 and Hb_4O_6, but Hb_4O_4 should be an observable intermediate as illustrated in Figure 4.7(a). The illustration in Figure 4.7(b), allowing for interaction between the two $\alpha\beta$ dimers in the tetramer, is probably a better approximation of the actual situation.

With the complete three-dimensional structure of hemoglobin available, it should be possible to explain the subunit interaction in terms of the detailed structure of the individual subunits. This goal has not yet been reached. It is evident, however, that both the interaction of subunits and the binding of oxygen are associated with a substantial change in conformation. The most dramatic demonstration of such a conformational change is the fact that crystals of the nonoxygenated hemoglobin (analyses by X-ray crystallography were originally done with O_2-free Hb) crack and disintegrate when oxygenated. Preliminary data on the comparison of the structures as seen through X-ray crystallography of Hb_4 and Hb_4O_8 show that the oxygenated form has a

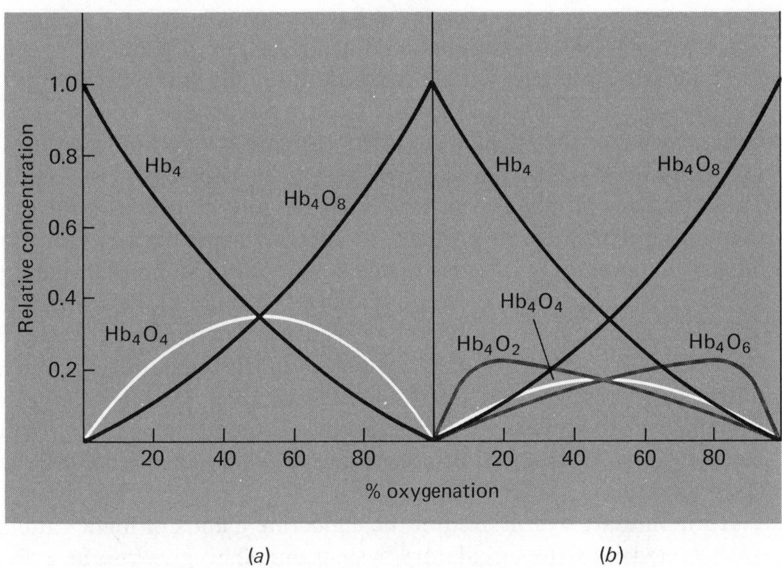

Figure 4.7 (*a*) A model for the titration of hemoglobin with O_2 assuming strong interaction between α-β subunits (between the first and second and between the third and fourth ligand binding). (*b*) A model assuming interaction both between α-β subunits and between α-β dimers. [See Coryell, *et al.*, *J. Phys. Chem.*, **43**, 825 (1939).]

tighter structure with the two β chains closer both to the α chains and to each other; it has also been known for a long time that several physical properties of Hb_4 and Hb_4O_8 reflect a difference in structure between the two. It is interesting to note that the C-terminal end of each subunit is close to the heme and thus could affect the O_2 binding. However, rather than implicating the C-terminal ends as direct affectors of O_2-heme interaction it seems more reasonable to propose that the conformational change leading to both the cooperative binding and the Bohr effect is dependent upon an intact C-terminal sequence, at least in the β chains.*

Before we leave the subject of hemoglobin, a physiologically significant detail must be mentioned. The toxic effect of carbon monoxide (CO) is based on the high binding affinity of Hb for this ligand. It is well established that the binding site for CO is the same as that for O_2 and indeed that CO binding shows cooperativity and Bohr effect very similar to those of O_2 binding. The affinity of Hb for CO is higher than its affinity for O_2 and, because of the direct competition for the same site, low concentrations of CO can effectively block the transport of the vital O_2 to the different tissues.

* For a recent discussion of these aspects, see M. F. Perutz *et al.*, *Nature*, **219**, 131 and 902 (1968).

4.3 INSULIN—REGULATORY PROTEINS

Since we included in the functional classification of proteins a separate class of "regulatory proteins," it is necessary to qualify this terminology somewhat since in the broadest sense all proteins have a regulatory role. For the purpose of this discussion, however, it seems appropriate to single out a group of proteins for which *no measurable function other than regulation* has been observed. These proteins do not appear to be enzymes and do not fit in the categories of immune proteins, carrier proteins, or contractile proteins but they profoundly affect the functioning of other systems. We shall assume that these effects are based on specific recognition and binding. The regulatory proteins represent a very heterogeneous group including regulatory subunits in allosteric enzymes (see Section 3.5), proteins directly involved in the regulation of gene expression (histones, repressors, and interferon; see Chapter 8), and a large number of proteins whose specific site of action is yet undetermined. Among this latter category are the protein hormones, one of which, insulin, will be briefly discussed here.

Hormones are synthesized in the endocrine glands of higher animals. They are secreted into the circulatory system and they give rise at extremely low concentrations to profound physiological and biochemical effects in the intact animal. Because of their unique features—acting at low concentrations, affecting several apparently unrelated properties of the organism, and showing relatively insignificant effects on individual, isolated, in vitro systems—it has been difficult to define hormone action at the molecular level. The problem is not so much to design in vitro assays that duplicate some physiological effect as it is to interpret the observed hormone-induced changes as being either direct, primary effects of the hormone, or secondary effects far removed from the hormone's true molecular site of action. It must be kept in mind that the protein (or polypeptide) hormones make up only a small subgroup of hormones and that a general discussion of hormone function should also apply to other hormones of low molecular weight, such as steroids, adrenaline, and thyroxine.

Structure

Insulin was the first protein molecule for which the complete structure was determined by chemical means. The primary structure of insulin, given in Figure 4.8 was the result of some 10 yr of careful and meticulous studies. It is worthwhile to stress that this outstanding achievement in protein-structure analysis was completed less than 20 yr ago just to indicate the impetus it gave to the study of protein structure and function as shown by the developments in the interim years. It is interesting to ponder whether scientific advances are more generally made in this stepwise or wavelike fashion. Progress is steady and perhaps not too impressive for long periods of time, and then suddenly a big discovery, achievement, or idea, the "breakthrough," comes along to open new avenues of study making possible what was thought impossible and often

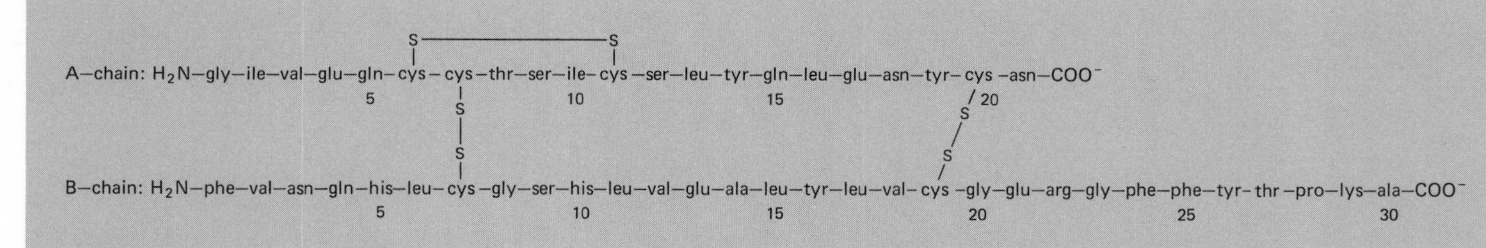

Figure 4.8 The primary structure of bovine insulin [from Sanger, F. and co-workers, *Biochem. J.*, **49**, 481 (1951); **53**, 366 (1953); **60**, 541 (1955)].

initiating a whole new field of thought and accomplishments. It may also be well to note that these breakthroughs are seldom just an inspired idea or a chance discovery but most often come as the culmination of the combination of good ideas, alert recording, and use of the unexpected findings with hard, conscientious, and patient work.

Although our main purpose is to attempt to correlate the structure of insulin with its functional properties, there is one aspect of this simple two-chain structure that should be examined first. In the case of ribonuclease (see Section 3.1) it was shown that the four S—S bonds could be reduced to give an unfolded polypeptide chain with all the natural covalent cross-links (S—S bonds) cleaved. Upon reoxidation the original S—S bonds reformed and an excellent yield of the original three-dimensional, active structure could be obtained. Can the same experiment be done with insulin?

In this case a bimolecular reaction is involved in the recombination and reoxidation of a mixture of reduced A and B chains and in addition several side reactions (formation of A_2, B_2, and both monomeric and polymeric forms) are possible. The experiment has been carried out both with natural A and B chains obtained by reduction of the S—S bonds in insulin and with chemically synthesized, reduced A and B chains* (the latter system represents another remarkable achievement—the stepwise chemical synthesis of two polypeptide chains of this size and complexity). In either case the yield of biologically active insulin was, perhaps not unexpectedly, dismally low. In the case of ribonuclease, the good yield of reoxidized enzyme was used as an important argument that the enzyme could be synthesized in the reduced form and then upon completion would spontaneously fold into the proper conformation for "correct" S—S pairing. Clearly the same argument cannot be used for insulin. If chains A and B were indeed synthesized separately in the reduced form, all the in vitro experiments predict that very poor yields of active hormone would result. The fact that one of the enzymes that destroy insulin in the tissues, insulinase, acts by cleaving one of the S—S bonds also argues against biosynthesis of individual chains followed by recombination and oxidation. The resolution of this dilemma appears to lie in the recent discovery of proinsulin,† a single-polypeptide-chain protein having a molecular weight of 10,000. Proinsulin can be isolated and activated by trypsin to produce insulin and a peptide having a molecular weight of 4,000. [Figure 4.9]. It appears that insulin, like several of the proteolytic enzymes, is synthesized in the pancreas as a single-chain inactive prohormone. This prohormone, as shown by the experiments in Figure 4.9, can be reduced and oxidized in good yield, and upon peptide-bond cleavage the reoxidized proinsulin is converted to active insulin.

It is interesting to speculate a little further about this phenomenon of reoxidation of —SH groups in a single polypeptide chain in contrast to that in multichain structures. In this chapter we have considered another multichain

* P. G. Katsoyannis, *Recent Progr. Hormone Res.* **23**, 505 (1967), and H. Zahn, *Ann. Endocrinol.*, **27**, 575 (1966).

† D. F. Steiner, *et al.*, *Proc. Nat. Acad. Sci.*, **57**, 473 (1967), and *Science*, **157**, 697 (1967).

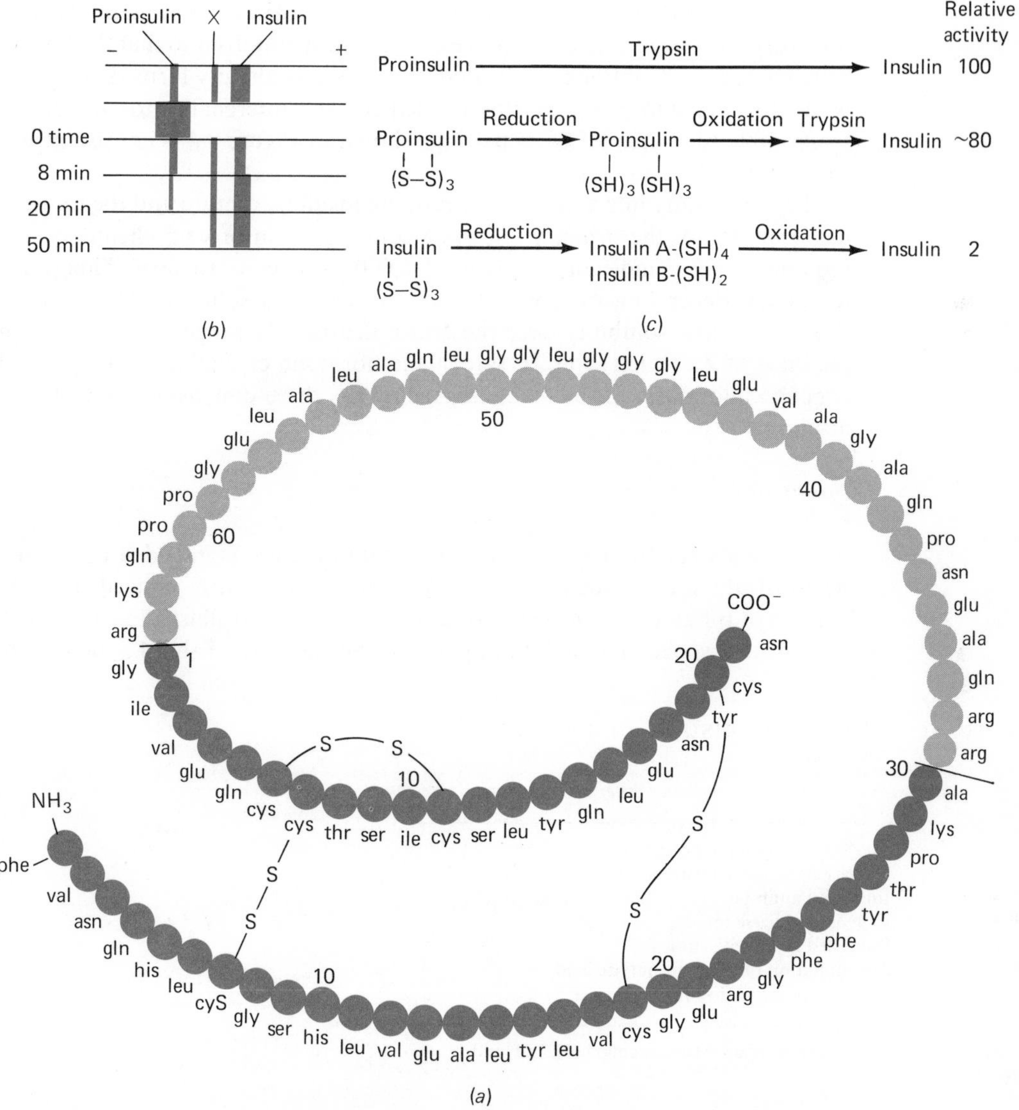

Figure 4.9 (*a*) The primary structure of porcine proinsulin. [From Chance, R., *et al.*, *Science*, **161**, 165 (1968) with permission.] (*b*) The conversion of proinsulin to insulin by trypsin digestion followed by gel electrophoresis. The tryptic activation involves several cleavages, but eventually leads to the removal of the segment 31–63. The nature of product X is not known. (*c*) The reduction and reoxidation of proinsulin and insulin.

molecule held together by interchain S—S bonds, namely γ-globulin. In that case, good yields of either half-molecules from individual light and heavy chains or intact γ-globulin from reduced halves were obtained. It is significant, however, that in the case of γ-globulin, partial reduction of only one (or at the most a few) S—S bond(s) is a prerequisite for recombination and proper reoxidation. Perhaps this implies that individual heavy chains and light chains, because of their size, fold and oxidize as they are synthesized so that the finished product has all its intrachain disulfide S—S bonds already formed, and is left with only the odd —SH groups needed for the interchain cross-links. The remarkable regularity of S—S pairings shown in Figure 4.4 is consistent with such a picture.

There are many interesting features of the insulin structure and the molecule has been one of the primary reference compounds in protein chemistry. The monomer has a molecular weight of 6,000 and tends to form dimers and tetramers (depending on concentration) in aqueous solution. The hormone shows very low solubility over the entire neutral pH range, especially in the presence of Zn^{2+} ion; the Zn salt of the hormone crystallizes readily. Little concrete information is available regarding the three-dimensional structure of insulin.

Function

As already stated, the discussion of insulin function is included here not to demonstrate an obvious structure-function relationship—indeed no such relationship has yet been demonstrated—but rather to illustrate some of the difficulties associated with the study of hormone activity. Table 4.5 shows some

TABLE 4.5 EFFECTS OF INSULIN IN THE INTACT ANIMAL

Tissue[a]	*Observed Effects*
M	Increased membrane potential
M	Increased oxygen consumption
M, A, L	Increased amino acid incorporation into proteins
M, A, Ma, L	Increased glucose uptake
M, L	Increased glycogen synthesis
M	Accumulation of α-amino butyric acid
A, Ma, L	Increased fatty acid synthesis

[a] M = muscle, A = adipose tissue, Ma = mammary gland, and L = liver.

examples of the types of observations one might make after administering insulin to an animal. The picture is complex indeed; although it may at first sight look as though several of the observed effects are closely related, more careful examination has shown, in some instances at least, that apparently related effects are not linked at all. Thus the insulin effect on glycogen synthesis can be eliminated (by administration of puromycin) without any effect on insulin-stimulated glucose uptake and, conversely, the insulin effect on glucose uptake can be blocked (by administration of N-ethylmaleimide) without affecting glycogen synthesis. In view of this it must be concluded that insulin has a large number of target tissues and that in each tissue there are several different sites at which the hormone's regulatory function is exerted.

Perhaps the best-defined system in which an insulin effect can be documented at the molecular level is the glucosyl transferase system. This enzyme, responsible for glycogen synthesis, exists in two forms: One requires glucose-6-phosphate as an activator and the other is independent of glucose-6-phosphate. After insulin treatment, the level of concentration of the independent form increases in both muscle and liver; it thus appears that the glycogen-synthesis effect can be explained by an activation of the proper enzyme, (For details, see Larner in this series).

Even with this precise a definition of the functional site, however, the main problem still remains. The effect of insulin is only observed when insulin is added to the intact animal or to isolated intact tissues. For this reason it becomes impossible to conclude that insulin is the primary effector in each of the different functional changes observed. The alternative explanation—that insulin has a single effector site where it regulates the production of some unknown compound (X), which in turn is the effector in each of the functions regulated by insulin—is just as valid as the first one. In fact the second mechanism has been favored in most hormone system models because it allows for the multiplicity of effects at very low hormone concentrations. This mechanism is the basis for the so-called "cascade effect" (see Chapter 8), in which low concentrations of a hormone activating the enzyme responsible for the synthesis of the true effector substance (X), are thought to cause the synthesis of a sufficiently high concentration of the effector to permit activation or inhibition at all the sites of action in all the different target tissues.

The hormone could thus be considered as the initial trigger signal, the primary message. As the activator of an enzyme that catalyzes production of the secondary message (compound X, the true effector), the small trigger signal (low concentration of hormone) is thus amplified into a strong signal (high concentration of X). Recent work in the field of molecular endocrinology has uncovered at least one potential candidate for the role of secondary message or X, namely adenosine-*cyclic*-3′,5′-monophosphate, or *cyclic*-AMP.* The enzyme adenyl cyclase, responsible for the synthesis of this compound, is activated by at least two hormones, glucagon and epinephrin. Without specific knowledge of the enzyme involvement, it has further been established that the intracellular levels of *cyclic*-AMP are sensitive to a large number of hormones, some of which cause increased levels while others cause decreased levels. Among the latter is insulin; thus a possible primary effect of insulin may be to decrease the levels of the true effector, *cyclic*-AMP. Whether this is accomplished by inhibition of adenyl cyclase, by activation of the hydrolytic step (see Figure 4.10), or by some

* G. A. Robinson, *et al.*, *Ann. Revs. Biochem.*, **37**, 665 (1968).

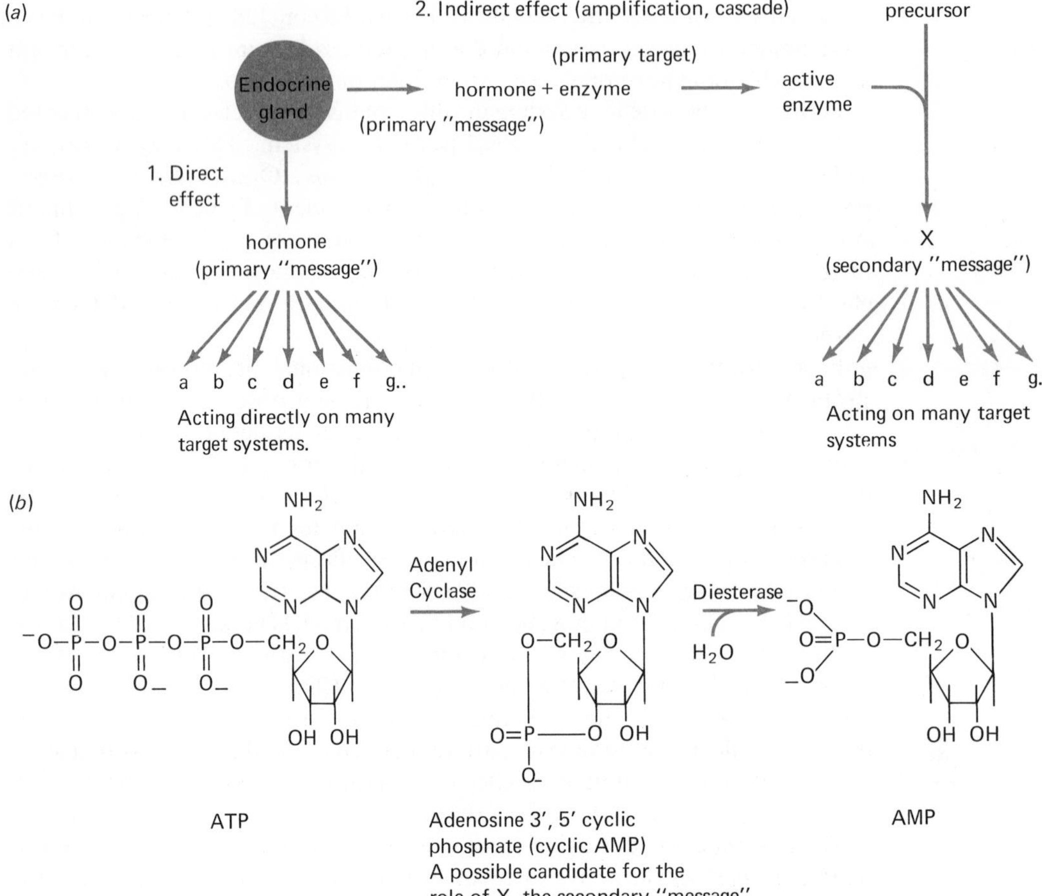

Figure 4.10 (*a*) Hypothetical models for hormone action, distinguishing between the direct effect of a given hormone on many different target systems and the direct effect on a single target system causing an increase or decrease in the levels of a secondary "message" or regulatory substance. The target systems are most likely to be individual enzymes or transport proteins, and the effect can be either an activation or an inactivation. (*b*) Cyclic AMP may be a secondary "message". It is produced from ATP through the action of the enzyme adenyl cyclase and is destroyed by hydrolysis catalyzed by a diesterase. Regulatory substances which lead to increased levels of cyclic AMP (glucagon, catecholamines, ACTH, leutenizing hormone, thyroid-stimulating hormone, angiotensin, gastrin, histamine, serotomin) could act directly (or indirectly) either by activating the cyclase or by inhibiting the diesterase. Similarly, substances which cause decreased levels of cyclic AMP (insulin, melatonin, prostaglandins) could act by inhibiting the cyclase or activating the diesterase. A large number of widely different processes and enzymes are activated or inhibited by cyclic AMP; so cyclic AMP could explain the broad spectrum of effects of a given hormone (see Table 4.5).

completely different mechanism is only speculation at present. The scheme in Figure 4.10 is, however, representative of the type of mechanisms that are currently considered in hormonal regulation.

The problem of in vivo regulation, illustrated here for insulin action, is not at all unique to hormones. In fact, as biochemists turn more and more to studies of intact cells, tissues, and organisms, the problem of elucidating all the steps of events between the known primary effector substance and the observed effect will become a very general one.

4.4 MUSCLE CONTRACTION—CONTRACTILE PROTEINS

In this section muscle contraction will be considered from the point of view of the proposed model—that proteins function through a single mechanism involving ligand recognition and binding. Several other systems of contraction and motility have been studied. Flagella and ciliary appendages in cells of higher animals as well as in bacteria have been defined in terms of chemical composition and microscopic structure; mitochondrial contraction has been investigated in both plants and animals. The muscle system is undoubtedly the one that is best defined and elucidation of the mechanism of muscle contraction will certainly shed light on the mechanisms of the other systems as well.

There are three general types of muscle in mammals. Smooth muscle lines the ducts and vessels of the digestive, respiratory, and circulatory systems; it is uniform in appearance and is not enclosed in a distinctive sheath. Cardiac muscle, the muscle of the heart, shows striation and nonuniform structure and the muscle fibers are surrounded by thin sheaths. Skeletal muscle is the major type of muscle, showing marked striation and having a distinct sheath. Each muscle can be seen to be built of several muscle fibers 0.05 to 0.1 mm (millimeters) thick, each encased in a membrane. Each fiber, in turn, is made up of several fibrils 10,000 Å (angstroms) thick (see Figure 4.12 below); this *myofibril* is the unit of structure and function in muscle contraction.

Structure

The major protein components of striated muscle are actin and myosin. The properties of these proteins are described briefly below.

Myosin. Myosin, as extracted from muscle, has a molecular weight of 600,000*; after treatment with guanidinium chloride or urea, a subunit having a molecular weight of 200,000 is found. The myosin molecule is postulated to be a trimer with the structure given in Figure 4.11. This is based primarily on

* Considerable disagreement exists as to the exact values of molecular weight and thus the number of subunits in myosin. Although there does not appear at this time to be any clear basis on which to select one particular set of values, one set of data was chosen quite arbitrarily for this discussion.

several observations: (1) the head and tail shape of this dimension in the electron microscope: (2) the limited susceptibility of the molecule to proteolytic digestion reproducibly yielding two pieces, light meromyosin (LMM) and heavy meromyosin (HMM); and (3) the properties of isolated LMM (fibrous, with a molecular weight of 260,000) and HMM (globular, with a molecular weight of 360,000), which are consistent with this picture. Myosin is also an enzyme (ATPase), catalyzing the hydrolysis of ATP, and the ATPase activity is stimulated by Ca^{2+} and inhibited by Mg^{2+}. The enzyme is very sensitive to pH. After proteolysis, only HMM retains ATPase activity. Myosin tends to aggregate and reproducibly form a uniform complex 6,000 to 7,000 Å long, with a diameter of 100 to 200 Å and a particle weight of 63×10^6.

Actin. Actin exists in two forms, the globular monomer, *G*-actin, and the fibrous polymer with a very high molecular weight, *F*-actin (Figure 4.11). The *G*-actin has a molecular weight of about 57,000 and contains one mole each of ATP and Ca^{2+} (or Mg^{2+}) per mole of *G*-actin. Removal of ATP or Ca^{2+} (with chelating agents such as ethylenediamine tetra-acetic acid) leads to "inactivation" and dimerization, but no further aggregation occurs. If "active" *G*-actin is exposed to 0.1 *M* salts containing Mg^{2+}, it very rapidly polymerizes to *F*-actin; in the process, ATP is hydrolyzed to ADP + P_i. Since the ADP in *F*-actin does not readily exchange with ATP, one cannot reasonably consider actin as an ATPase. (With sonication or other means of mechanical disruption in the presence of ATP, the reaction can be reversed to give *G*-actin–ATP + ADP from *F*-actin–ADP + ATP, and under these artificial conditions *F*-actin may be considered to be an ATPase.) It is not clear that hydrolysis of ATP is obligatory for polymerization of *G*-actin to *F*-actin; with artificially produced *G*-actin–ADP, polymerization still takes place. The structure of *F*-actin appears to be best described by a double-helical array of *G*-actin molecules, with 13 to 15 subunits per turn of the helix (Figure 4.11).

Actomyosin. Actomyosin is the molecular complex of actin and myosin. It is obtained by extraction from skeletal muscle, first with water to remove glycolytic enzymes and then with 0.6 *M* KCl. After extensive extraction with the second solvent a highly viscous solution of quite pure actomyosin is obtained. When this solution is squirted through a narrow orifice into dilute salt solutions, the actomyosin precipitates out in the form of long threads. These threads, when exposed to ATP and in the presence of K^+ and Mg^{2+}, contract in the most dramatic fashion while ATP is hydrolyzed to ADP and P_i.

The Contractile Apparatus

In trying to formulate a molecular mechanism for contraction, observations on the intact muscle or on the actomyosin-model system have both been used. It may be useful to consider the contractile process in terms of two models, the structural one and the biochemical one. The structural model is derived from

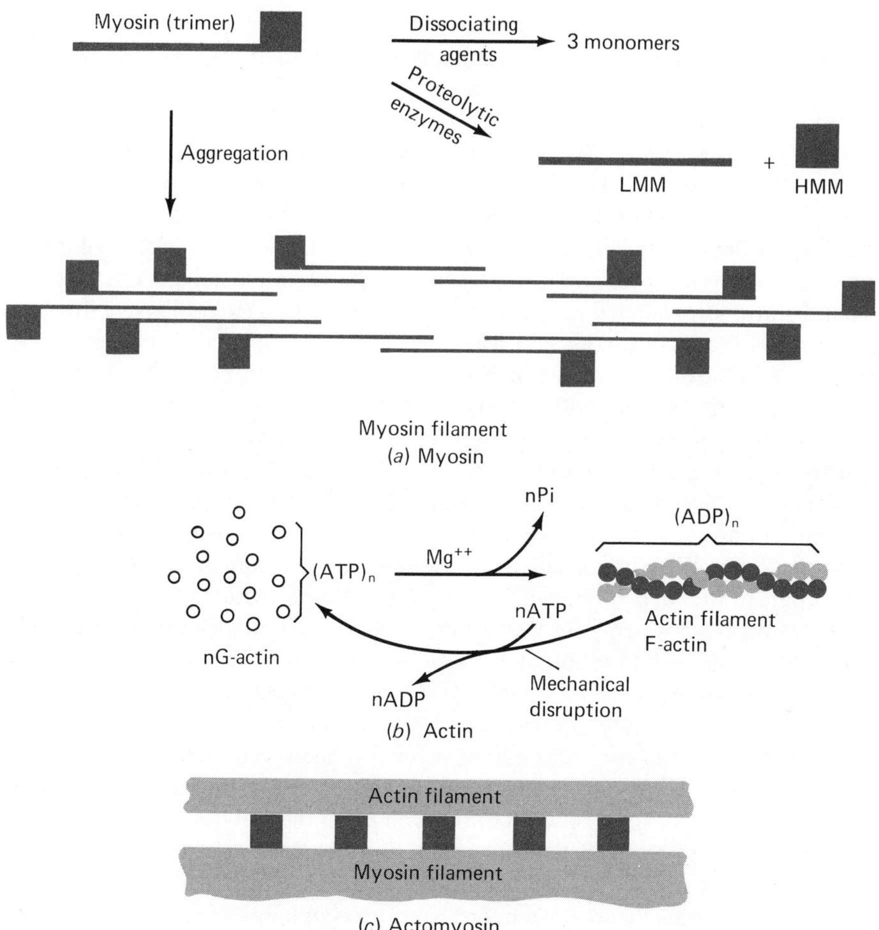

Figure 4.11 Models of myosin, actin, and actomyosin. (*a*) Myosin is a trimer which can be dissociated into monomer units. Proteolytic digestion reproducibly cleaves only a limited number of peptide bonds to give two separate fractions: the fibrous part, light meromyosin (LMM), and the globular part, heavy meromyosin (HMM). The ATPase activity of myosin resides in the HMM fraction. Myosin aggregates readily in 0.1 *M* KCl to give myosin filaments. The observations of these filaments suggest that they are put together through interaction of the LMM part of the molecule, reversing polarity of the molecular arrangement in the middle. The model is presented only in two dimensions, but the real filament must be a cylindrical unit with the globular projections (the cross-bridges) protruding in three dimensions. (*b*) Actin can be studied in two forms, globular (G-actin) or fibrous (F-actin). The latter is a double helical arrangement of very highly aggregated G-actin monomers. (*c*) The minimum contractile unit, actomyosin, is an association of actin and myosin filaments. It is proposed that the globular projections of myosin provide the attachment between the filaments. (Adapted primarily from Huxley, H. E., *Science*, **164**, 1356 (1969).]

very thorough analyses by X-ray diffraction and equally thorough examination in the electron microscope, and has been supplemented with immunochemical evidence. The model is illustrated in Figure 4.12 at different levels of observation and model building. Both electron-microscopic studies comparing the dimensions of the high-molecular-weight aggregates of myosin and actin with the dimensions in the myofibrils, and more recent immunochemical studies have made it possible to identify the various regions of the fibril in terms of their molecular constituents. Thus antibodies prepared toward pure myosin become associated with the *A* bands while antibodies toward actin are associated with the *A* and *I* bands; antibodies toward HMM concentrate in the center of the *A* band and antibodies toward LMM concentrate in the distal portions of the *A* band (antibodies labeled with fluorescent dye were used to readily visualize the process). This model involves at least two different kinds of specific protein-protein interactions—first, the myosin-myosin interactions, which form the myosin filaments, and second, the myosin-actin interaction, which is proposed to be the basis for the actual contraction. The model in Figure 4.12 suggests that the myosin filament is stabilized through interaction of the LMM part of the myosin molecule, leaving the globular HMM parts as projections on the filament surface. When isolated LMM is allowed to aggregate in 0.1 *M* salt, completely smooth, long, needlelike filaments form, demonstrating the ability of this part of the myosin molecule to interact. No projections can be found in these aggregates. When, on the other hand, intact myosin is aggregated, typical filaments with the characteristic projections are observed. Both observations are consistent with the model. The HMM projections are the so-called cross bridges, which are the proposed sites responsible for the interaction between myosin and actin. The evidence for this particular interaction is based primarily on the electron micrographs showing the cross bridges as the most obvious points of attachment to actin, as well as on the fact that there is a direct linear relationship between the number of cross bridges overlapping the actin filament and the muscle tension generated at different stages of contraction or relaxation. In other words each cross-bridge contact to actin appears to generate a constant amount of muscle tension. The biochemical model of muscle contraction is yet very speculative. In addition to actin and myosin structure, this model must consider the role of the ATPase activity, the metal-ion effects, and the molecular mechanisms that trigger the contraction, as well as explain the sliding motion of the filaments. It has been estimated that the free energy released in the hydrolysis of one ATP molecule at each cross bridge is sufficient for a sliding movement of 50 to 100 Å per sarcomere; the key to contraction must thus be to elucidate the ATPase mechanism relative to myosin-actin interaction in the intact muscle. The following observations on the ATPase activity must be relevant as a beginning toward this elucidation. Pure myosin in the presence of physiological concentrations of Ca^{2+} and Mg^{2+} shows relatively low ATPase activity but addition of actin under conditions in which actomyosin can form greatly stimulates the activity. Unpurified actomyosin similarly shows low ATPase activity at low (10^{-7} *M*) Ca^{2+} concentration,

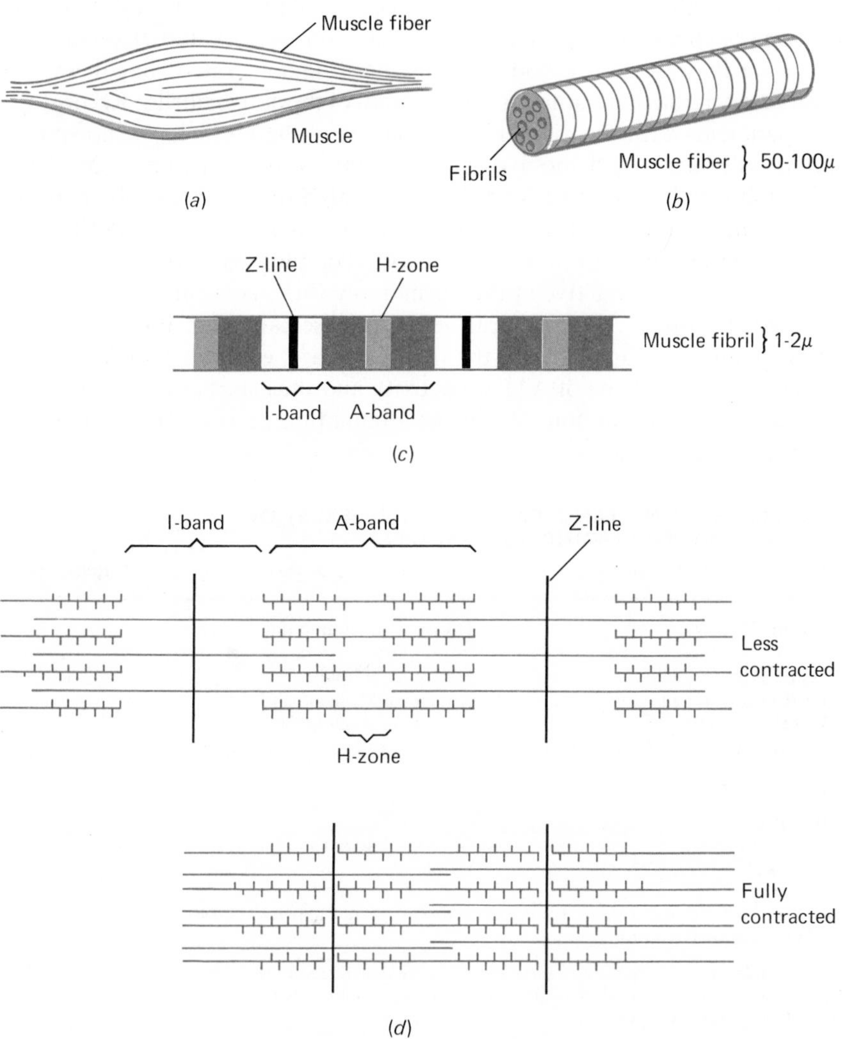

Figure 4.12 Different levels of organization of the contractile apparatus of striated muscle. (*a*) Intact muscle, (*b*) muscle fiber, (*c*) muscle fibril showing the characteristic bands which are readily observed in the microscope. In the contracted muscle the I-band and H-zone are diminished or absent. (*d*) A proposed model for the molecular arrangement in the muscle fibril incorporating the known properties of the actin and myosin filaments and the observed microscopic banding. This model is the basis for the so-called "sliding filament" theory, explaining the shortening and lengthening of the muscle on the basis of increasing and decreasing actin–myosin overlap and interaction accomplished by a sliding movement of the two filaments relatively to one another. [From Huxley, H. E., *Science*, **164**, 1356 (1969).]

but has high activity at 10^{-5} M Ca^{2+} concentration. If the actomyosin complex is purified, this sensitivity to Ca^{2+} is abolished and the ATPase activity is normal at low Ca^{2+} concentrations. When a protein fraction is added back to the purified actomyosin, sensitivity to Ca^{2+} is restored. The main components of this protein fraction are *tropomyosin* and *troponin* and, again using fluorescent antibodies, it has been demonstrated that these two proteins exist in the thin filaments (the H band) of the native fibrils. Reconstitution experiments with the different components of the fibrils have shown that troponin will bind to a tropomyosin-actin complex, but not to myosin; it has also been shown that troponin binds Ca^{2+} with high affinity. These observations represent the foundation for the beginning of a biochemical model for the events in muscle contraction. The actomyosin complex, with tropomyosin and troponin attached, is inactive at the normal low Ca^{2+} concentration found in unstimulated muscle. The electrical nerve impulse causes an influx of Ca^{2+}, which is bound by troponin, and the troponin-Ca^{2+} complex (not free Ca^{2+}) now acts as the activator of ATPase activity and thus starts the sliding of the fibers leading to contraction. When the stimulation ceases, Ca^{2+} is pumped out and the system relaxes.

TABLE 4.6 SOME OTHER PROTEINS (AND PROTEIN COMPLEXES) OF INTEREST IN STRUCTURE–FUNCTION STUDIES[a]

Class	*Protein*	*Function*	*Reference*
Carrier proteins	Serum albumin	Transport of compounds of low solubility	1
	α- and β-lipoproteins	Transport of lipids	1
	Ferritin-transferrin	Storage and transport of iron	2
Contractile proteins	Flagella and cilia	Motility in single cells	3
Regulatory proteins	Histones and other nuclear proteins	Regulation of gene (see Chapters 5–8)	4
	Protein hormones		5
Chemoreceptors[b]	Bacterial	Chemotaxis	6
	Taste protein	"Bitter" taste sensing	7
Photoreceptors[b]	Opsins	Vision	8

[a] See the classification in Table 2.3 and also the discussion in Chapters 9 and 10 on membrane-associated proteins.

[b] These could probably in the broadest sense best be considered in the group of regulatory proteins in that they mediate a signal-response linkage by turning on and off the response reactions.

1. F. W. Putnam, in *The Proteins, Second edition*, Vol. 3, H. Neurath, (Ed.), Academic Press, New York, 1965, p. 153.
2. P. M. Harrison, in *Iron Metabolism*, F. Gross (Ed.), Springer-Verlag, Berlin, 1964, p. 40.
3. I. R. Gibbons, *Ann. Rev. Biochem.*, **37**, 521 (1968).
4. R. H. Stellwagen and R. D. Cole, *Ann. Rev. Biochem.*, **38**, 951 (1969).
5. E. Frieden and H. Lipner, *Biochemical Endocrinology of the Vertebrates*, Prentice-Hall, Englewood Cliffs, N.J., 1971.
6. J. Adler, *Science* **166**, 1588 (1969).
7. F. R. Dastoli, *et al.*, *Nature*, **218**, 884 (1968).
8. G. Wald, *Science*, **162**, 230 (1968).

This very oversimplified model obviously leaves many questions unanswered but it does introduce some ways to visualize a highly sophisticated protein complex in which an enzyme, two regulatory proteins, and an interacting pair of fibrous proteins can translate a specific chemical reaction into mechanical work.

4.5 CONCLUSION

The fundamental idea of this chapter has been to consider some noncatalytic proteins of very different functions with the aims of (1) verifying the proposal that proteins carry out their function through ligand recognition and binding, and (2) focusing attention on some proteins whose functions are expressed only by extremely complex structural units. Obviously the selection of the examples is again quite arbitrary; many other functional proteins that could have been discussed are outlined in Table 4.6.

REFERENCES

Immunochemistry:

"Antibodies," *Cold Spring Harbor Symp. Quant. Biol.* **32** (1967).

Kabat, E. A., *Structural Concepts in Immunology and Immunochemistry*, Holt, Rinehart and Winston, New York, 1968.

Metzger, H., "The Antigen Receptor Problem," *Ann. Rev. Biochem.*, **39**, 889 (1970).

Müller-Eberhard, H. J., "Complement," *Ann. Rev. Biochem.*, **38**, 389 (1969).

Porter, R. R., "The Structure of Immunoglobulins," *Assays in Biochemistry*, **3**, 1 (1967).

Williams, C. A., and M. W. Chase (Eds.), *Methods in Immunology and Immunochemistry*, Academic Press, New York, 1967.

Hemoglobin—recent reviews:

Antonini, E., and M. Brunori, *Ann. Rev. Biochem.*, **39**, 977 (1970).

Braunitzer, G., *J. Cell. Physiol.*, Suppl. 1, 1 (1966).

General reading:

Perutz, M. F., "The Hemoglobin Molecule," *Scientific American*, **211**, November (1964).

Insulin review:

Behrens, O. K. and E. L. Grinnan, "Polypeptide Hormones," *Ann. Rev. Biochem.*, **38**, 83 (1969).

Hales, C. N., "Some Actions of Hormones in the Regulation of Glucose Metabolism," *Essays in Biochemistry*, **3**, 73 (1967).

Muscle Contraction:

Davies, R. E., "On the Mechanism of Muscular Contraction," *Assays in Biochemistry*, **1**, 29 (1965).

Hoyle, G., "How is Muscle Turned On and Off?" *Scientific American* **222**, April (1970).

Huxley, H. E., "The Mechanism of Muscular Contraction," in Brachet, J., and A. E. Mirsky (Eds.), *The Cell*, Vol. IV, Academic Press, New York, 1968.

Young, M., "The Molecular Basis of Muscle Contraction," *Ann. Rev. Biochem.*, **38**, 913 (1969).

FIVE NUCLEIC ACIDS: SYSTEMS AND METHODS

In most scientific advances, information is of necessity sought at levels of progressively increasing detail. The short history of biochemical genetics illustrates this progressive advance very well in one area of very rapid development. Knowledge of physical and chemical properties of the nucleic acids and the biological basis of genetics were the minimum prerequisites for stating the first proper questions regarding biochemical genetics. Once the model had been proposed that information was stored in the DNA structure in a "language" made up of nucleotide "words" and that this information could be translated into specific protein structures in a language of amino acid words, the next step was to establish the flow of events in information storage and transfer and obtain sound experimental evidence for the model. Finally, when the model had thus been confirmed, the more detailed questions about the molecular mechanisms of each individual step in the process could be stated and the answers sought.

In discussing this exciting area of research a deliberate attempt has been made to present the material according to such a stepwise progression of model building and model refinement. The procedure has the obvious advantage of following a fairly logical sequence of events, not necessarily in chronological order but in an order which in retrospect should have been or could have been a logical chronological progression. (This description is formulated by a person who watched the play from the gallery, however, and it probably differs significantly from the one the actors themselves would set down.) The procedure is not unlike the exploration of the moon, first through telescopes, then from orbiting spaceships, and finally by landing and collecting samples for detailed analysis. It may seem unnecessary for the person born in the space age to concern himself with the poor resolution and the difficult interpretation of the earliest data. On the other hand can a person really evaluate and appreciate the detailed chemical analysis of a moon rock without having knowledge of other rock analyses for comparison or without having observed the full moon 250,000 miles away and pondered its relationship to the solar system, the earth, and the observer himself?

This next section of the book, then, concerns itself with the informational molecules at three levels. In Chapter 5 the chemical properties and the analytical tools available for studying both structure and function are reviewed and in Chapter 6, the different steps involved in the storage and transfer of genetic information are established. Finally, in Chapter 7 the molecular mechanisms of each step are explored. Chapters 6 and 7 thus essentially cover exactly the same topics but at different levels. With the extensive coverage these topics have received in our times, a large number of students will undoubtedly and justifiably feel that they have "observed the moon" sufficiently long and would like to get on with the rock analyses; they are urged to move directly to Chapter 7. Chapters 5 and 6 will hopefully help those who feel they need some of the background material to see the whole forest before starting to study the individual trees.

5.1 INTRODUCTION

The nucleic acids are the central elements for information storage and transfer. This chapter introduces the main groups of DNA and RNA and reviews their basic structural properties and some of the experimental tools available for structural analysis. The methods available for characterizing their functional properties, based either on the direct influence of the nucleic acids on the synthesis of proteins or on the more indirect studies of nucleic acid interactions, are also briefly outlined.

The constancy of expression of living forms necessitates postulating genetic material. Based on our current knowledge of in vivo catalysis and intermediary metabolism, it is clear that all the characteristics of any living organism—size, shape, color, metabolic requirements, and end products—depend upon its possessing a characteristic complement of enzymes. The ability to make these enzymes and just these enzymes is passed on from parents to progeny with (almost) faultless precision and this fact can only be explained on the basis of the existence of a genetic material, blueprints for the production of that particular organism. The nature of this blueprint must be such that it is self-replicating, so that an exact copy can be transmitted from parents to progeny.

It must also be such that its information can be translated into the characteristic complement of protein structures.

Perhaps no area of science has ever undergone such an exciting and explosive development as biochemical genetics. From the first demonstration of DNA as the carrier of genetic information in 1944* through its careful chemical characterization† to the establishment of the double-stranded structure of DNA in 1953‡, the concept of a chemical language made up of nucleotides that spell out the molecular characteristics of each living cell was confirmed. The next steps—the formulation of the mechanisms to explain how the fundamental DNA language could be transcribed to RNA language and subsequently translated into amino acid sequences—were highlighted by the discovery of the unstable messenger RNA (mRNA) in 1956§ and the deciphering of the first word of the genetic vocabulary in 1961.|| In a span of well under 30 yr, the seed of a biological idea has blossomed into a molecular model, the potential and consequence of which are certainly not yet fully understood and appreciated.

In attacking the problem of correlating structure and function of the informational molecules, it is again necessary to examine the fundamentals as expressed by the following questions: What kinds of molecules and structures are available for study? How can these structures be examined and described? How can their functions be studied, described, and quantitatively measured? This chapter, then, will be addressed to answering these three questions.

5.2 CLASSIFICATION

The molecules and structures involved in basic information storage and transmission are the nucleic acids, free or in the different characteristic nucleoprotein complexes. Thus we are here interested in cellular DNA, the three functional groups of cellular RNA, and also the DNA and RNA of viruses; the last represent relatively simple and biologically single-purpose informational structures (Table 5.1).

Deoxyribonucleic Acid

Most of the DNA in nature exists in the molecular-weight range from 10^7 to 10^9 daltons; a few also fall within one order of magnitude on either side of this range. Because of the enormous size of the linear DNA polymers, the precise determination of their molecular weight has been a persistent problem in biochemistry. Simple manipulation of DNA solutions sets up sufficient

* O. T. Avery, C. M. MacLeod, and M. McCarty, *J. Exp. Med.*, **79**, 137 (1944).

† See E. Chargaff, "Essays on Nucleic Acids," Elsevier, New York, (1962).

‡ J. D. Watson and F. H. C. Crick, *Nature*, **171**, 737, 964 (1953).

§ E. Volkin and L. Astrachan, *Virology*, **2**, 149, 433 (1956).

|| M. W. Nirenberg and J. H. Matthaei, *Proc. Nat. Acad. Sci.*, **47**, 1558 (1961).

TABLE 5.1 SUMMARY CLASSIFICATION OF NUCLEIC ACIDS

	Molecular Weight (in 10^3 Daltons)	*Number of Different Molecules in Any One Cell (Organism, Virus)*
DNA (deoxyribonucleic acid)		
Chromosomal (nuclear) DNA	1,000,000	1 to several thousand
Viral (DNA)	2,000–200,000	1
Mitochondrial (DNA)[a]	10,000	1 to 10
RNA (ribonucleic acid)		
Transfer RNA	25	20 isolated, probably 50 to 60
Messenger RNA	100–4,000	Several thousand
Viral RNA	500–10,000	1
Ribosomal RNA	40–1,600	At least 3

"Counterions"—Basic Compounds Found Associated with Nucleic Acids[b]

Compound	*Structural Characteristics*	*Source*	*Nucleic Acids*
Histones[c]	10–20,000 MW proteins, high in arginine and lysine	Animals and Plants	DNA
Lysine rich (up to 30% lysine)	Contains no cystine and very little methionine		
Arginine rich (up to 13% arginine)	Contains no cystine and very little methionine		
Protamine (MW = 5000)	Similar to arginine-rich histones, no sulfur, and no tyrosine or tryptophan	Fish sperm cells	DNA
Spermine	$H_3\overset{+}{N}{-}(CH_2)_3{-}\overset{+}{N}H_2{-}(CH_2)_4{-}\overset{+}{N}H_2{-}(CH_2)_3{-}\overset{+}{N}H_3$	Bacterial	RNA and DNA
Spermidine	$H_3\overset{+}{N}{-}(CH_2)_3{-}\overset{+}{N}H_2{-}(CH_2)_4{-}\overset{+}{N}H_3$	Viruses	
Bis-(3-aminopropyl)-amine	$H_3\overset{+}{N}{-}(CH_2)_3{-}\overset{+}{N}H_2{-}(CH_2)_3{-}\overset{+}{N}H_3$	Plant viruses	
Ribosomal proteins	Weakly basic, heterogeneous, and quite similar in widely different species	All forms	RNA

[a] See Y. Suyama and K. Miura, *Proc. Nat. Acad. Sci.*, **60**, 235 (1968) and S. Gramick and A. Gibor, *Progr. Nucleic Acid Res.*, **6**, 143 (1967).

[b] From L. S. Hnilica, *Progr. Nucleic Acid Res.*, **7**, 25 (1967).

[c] Histones are also found acetylated and phosphorylated.

shear forces to fragment the polymer and even now several recorded DNA molecular weights must be considered as tentative. With reference to the summary in Chapter 1, DNA in nucleated cells (eucaryotic cells) is found almost exclusively in the nucleus, where, in association with basic proteins (histones), it makes up the chromatin. The chromatin can be observed to exist in the nucleus as distinct bodies, the chromosomes. Each chromosome contains at least one molecule of DNA but the complete chromosome structure continues to be a riddle. In bacterial cells (procaryotic cells), where no nucleus exists, the DNA is found throughout the cell, probably free of histones and attached to the plasma membrane in at least two points. In viruses, as indicated in Chapter 1, the DNA is generally found as the virus core, tightly surrounded by protein. In all cases we refer to this DNA as chromosomal or "nuclear" DNA, and it appears that the chromosomal DNA of any cell or virus is a constant and unique single molecule or set of molecules (amount of DNA increases with organism complexity). The total chromosomal DNA of any cell or virus is referred to as its *genome*. It is important that a small but significant amount of DNA different from the nuclear DNA is always found in the mitochondria (and in the chloroplasts of photosynthetic organisms). The molecular weight of mitochondrial DNA is about 10^7. In some cell types the "small" circular DNA units appear to be interlinked in "daisy chains" containing up to 12 or 14 rings per mitochondrion. Mitochondrial DNA may be quite similar in different species. Structurally the giant DNA molecules are generally double-stranded helices; the complementary strands are stabilized by the hydrogen bonding between A–T and G–C pairs and by base stacking and hydrophobic interactions. The two strands are antiparallel (see below). Some DNA molecules, so far found only in viruses and not in higher cells, are single stranded and some, both single stranded and double stranded, are circular (the two ends are covalently closed with a phosphatediester link). For our purpose we will consider that all DNA, whether it is chromosomal, extrachromosomal, or viral DNA, at some stage in the life cycle exists as a double-stranded linear helical molecule. Functionally the DNA is proposed as the basic storage molecule for genetic information.

Ribonucleic Acid

Three classes of RNA are found in living cells, transfer RNA (tRNA), messenger-RNA (mRNA), and ribosomal RNA (rRNA). This system of subdivision and nomenclature is based on the functional and structural features of the different RNA's, which will be discussed in the next chapter.

Transfer RNA has a molecular weight of 25,000 and is found free in the cytoplasm of all cells as the smallest nucleic acid molecule. It is unique in that it contains a large number of unusual bases not found in the other nucleic acids (C- and N-methylated bases, pseudouridine, hypoxanthine, and so forth). It is a single-stranded molecule with some complementary-structured sequences. Approximately 20 different tRNA molecules have been isolated but the total number in each cell is probably greater than this (see Chapter 6). The complete nucleotide sequence has been determined for several tRNA molecules; the large number of unusual bases greatly facilitated this accomplishment. It appears that all tRNA molecules have a common C–C–A 3'-terminal sequence.

Messenger RNA is a larger molecule than tRNA, apparently containing only the conventional RNA bases A, G, U, and C. In general it is assumed that most cellular mRNA's fall in a fairly narrow molecular-weight range of about 10^6 (this is only a generalization, however, and especially if viral RNA is included in this functional class of RNA, a much wider range of sizes must be considered). The mRNA is the most elusive of the RNA types, present at any time only at low concentrations. In most procaryotic cells the mRNA has a very short half-life whereas there is strong evidence for quite stable mRNA molecules in higher cells. The cellular mRNA is always single stranded whereas viral RNA can exist in a double-stranded form (see Section 6.7). Messenger RNA, as shall be seen, can be found free in the cytoplasn or associated with the nucleus and with ribosomes. Although there is no direct evidence for a large number of different kinds of mRNA in any one cell, our current functional model predicts that the number of different mRNA molecules in any one cell should approach the number of different polypeptide chains in that cell.

Ribosomal RNA, the third type of RNA, is found as a nucleoprotein complex, the ribosome. Ribosomes are scattered throughout the cytoplasm as well as attached to the plasma membrane in bacterial cells and to specialized membranes (the endoplasmic reticulum) in cells of higher organisms. Bacterial ribosomes are in general readily prepared by fractional centrifugation of cell homogenates, whereas ribosomes from higher cells are quite difficult to obtain free from the lipoprotein membrane material, which together with ribosomes make up the microsomal fraction. Figure 5.1 illustrates the makeup of the *Escherichia coli* ribosomes as a typical example of the ribosome structure. It should be noted that the stability of 100-S, 70-S, 50-S, and 30-s ribosomes is a function of Mg^{2+} concentration. The ribosomes contain two different kinds of high-molecular-weight RNA, one with a molecular weight of 600,000 from the 30-s ribosomes, and one with a molecular weight of 1,200,000 from the 50-s ribosomes; in addition a 5-s RNA with a low molecular weight (40,000) has been found associated with the 50-s ribosomes. It appears that there is only one, or at the most a few, types of each of these in any one cell. The ribosomal proteins have all been separated and some information about their properties is also given in Figure 5.1. Table 5.2 summarizes some basic properties of the nucleic acids and some conventions of nomenclature and structural representations.

5.3 SUMMARY OF STRUCTURE: STRUCTURAL PARAMETERS

In the introductory remarks above, we have referred to several structural forms and terms that are all founded on chemical and physical observations. These are discussed by Barker and Van Holde in this series and are briefly reviewed in Tables 5.2 and 5.3. Most of the methods will be considered in the actual experiments discussed in the next chapters. It must be kept in mind, however, that the specific structural information available for nucleic acids is very limited. Both the size of the molecules and the small number of monomer building blocks involved in the polymers have complicated the covalent-structure determination (nucleotide-sequence determination). With the

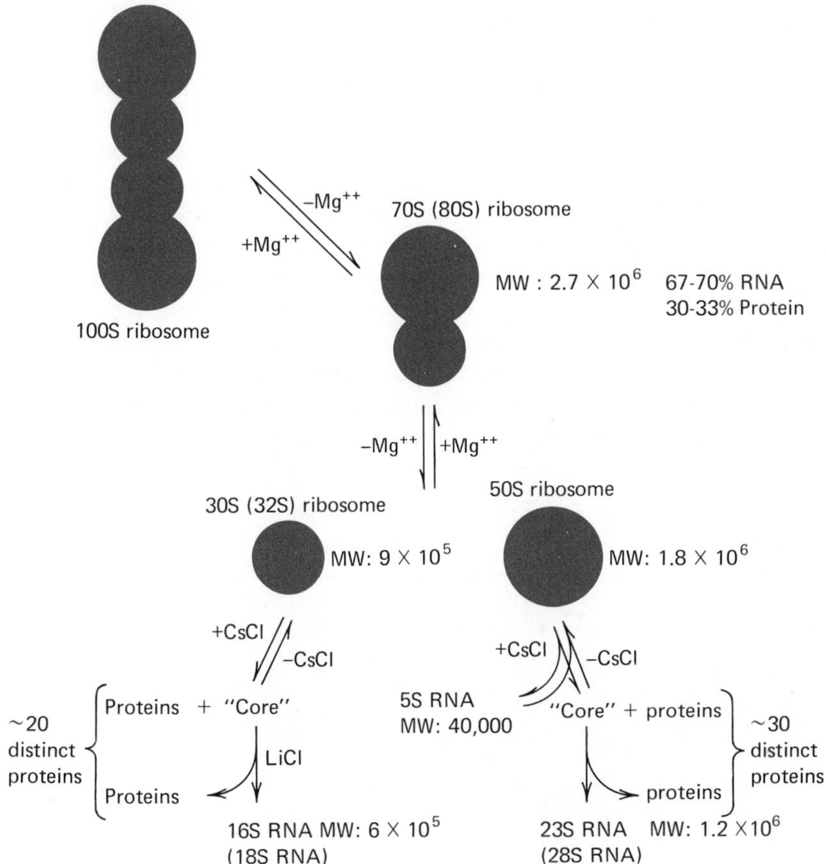

Figure 5.1 The architecture of *Escherichia coli* ribosomes. The figure illustrates the Mg^{++} and salt-induced dissociation and association steps. Three distinct types of RNA are found. The 5S RNA contains unusual bases (like tRNA), and its complete nucleotide sequence has been determined [Brownlee, G. G., *et al.*, *Nature*, **215**, 735 (1967)]. The approximately 50 different proteins have been separated and partially characterized. There appears to be one molecule of each per ribosome and they are all immunochemically different. Their molecular weights range from about 10,000 to 50,000. The numbers given in parentheses in the figure refer to guinea pig ribosomes. The mammalian ribosomes are slightly larger than the *E. coli* ones, but appear to be very similar in general structural properties.

exception of some of the smaller nucleic acids (notably tRNA), very little is known about nucleotide sequences. Nevertheless, certain aspects of structure can be described quite precisely once the limitations and the terminology are defined. What we are primarily concerned with here can best be stated by asking the following questions. How can we separate or otherwise distinguish between the different nucleic acids, both free and in their natural nucleoprotein complexes? How can we define a native and a denatured state of the isolated molecules in structural terms? How can we compare nucleotide sequence in the different molecules without knowing the actual sequence of any of the molecules under study?

It should be emphasized that most of the work in this area has been carried out on a rather unusual scale. Several procedures are based on experiments in which no product can be obtained in quantities that can be either seen or weighed. The experimental techniques have nevertheless led to unequivocal results and are often ingeniously simple. The high UV absorption of nucleic acids provides a convenient method of detection and quantitation at the μg level but it does not satisfactorily distinguish between RNA and DNA. The most important quantitative technique involves the use of radioactive isotopes. By using compounds labeled with isotopes of high enough specific activity (^{32}P, ^{14}C, and ^{3}H can all readily be introduced into the covalent nucleic acid structure), precise quantitative measurements can be made at the pg (picogram) level of quantities of nucleotides and nucleic acids (as well as proteins).

TABLE 5.2 STRUCTURAL FEATURES OF NUCLEIC ACIDS

A. Building Blocks

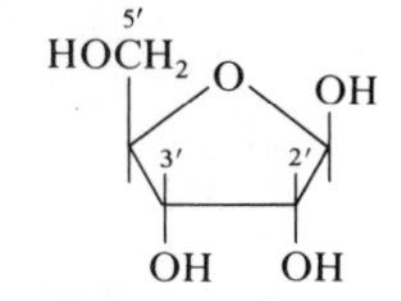

5′ HOCH2 O OH 3′ 2′ OH OH

β-D-ribofuranose

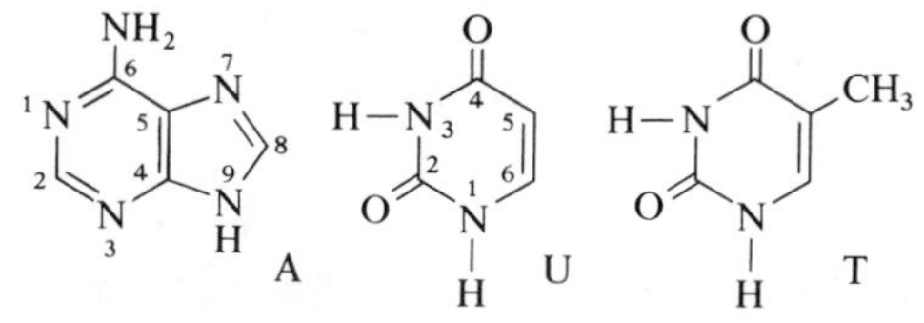

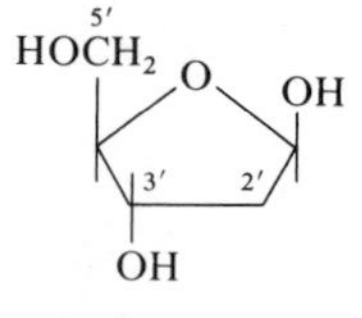

NH2 6 7 1 N 5 N 8 2 4 9 N N 3 H A

O H—N3 4 5 2 1 6 O N H U

O H—N CH3 O N H T

5′ HOCH2 O OH 3′ 2′ OH

2-deoxy-β-D-ribofuranose

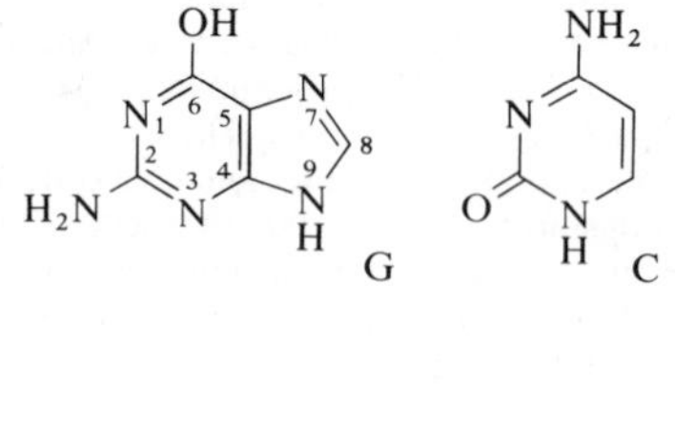

OH N1 6 5 N 7 8 2 3 4 9 H2N N N H G

NH2 N O N H C

1. Common purine and pyrimidine bases:

Name	*Abbreviation*
Adenine	A
Guanine	G
Uracil	U
Thymine (5-methyluracil)	T
Cytosine	C

2. Unusual bases:

Name	*Abbreviation*
6-Methylaminopurine	AM
6,6-Dimethylaminopurine	AM_2
Hypoxanthine (6-oxypurine)	I
2-Methylamino-6-oxypurine	GM
2,2-Dimethylamino-6-oxypurine	GM_2
1-Methylhypoxanthine	IM
6 O-Methylguanosine	GOM
5-Hydroxymethyluracil	HMU
5-Methylcytosine	CM
5-Hydroxymethyl cytosine	HMC
N(4)-Acetylcytosine	CAc
6-Aminoisopentenylpurine	A-isop (AC_5)
5,6-Dihydrouracil	UH_2

3. Nucleosides: The bases are linked to C1 of ribose or deoxyribose (β-glycosidic bond) through N9 of purines and N1 of pyrimidine. The only exception is pseudouridine (5-ribosyluracil), whose symbol is ψ. Ribonucleosides are abbreviated with the same symbol as the base, for deoxyribonucleosides the letter "d" is added as a prefix.

4. Nucleotides: These have a phosphate on either of the free —OH groups of the sugar, indicated by primed numbers.

B. Polymers:

1. Sequence. All nucleic acids are polymers of nucleotides linked by a phosphate diester bond from the 3′-OH of one to the 5′-OH of the next. In any representation it is important to indicate the 5′-end (head) and the 3′-end tail). Different representations are used: The tetranucleotide represented below can be described in shorthand as A-U-G-C, or as pApUpGpC or as the "stick" representation:

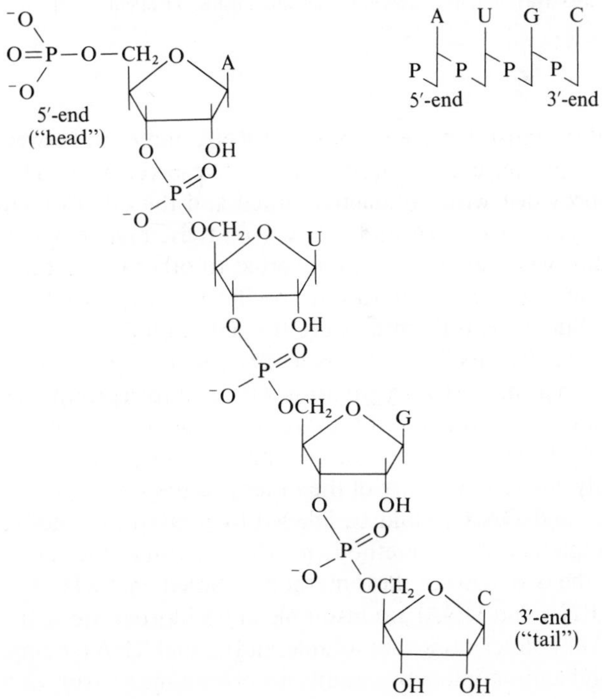

TABLE 5.2 STRUCTURAL FEATURES OF NUCLEIC ACIDS *(cont.)*

In the "stick-representation" the ribose is represented as a line with C1 (attached to the base) on top and C5 on the bottom. If the tetranucleotide contained 2-deoxyribose, the shorthand representations dA-dT-dG-dC, pdApdTpdGpdC or d(A-T-G-C) would be used. (Note that dU never occurs in a DNA sequence, only the 5-methylated derivative thymine is found, so T is substituted for U in the hypothetical tetranucleotide.)

2. Noncovalent structure: The basic structural determinant is hydrogen bonding between base pairs:

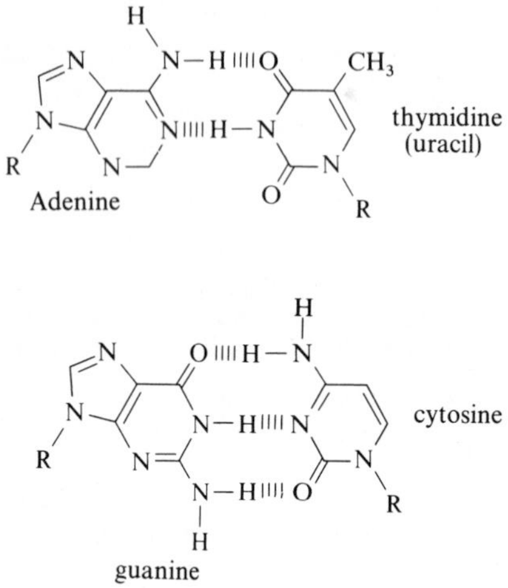

Adenine — thymidine (uracil)

guanine — cytosine

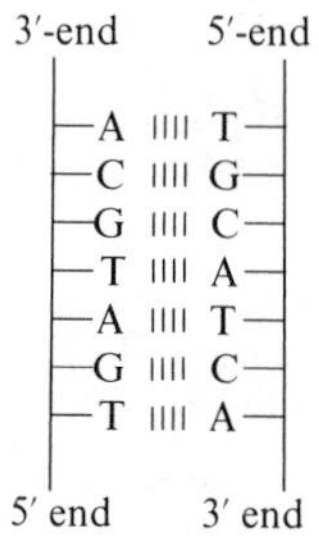

3′-end		5′-end
A	IIII	T
C	IIII	G
G	IIII	C
T	IIII	A
A	IIII	T
G	IIII	C
T	IIII	A
5′ end		3′ end

antiparallel complementary (A = T, G = C) sequences of a double-stranded segment of DNA

3. Higher orders of structure (associated forks of nucleic acids). Nucleoproteins: DNA-histones, RNA-protein (ribosomes), other polycationic counterions are found in viruses and microorganisms. Divalent metal ions (Mg^{2+}) strongly associate with nucleic acids.

It is possible to distinguish between DNA and RNA and even between different types of RNA by the use of specific radioactive precursors. Thus if a growing culture is provided with radioactive uracil and the other (unlabeled) bases, the cells will synthesize RNA and DNA from these precursors. Under the proper conditions, when uracil is not converted to other bases, the radioactivity will remain in uracil and will appear specifically in RNA (DNA does not contain uracil). Similarly, if thymine were the only radioactive precursor, DNA would be specifically labelled. This is a very general approach in biochemistry, applicable in principle to any situation where one of a group of similar molecules has a unique constituent and the precursor for this particular constituent can be supplied under such conditions that it can be taken up by the cell and is not rapidly converted to any of the other precursors.

Even if both RNA and DNA should be labeled by a given radioactive precursor, there are simple and direct methods for distinguishing the rate or the extent of labeling of the two. One of these methods is based on the facts (1) that the polymers (both RNA and DNA) are insoluble in trichloroacetic acid, while monomers and short oligonucleotides are soluble, and (2) that RNA is completely degraded to acid soluble monomers by conditions of alkaline hydrolysis which will not cause any degradation of DNA (see Barker in this series). Consider an experiment in which a soluble radioactive precursor labels both RNA and DNA. The acid insoluble radioactivity in an aliquot of this incubation mixture is a direct measure of total incorporation into both RNA and DNA. In an identical aliquot treated first with alkali (to hydrolyze RNA) and then with trichloroacetic acid, only DNA precipitates, and the extent of incorporation into DNA is thus measured directly. The extent of incorporation into RNA is next obtained simply from the difference between the two aliquots.

Nucleic acids can be purified by a number of standard techniques of chromatography and electrophoresis and, to insure that one indeed can work with isolated individual compounds, several unique systems have been developed specifically for the nucleic acid field (separation on methylated albumin, benzoylated diethylaminoethyl cellulose,* as well as reverse-phase chromatography.†) One of the most useful methods is based on the characteristic density differences between native DNA, denatured (single-stranded) DNA, RNA, ribosomes, and so forth. Sedimentation in a density gradient allows very sharp separation of the different molecules (Table 5.3).

For the problem of establishing nucleotide sequences (see Barker in this series) there are at present four possible solutions: (1) nearest-neighbor analysis; (2) 3′- and 5′-terminal analysis, partial degradation, and overlaps; (3) fine-structure genetic mapping (discussed under mutations in Chapter 7); and (4) biosynthesis with very short pulses of radioactivity. The last two methods have been amazingly powerful in deducing long nucleotide sequences in both DNA and RNA.

5.4 DETERMINATION OF FUNCTION: FUNCTIONAL PARAMETERS

In studying the biological role of the information molecules, one immediately notices a new degree of complexity in comparison to proteins. With an enzyme we could assay for biological activity by studying a simple precursor-product (substrate-product) conversion, observed upon the addition of the enzyme. For the informational molecules, no such assay can be designed. The term "informational molecule" implies that these molecules determine the nature of a product. However, in order for this product to be formed, not only the information but also the whole complicated synthetic apparatus must be present. In the minimum assay system in vitro, therefore, in addition to all the informational molecules we must have all the precursors, coenzymes, and enzymes required for making product (protein) and we must be able to characterize the product in order to evaluate the information supplied. Because of the complexity and instability of the isolated synthetic apparatus, it is difficult to carry out in vitro assays and it is often necessary to leave the cell intact and try

* G. M. Tener, *et al.*, *Fed. Proc.*, **25**, 519 (1966).

† For a review see G. D. Novelli, *Ann. Revs. Biochem.*, **36**, 449 (1967).

TABLE 5.3 EXPERIMENTAL APPROACHES TO STRUCTURAL DEFINITION

Method, Probe	*Information*
Base composition	Characteristic base ratio, of some use for identification of a certain nucleic acid, important in exploring whether the molecule is double stranded (A = T(or U), G = C) or single stranded (A ≠ T, G ≠ C)
Nearest-neighbor analysis	A first approximation to sequence analysis for further characterization. Also demonstration of the antiparallel nature of double stranded DNA. $A \longrightarrow G = C \longrightarrow T, G \longrightarrow T = A \longrightarrow C$, etc. (Rather than $A \longrightarrow G = T \longrightarrow C$ and $G \longrightarrow T = C \longrightarrow A$, which would be expected for parallel chains).
UV absorbance	High UV absorbance (usually read at a wavelength of 260 nm) is characteristic of all nucleic acids. In structured polymers base–base interactions and hydrogen bonding lead to a reduction in the absorbance calculated for the sum of the absorption of the component bases (hypochromic effect). Nonstructured (random coil) nucleic acids have absorbances closer to the theoretical. Thus the transition between structured and nonstructured (native and denatured DNA, hydrogen bonded and nonhydrogen bonded, double helix and random coil) can be followed by UV-absorption changes. Because of the very sharp temperature transition for helix-coil conversion of pure DNA, this is often referred to as melting point[a] for thermal denaturation.
Hydrodynamics	Sedimentation, diffusion, viscosity are functions of the size and shape. Changes in shape (for example the helix-coil transition for DNA) are conveniently detected by hydrodynamic measurements.
Density	Molecules sedimenting in a density gradient, will sediment to a point at which the solvent density matches the buoyant density of the molecule in that solvent (the molecule floats). If the characteristics of the density gradient are known, the position of a compound in that gradient accurately determines its density. Because of large differences in the density of macromolecules, and even in native and denatured states of any given molecule (approximate values: protein 1.3; native DNA, 1.66–1.76; denatured DNA, 0.02 higher than native; RNA, 2.0), density measurements on RNA, DNA, and protein mixtures have become very useful.[a]
Enzymatic attack	Specific hydrolytic enzymes (nucleases = diesterases) can be used as structural probes, some attacking only DNA or RNA, some only double-stranded structures, some only single-stranded ones. Specificity of cleavage (exonucleases versus endonucleases, 5′-side versus 3′-side) is important in sequence determinations.

[a] The melting temperature of a double helix is related to the stability of the double helix. Since G—C hydrogen bonds contribute most to stability, there should be a linear relationship between melting temperature and G + C content of DNA. This has been found to be true. G + C also contribute most to the densities of the polymer, and so there is also a linear relationship between density and G—C content. With proper calibrations and controls, one can actually estimate very accurately the G + C content of an unknown DNA from its melting temperature or its density, provided that it contains only the four normal bases.

to develop in vivo "activity" assays. This is one reason why the viruses have become such extremely important tools: They can be isolated, and both their nucleic acid and their protein can be characterized. When a virus infects the host cells, it essentially carries out the most difficult experimental step of any in vivo work, the injection or introduction of a new compound into the cell without destruction of the cell's synthetic apparatus. Thus it is possible, however complicated, to study the direct effect of DNA or RNA on the enzyme-catalyzed formation of specific peptide-protein products both in vitro and in vivo.

Simpler and more convenient functional handles have been defined on the basis of direct nucleic acid interactions. It is now well established experimentally that functionally related nucleic acids are also structurally related either as copies or as complements* of one another (see Table 6.1). Since complementary sequences form complexes (a double helix in the case of DNA), the direct study of nucleic acid interactions has become a useful tool in establishing functional relationships. Experimentally, the presence of complementary nucleic strands can be established in a number of ways, some of which have already been mentioned. Three types of double-stranded structures are involved, DNA-DNA double helices (*duplex*), RNA-RNA duplex, and DNA-RNA hybrids. Again, separation based on density differences offers a convenient method for distinguishing between these types of interacting nucleic acids. Another important approach to the assay of nucleic acid interactions is based on the procedure outlined in Figure 5.2. The principle of immobilizing one partner in a complex is a very general one that has been used successfully in the study of different types of interactions. Several specific examples of these methods will be considered in the next chapters.

5.5 ENZYMES USED IN NUCLEIC ACID CHARACTERIZATION

Before ending the survey of the approach to structural and functional characterization of nucleic acids, it is necessary to list the various enzymes that are involved both in the biosynthesis and biodegradation of nucleic acids. These enzymes will be encountered in several different sections of the next chapters as well as in Barker and Larner in this series, but it is probably convenient to attempt a tabular listing at this point. The list in Table 5.4 is not exhaustive; new specific enzymes are being found continuously. Even a cursory examination should, however, give an impression of the very broad spectrum of highly specific enzymes available in nature. There are two important aspects of this list of enzymes. First, they represent chemical reagents of inestimable importance to the structural analysis of nucleic acids. Once the specificity of a given enzyme has been established, its action on an unknown nucleic acid is as characteristic as that of any reagent in qualitative analysis; the results of enzymatic attack on nucleic acids can thus be interpreted quite unequivocally in terms of specific structural features of the unknown substrate molecule. Many ingenious analytical procedures have been developed entirely upon the action of specific enzymes; several, such as nearest-neighbor analysis and end-group analysis, have already been mentioned. The second important aspect of

* Complementary sequences are related by a faithful matching of the proper base pairs, A to T or U, G to C. Thus the sequences A–G–G–T–C and T–C–C–A–G are complementary.

the enzymes in Table 5.4 is the concept that if they are present, it is for a purpose. Thus the existence of a given specific catalytic activity in a cell should be accounted for in any model attempting to describe how that cell functions. On these premises, if one predicts a certain chemical reaction as part of a model or scheme, then an important part of the proof for that reaction is a demonstration of the existence of the enzyme that catalyzes the chemical reaction. In the next chapters, several cases will be discussed in which either the existence of an enzyme has been predicted by the model and the subsequent discovery of that enzyme has thus given support to the model, or the accidental discovery of a new enzyme has led to new concepts and refinements in the model.

TABLE 5.4 ENZYMES INVOLVED IN NUCLEIC ACID DEGRADATION AND SYNTHESIS[a]

A. Degrading enzymes

	Type of Cleavage[b]			*Specificity*[c]				
	a	*b*	*Double*	*Single*	*U(T)*	*C*	*A*	*G*
I. RNA and DNA								
1. Diesterases (exo-)								
Spleen diesterase		+			+	+	+	+
Snake venom diesterase	+			+	+	+	+	+
Bacilus subtilis diesterase (Ca-dependent)		+	+	+	+	+	+	+
2. Diesterases (endo-)								
Micrococcal diesterase		+	+					
Venom diesterase		+						
3. Monoesterases (phosphatases)								
Alkaline phosphatase (nonspecific)								
5′-esterases								
3′-esterases								
II. RNA								
1. Diesterases (endo-)								
Pancreatic ribonuclease		+		+	+	+		
Plant ribonuclease		+		+	+	+	+	+
B. subtilis ribonuclease		+		+			+	+
Takadiastase RNase T_1		+		+				+
Takadiastase RNase T_2		+		+				+
E. coli ribonuclease II	+							

A. Degrading enzymes

	Type of Cleavage[b]			*Specificity*[c]				
	a	*b*	*Double*	*Single*	*U(T)*	*C*	*A*	*G*
III. DNA								
1. Diesterases (exo-)								
E. coli exonuclease I	+			+				
E. coli exonuclease II	+		+	+				
E. coli exonuclease III	+		+					
E. coli exonuclease IV	+			+ (low MW only)				
2. Diesterases (endo-)								
(Pancreatic) DNase I	+		+					
(Thymus or spleen) DNase II		+	+	(+) double stranded cleavage				
E. coli endonuclease I	+		+	+ (inhibited by tRNA)				
E. coli endonuclease II								
E. coli endonuclease III								

B. Synthetic Enzymes

Enzyme	*Source*	*Substrates*	*Primer*	*Template*	*Product*
1. Polynucleotide[d] (Phosphorylase)	Microorganisms	Nucleoside diphosphates	+	—	Polyribonucleotides
2. DNA-polymerase (DNA-replicase)	*E. coli* and mammalian tissues	dATP + dTTP + dGTP + dCTP	DNA (oligodeoxynucleotides)	DNA	DNA
3. RNA-polymerase (transcriptase)	Microorganisms, plants, animals	ATP + UTP + GTP + CTP	—	DNA	RNA
4. RNA-polymerase (RNA-replicase)	Virus-infected microorganisms	ATP + UTP + GTP + CTP	—	RNA	RNA
5. Joining enzymes (ligases)	*E. coli* and virus-infected *E. coli*	Single-stranded DNA pieces (2 polydeoxynucleotide ends)	(DNA) (also requires coenzymes: NAD or ATP)	(DNA)	End-to-end condensation product (circular or linear)

[a] The summary in this table is based on the work from many laboratories and is taken in part from: F. Egami, K. Takahashi, and T. Uchida, *Progr. in Nucleic Acid Res.*, **3**, 59 (1964); M. Laskowski, *Advance. Enzymol.*, **29**, 165 (1967); and J. F. Koerner, *Ann. Rev. Biochem.*, **39**, 291 (1970).

[b] Type of cleavage to give either (a) 5′-phosphate or (b) 3′-phosphate:

X Y
P
a b

a ⟶ 5′-phosphate, 3′-OH

b ⟶ 3′-phosphate, 5′-OH

[c] Specificity for either single-stranded or double-stranded nucleic acid and for cleaving the diester linkage (in the direction from head to tail) *after* the nucleotides indicated. A plus is indicated only in cases where the specificity is well established.

[d] The in vivo function of this enzyme is undoubtedly primarily to degrade polynucleotides. Because of its extensive use in in vitro synthesis of "artificial" polymers, it has been included under synthetic enzymes.

1. A specific nucleic acid is fixed (covalently or noncovalently) to an insoluble matrix.

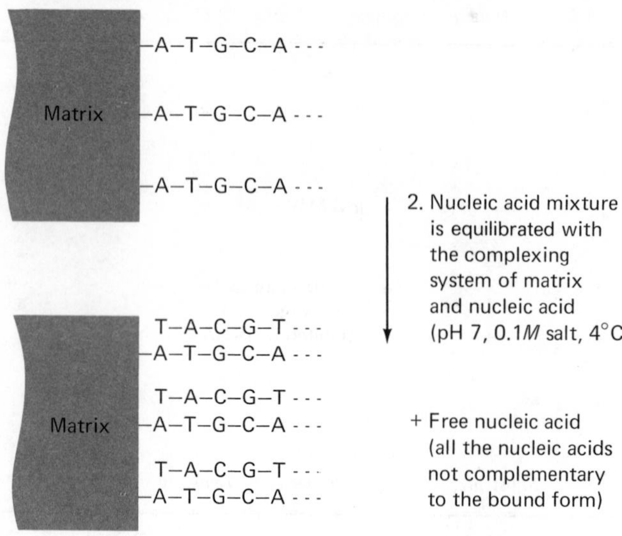

3. Matrix is washed free of all unbound nucleic acid.

4. Matrix is washed under conditions that disrupt hydrogen bonds (high salt concentration, high temperature)

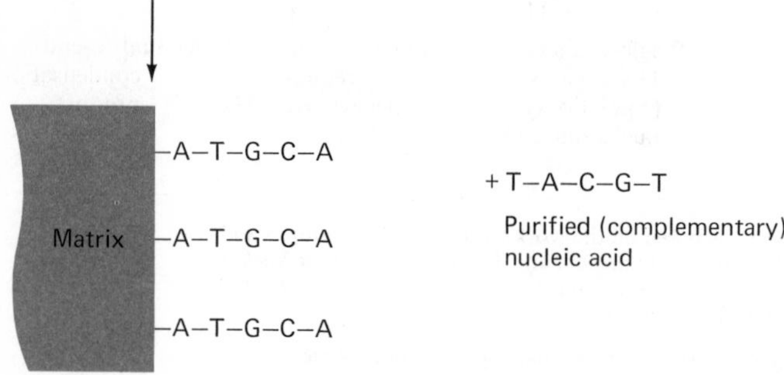

Figure 5.2 Determination of nucleic acid interactions on insoluble matrices. The basic principle involved is the fact that complementary sequences of nucleic acids interact and form strong complexes at low temperature, neutral pH, and low ionic strength (binding conditions). By increasing either temperature, the hydrogen ion concentration or the salt concentration (dissociating conditions), the weak interaction of the hydrogen bonds can be disrupted, and the partners in the complex will separate. By attaching one of the partners in a complexing system to an insoluble, and ideally, inert matrix, and making the other partner radioactive, complex formation between the two can be detected very sensitively in terms of the amount of radioactivity held by the nucleic acid-matrix under binding conditions and subsequently released under dissociating conditions. A typical example is shown above.

REFERENCES

General (multivolume) reference sources:

Chargaff, E. and J. N. Davidson (Eds.), *The Nucleic Acids*, Academic Press, New York, 1960.

Davidson, J. N. and W. E. Cohn (Eds.), *Progress in Nucleic Acid Research and Molecular Biology*, Academic Press, New York, 1963.

Other reference books:

Michelson, A. M., *The Chemistry of Nucleosides and Nucleotides*, Academic Press, New York, 1963.

Spirin, A. S., *Macromolecular Structure of Ribonucleic Acids*, Reinhold Publishing Co., New York, 1964.

Experimental techniques:

Grossman, L. and K. Moldave (Eds.), *Nucleic Acids* (Parts A and B), *Methods in Enzymology*, Vol. XII, Academic Press, New York, 1968.

Chemical synthesis of polynucleotides:

Khorana, H. G., *Some Recent Developments in the Chemistry of Phosphate Esters of Biological Interest*, John Wiley and Sons, Inc., New York, 1961.

Khorana, H. G., *et al.*, "Polynucleotide Synthesis and the Genetic Code," *Cold Spring Harbor Symp. Quant. Biol.*, **31**, 39 (1966).

SIX THE MODEL AND ITS EXPERIMENTAL FOUNDATION

As stated in Chapter 5, the following chapter presents the model for storage and transmission of genetic information. To make the model as inclusive as possible, the steps involved in the specialized cases of virus replication have also been included. We shall therefore consider first the storage and replication of the total genetic information (the genome) of eucaryotic cells (complex chromosomes), procaryotic cells (simple double-stranded DNA chromosomes), DNA viruses (double-stranded or single-stranded DNA chromosomes) or RNA viruses (single-stranded RNA chromosomes). Next we shall discuss the transcription of the DNA language into the RNA language and the translation of the RNA language into the amino acid language (sequence) of proteins. Finally we shall establish the chemical nature of the "dictionary" of what appears to be the only universal language on this planet. Throughout this chapter the focus will be on the reasoning and the experimental evidence that provide the basis for a qualitative or descriptive model. As with any good model, this one should make predictions that can be tested and from the tests it should be possible to refine the model; these refinements lead in turn to more detailed predictions that can be further subjected to experimental test. These more detailed mechanistic aspects of the model will then be examined in Chapter 7.

6.1 THE MODEL

A simple representation of the model is given below, illustrating the components and steps that occur inside all cells as well as the extracellular genomes of viruses, which are replicated, transcribed, and translated only after they have entered the host cell.

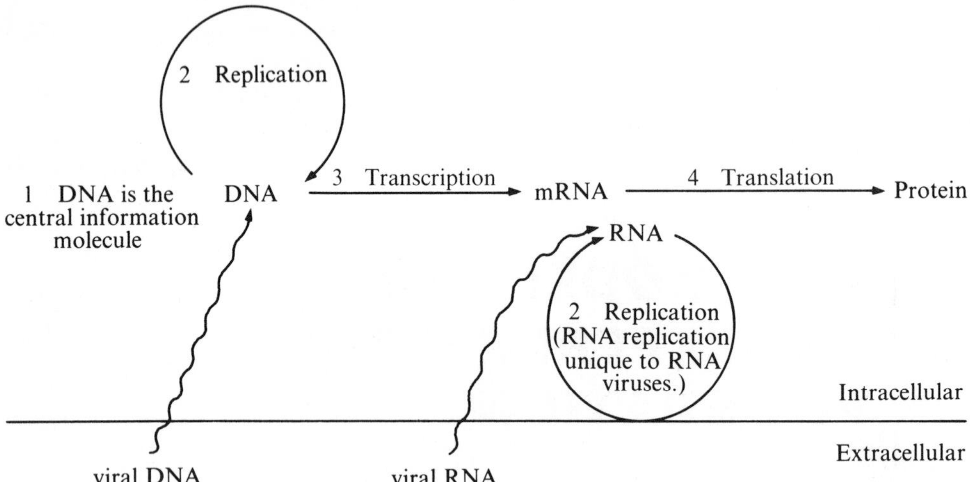

It is convenient to discuss this model according to the following outline, after which we will examine the evidence on which it is founded.

1. DNA is the central informational molecule, containing all the "blueprints" for the complete duplication of the cell (in a few viruses the genetic material is RNA).
2. The replication of the DNA (or RNA) genome—the process in which the original parent genome is duplicated to give an identical genome to each of the daughter cells—is the basis for the genetic constancy essential to the preservation of species characteristics from parents to progeny.
3. The language of DNA nucleotide sequence is transcribed into the language of RNA nucleotide sequence. The product of the transcription, the RNA, carries the information of the transcribed DNA segment to the synthetic apparatus.
4. The RNA nucleotide language is translated to the language of amino acid sequence by the synthetic apparatus, consisting of the ribosomes, tRNA, and enzymes and cofactors.
5. For all of these sequences of events to be correct, there must be a precise chemical equivalence between the three languages such that a given sequence of deoxyribonucleotides corresponds to a given sequence of ribonucleotides, which again corresponds to a given amino acid. In other words there must be a chemical dictionary for the nucleotide language and the amino acid language.

6. Viral infection should according to this model simply constitute the introduction of foreign information either as DNA, which is then replicated, transcribed, and translated into new DNA viruses, or as mRNA, which is then replicated and translated into new RNA viruses.

6.2 THE CENTRAL ROLE OF DEOXYRIBONUCLEIC ACID

On what basis can one consider DNA as the central information-storage molecule? The first and most direct part of the answer is based on the experimental confirmation of the postulate that introduction of a foreign DNA into a cell or chemical modification of a cell's own DNA should lead to some predictable change in that cell's character. Several such experiments have been carried out, and a few will be discussed as illustrations of the type of data that can be obtained. First we will consider different ways by which new DNA can be introduced into a cell.

In *phage infection*, when a bacterial virus infects the host cell, it can be demonstrated that the only thing that enters the host is the phage DNA. Less than 10 per cent (and in some instances, none) of the phage protein accompanies the DNA. Some 20 to 30 min after the entry of the DNA, the host cell breaks open and 100 to 200 new phage particles, identical (most of the time) to the original virus, are released. Without an analysis of the specific mechanism, it is clear that the DNA from the phage carries information needed for the synthesis of the new phage DNA and proteins. (The phage proteins are all unique to the phage and are not made by the normal uninfected host cells.)

The *transforming principle* is shown in the following experiment illustrating the phenomenon of bacterial transformation; several similar cases are also known. Two strains of *Pneumococcus* (*Diplococcus pneumoniae*), *A* and *B*, can grow on mannose as the sole source of carbon. Strain *A* can also grow on mannitol and the presence of the enzyme mannitol dehydrogenase can be demonstrated in extracts of this organism. Strain *B* cannot grow on mannitol and has no dehydrogenase. It is thus predicted, if our model is correct, that strain *B* lacks a functional gene for dehydrogenase, cannot synthesize the active enzyme, and therefore has lost the ability to oxidize mannitol. Table 6.1 summarizes an experiment in which cell-free extracts of strain *A* added to a culture of strain *B* can transform the *B* cells to cells capable of oxidizing mannitol. Treatment of the crude cell extracts with specific degradative enzymes gives strong evidence for DNA as the carrier of the dehydrogenase gene, and the final purification of the DNA from strain *A* conclusively shows that the *transforming principle*, the compound transforming the dehydrogenaseless strain *B* to a dehydrogenase-synthesizing organism, is DNA. It is also highly significant (although the mechanism is not apparent; see Section 7.2) that the ability to synthesize dehydrogenase is transferred from the transformed cells to their progeny. (An interesting sideline of this experiment is the fact that the high-molecular-weight DNA can enter through the cell membrane of the strain *B* cells with no apparent difficulty. Generally one thinks of transport across

TABLE 6.1 THE DEMONSTRATION OF BACTERIAL TRANSFORMATION AND OF DNA AS THE TRANSFORMING PRINCIPLE

CH_2OH
HO—C—H
H—C—OH
HC—OH
HC—OH
CH_2OH

NAD^+ → $NADH + H^+$

OH
HO
HO H OH
O
CH^3OH

→ Energy, metabolic intermediates

A alone can do this
"Transformed" *B* acquires the ability

Both *A* and *B* can do this

Cell	*Addition*	*Growth on Mannitol*	*Mannitol Dehydrogenase*
A[a]	None	+	+
B[a]	None	−	−
B	Cell-free extract of *A*	+	+
B	Protease treated cell-free extract of *A*	+	+
B	RNase-treated cell-free extract of *A*	+	+
B	DNase-treated cell-free extract of *A*	−	−
B	+ Pure DNA from *A*	+	+
B	+ Extract of *A*, allow division, + wash, test progeny	+	+

[a] Strains *A* and *B* are two mutants of *Diplococcus pneumoniae*. Strain *A* is capable of growing on mannitol (and containing dehydrogenase), while strain *B* is incapable of growing on mannitol, but grows on mannose.

membranes as being relatively restrictive, limited to small molecules and ions. However, the nucleic acid molecules apparently have no difficulty in entering the cell at all.)

Transduction is a special form of transformation that is mediated by bacteriophages. Thus if an organism lacking a specific gene is infected by phage that originally multiplied in a host possessing this gene, the genetically deficient strain may be transduced to a competent one by the phage infection (the missing enzyme appears after infection). The explanation must be that the phage in some way carried the intact gene from the original competent host to the new deficient host. Again it can be demonstrated that purified transducing-phage DNA alone will transform the deficient strain, and that even fragments of the phage DNA are capable of transferring the missing genetic information. It should be emphasized here that this experiment does not work with phage that multiplies immediately (*lytic* virus) and causes lysis of the host. Only *lysogenic* (or *temperate*) viruses, which do not multiply immediately, can exist long enough in the host cell (as a *prophage*) to be incorporated into the bacterial chromosome. Once incorporated the viral chromosome is rarely released to yield a lytic cycle; it can be induced in a number of different ways, however.

Bacterial conjugation is another type of DNA transfer in unicellular organisms that is analogous to sexual mating in higher animals. In this case requiring direct contact (conjugation) between cells, it appears that most or even all the genetic information from the donor (male) can be transferred to the recipient (female). If the conjugation and thus the transfer is disrupted (usually mechanically, in a blender), it is possible to achieve a different extent of DNA transfer. Using a radioactive label in the DNA of the donor, one can then directly correlate amount of information transferred (number of deficient genes rendered competent) to the amount of DNA (amount of radioactivity incorporated) transferred; in all cases studied, there is a good linear correlation. Figure 6.1 illustrates the experimental procedure involved.

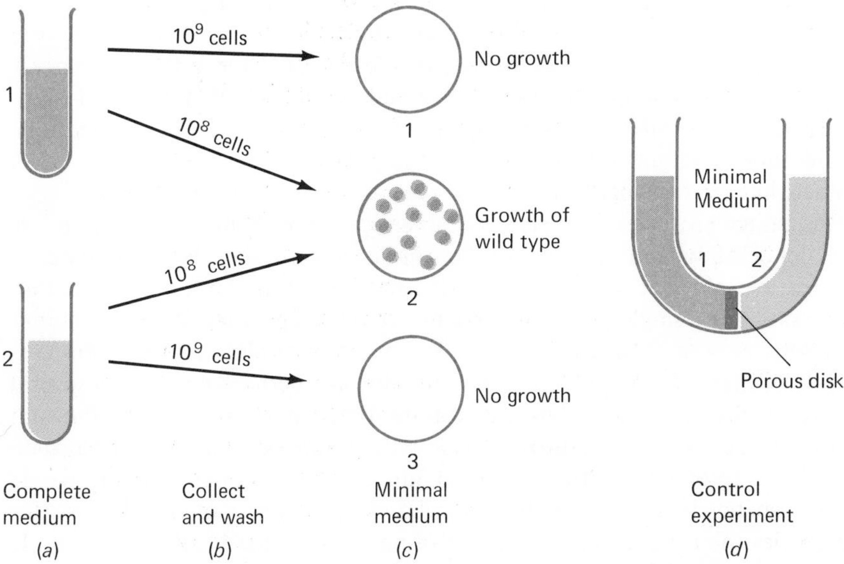

Figure 6.1 The demonstration of bacterial conjugation according to the initial experiment by J. Lederberg and E. L. Tatum. The success of this experiment was based on the use of double mutants (A^-B^- and C^-D^-) obtained by ultraviolet irradiation (see Chapter 7). Single mutants *revert* or back-mutate ($A^- \longrightarrow A^+$, $B^- \longrightarrow B^+$, etc.) at almost as high a frequency as that observed in the conjugation experiment. The probability of back-mutation of a double mutant ($A^-B^- \longrightarrow A^+B^+$), however, is very low, as observed by the absence of growth in plates (*a*) and (*c*). Cross feeding of the two strains (1 synthesizes B and C which become available to 2 through the medium) and transformation by a soluble extracellular component were eliminated by the control experiment (*d*).

Growth medium A: Complete medium, containing growth factors A, B, C, D.
Growth medium B: Minimal medium; A, B, C, D omitted.
Mutant 1 (strain A^-, B^-, C^+, D^+) requires A and B for growth.
Mutant 2 (strain A^+, B^+, C^-, D^-) requires C and D for growth.
(Wild type (A^+, B^+, C^+, D^+) grows on medium B.)

Mutation, another argument for DNA as the central information molecule, is based on chemical modification or alteration of DNA with an evaluation of the resulting effects on the proteins produced after the alteration. As this topic can be discussed more meaningfully in the next chapter, it will only be stated here that a number of different treatments (chemical reagents such as nitrous acid, hydroxylamine, or acridine dyes; high-energy irradiation such as UV light or γ irradiation; replacement of the normal DNA bases with analogues such as 5-bromouracil or 2-aminopurine in the biosynthesis of DNA) that are, in some instances at least, very specific for DNA cause faulty, modified DNA. Along with this chemical change or genotypic alteration* in the chromosome, one can observe characteristic phenotypic changes (as predicted if the genetic material is DNA). Moreover, the new phenotypic character is transferred to subsequent generations, again according to predictions.

The second part of the argument for DNA as the central genetic material is indirect. If such an information-storage molecule exists, it should satisfy certain requirements discussed below; the DNA molecule satisfies them all.

The total information stored in any one cell should be qualitatively constant during the life of the cell, independent of its age, nutrition, and metabolic activity, and it should be duplicated in the process of cell division in such a manner that all the progeny receive an exact copy of the parental DNA.

This statement requires some amplification. The term "information" is used deliberately and should not be confused with the chemical material carrying the information. It is now well established that the total information can be stored in a single DNA molecule in viruses and procaryotic cells, whereas eucaryotic cells may have a large number of different chromosomes (humans have 23 different chromosomes), each chromosome perhaps containing several hundred DNA molecules. Thus the total information as expressed by the sum of all unique DNA structures differs widely among species. Within each species the total information must remain constant, however, if the whole idea of genetics is correct. On the other hand the quantity of chemical matter that carries the total information must of necessity undergo changes if simple cell division or cell fusion (in sexual reproduction) is to lead to daughter cells with the same information content. This change in the quantity of chromosomal material is referred to as *ploidity*. At the vegetative (resting) state most of the procaryotic cells are haploid (containing a single copy of their information), whereas most of the eucaryotic cells are diploid (containing two copies of their information; a human somatic cell contains 23 pairs or 46 chromosomes.) Regardless of the ploidity of the cell, all the genetic material is doubled immediately prior to division so that, upon division, each daughter cell receives one copy of the original information. A haploid cell becomes diploid and then divides into identical haploid cells, whereas a diploid cell goes through a tetraploid state prior to division into two copies of the original diploid parent. This simple genetic doubling followed by division is a process common to all true cells and is known as *mitosis*.

In sexual reproduction in which a paternal and a maternal cell fuse to give the original daughter cell, it is necessary to halve the quantity of genetic material prior to sexual fusion, or the ploidity of each successive daughter would double.

* *Genotype* refers to the genetic makeup of the cell, whereas *phenotype* refers to the observed expression of this genetic makeup.

Such a halving process, known as *meiosis*, is also well established. The normal diploid cell, containing one set of maternal and one set of paternal chromosomes, duplicates its genetic material in a process very similar to mitosis and divides into two cells that each contain, like the parent cell, one paternal and one maternal chromosome. The next step in meiosis differs in detail in different organisms, but the key is a second division without prior duplication of the genetic material, after which each of the resulting four daughter cells has only one set of chromosomes (either the paternal or the maternal) and thus is haploid. These daughter cells are the germ cells or *gametes* (the ovum in the female and the sperm cell in the male) and the subsequent fusion of two of these haploid gametes yields a diploid *zygote* with the normal double set of chromosomes.

TABLE 6.2 A SURVEY OF DNA CONTENT AND COMPOSITION IN DIFFERENT SPECIES[a]

Organism	*DNA Content in pg Per Cell*[b]	*DNA Composition, Mole Percent G + C*[c]
Mammals	2.5–3.5[d]	39–42
Birds	0.7–1.5[d]	42–45
Fish	1 –2.5[d]	41–44
Amphibians, reptiles	3.3–7[d]	40–47
Arthropods	1.5–6[d]	35–45
Protozoa	0.3–15	27–57
Higher plants	5 –15 (root tip)	35–46
Bacteria and fungi	.002–0.1	25–75
Spirillum		26–30
Chlostridium		30–35
Bacillus		34–44
Lactobacillus		36–40
Neisseria		48–50
Corynebacteria		46–54
Escherichia		50–52
Aerobacter		52–56
Pseudomonads		55–70
Actinomyces		70–74
Viruses	.00001–.0005	28–73

[a] Taken in part from the extensive list in *Handbook of Biochemistry*, H. A. Sober (Ed.), The Chemical Rubber Co., Cleveland, 1968. For a more complete picture this reference should be consulted. The values given in this table have been selected to illustrate (1) the variation of DNA content with the complexity of the cells (or viruses), and (2) the relatively constant base ratio in closely related species, and at the same time the wide variety of base ratios found when all types of cells and viruses are considered (compare individual bacterial genera to the total range for all bacteria and fungi). The ranges given are only approximate. They cover the majority of species within each group, but omit several "exceptional" cases which do not "fit" the general model at our current level of knowledge.

[b] For higher organisms the DNA content of different cells and tissues varies considerably, and again the ranges given are only approximate.

[c] Given as G + C/total bases in mole percent.

[d] Ranges for sperm cells selected in an attempt to make the comparison as meaningful as possible.

The point of this grossly oversimplified picture of well-studied, well-understood, and complex cell processes is simply to emphasize the direct quantitative relationship between DNA and "genetic material" during the entire life cycle of different cells. Thus the amount of DNA in a haploid gamete is half that found in the somatic cell of the same individual, and the fertilization process leads to restoration of the normal quantity of DNA in the zygote. Polyploid cells also show the predicted proportional increase in DNA content.

The chemical composition of the information molecule of any one cell or organism should be characteristic of that cell or organism only. Another cell closely related to the first cell in its characteristics should have information molecules very similar to the first, while a cell very different in its properties should have information molecules different from the first (Table 6.2 and Figure 6.2). This argument depends in the strictest sense on knowledge of base sequence as well as composition, since two DNA's with completely different information content (nucleotide sequence) could fortuitously have identical base compositions.

The total quantity of information in an organism should bear a direct relationship to the complexity of the protein makeup of that organism; it must be large enough to code for all the proteins of the organism. This prediction is best tested by comparing both cells and viruses to allow a wide spectrum of complexity from the simplest single-protein nucleocapsid to the most complex cells (see Table 6.2).

It is now clear that the only molecule in living systems that universally satisfies all of these requirements is DNA.

6.3 REPLICATION OF DEOXYRIBONUCLEIC ACID

The second feature of the model that we stress is a very essential requirement for the molecules carrying the genetic information—*they must be capable of reproducing themselves* very faithfully. This process, replication, was first documented very elegantly by the now classical experiment (the Meselson-Stahl experiment) illustrated in Figure 6.3. The DNA of parent cells were labeled with heavy ^{15}N by growing the cells on ^{15}N-ammonium chloride. When these cells containing heavy (^{15}N–^{15}N) DNA were transferred to a normal ^{14}N-ammonium chloride medium and allowed to divide, the newly synthesized DNA should be made up of the "light" ^{14}N; because of the difference in density between ^{14}N–^{14}N DNA and ^{15}N–^{15}N DNA, it is possible to distinguish between parent and progeny DNA by density-gradient centrifugation. All the DNA molecules produced in each division are identical in all respects except for the ^{14}N and ^{15}N isotopes; they have the same base composition and apparently must contain the same information, as the cells continue to grow and divide as typical *Escherichia coli* cells. The experiment in Figure 6.3 not only establishes that a replicative process for DNA exists, but also demonstrates that the process takes place in a way that takes full advantage of the

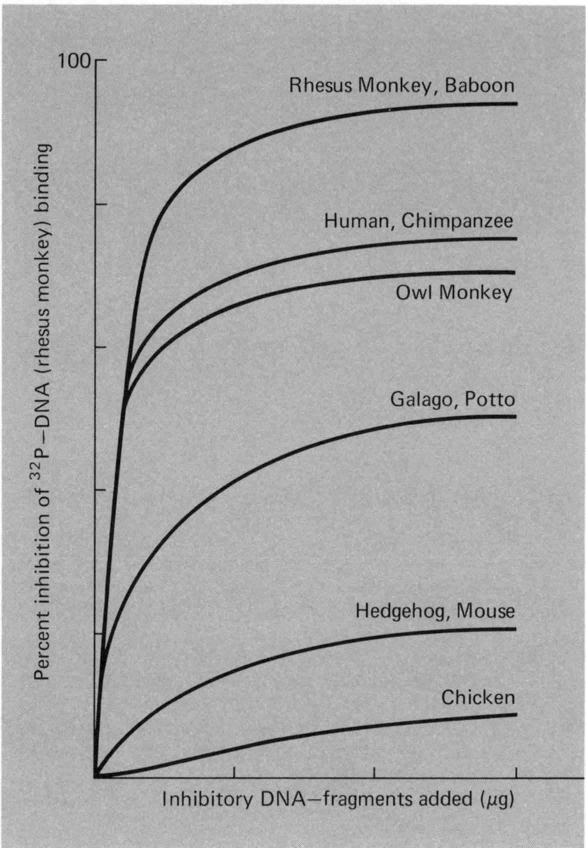

Figure 6.2 The study of species relatedness by direct comparison of their DNA. The method is based on interaction of complementary DNA sequences. Single-stranded DNA (obtained by heating) of species *a* is imbedded in agar. DNA of species *a* is also prepared with a ^{32}P-label. DNA from several species to be studied is prepared, and all these DNA's and the ^{32}P-labeled DNA_a are fragmented mechanically and separated into single strands by heating. In the control experiment the binding of ^{32}P-labeled DNA_a fragments to the homologous DNA_a agar system is determined and used as reference. The effect of the other DNA fragments from each of the other species is now compared as inhibitors of the homologous DNA binding. The more efficient a given mixture of DNA fragments is as inhibitor, the more similar it must be to the DNA of species *a*. In the figure, rhesus monkey DNA was imbedded in agar, and the binding of ^{32}P-labeled rhesus monkey DNA fragments was determined. Unlabeled rhesus monkey and baboon DNA were equally good as inhibitors; next came human and chimpanzee DNA, followed by owl monkey, galago and potto, hedgehog and mouse, and finally chicken DNA, which shows very little affinity for the imbedded DNA. The data are taken from Hoyer, B. H., *et al.*, *Evolving Genes and Proteins* (V. Bryson and H. J. Vogel, eds.), Academic Press, New York and London, 1965, p. 581. Clearly the information obtained in this type of study is more meaningful than the simple comparison base composition given in Table 6.2, and more in line with the direct comparison of amino acid sequences discussed in Chapter 3 (Figures 3.18, 3.19).

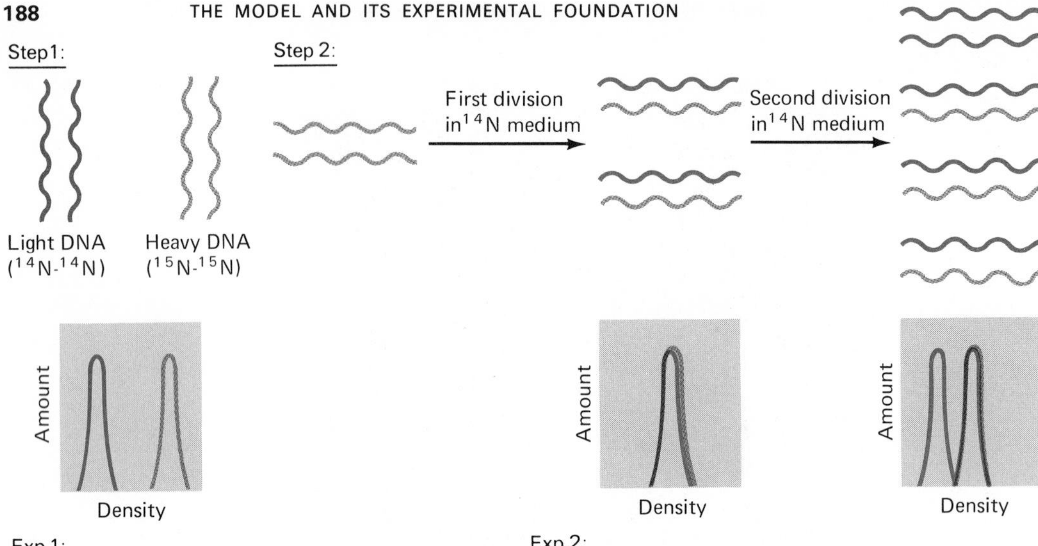

Figure 6.3 Demonstration of the replication of DNA according to the original experiment of Meselson and Stahl.

Step 1: *Calibration*: (*a*) Grow *E. coli* on $^{15}N–NH_3$, harvest DNA ("heavy DNA").
(*b*) Grow (in parallel) *E. coli* on $^{14}N–NH_3$, harvest DNA ("light DNA").

Step 2: *Replication*: Grow cells on $^{15}N–NH_3$ (as above) in a synchronous (all the cells divide at approximately the same time) culture, harvest the freshly divided cells, wash and
(*a*) transfer cells to a medium containing $^{14}N–NH_3$ and allow growth and 1 cell division: harvest DNA from part of the cells and
(*b*) allow remainder of cells to undergo a second division, and collect DNA.

Experiment 1: Set up density gradient centrifugation, use 1(*a*) and 1(*b*) for calibration and locate the equilibrium position of "heavy" (^{15}N) and "light" (^{14}N) DNA.

Experiment 2: Next run DNA from 2(*a*), 2(*b*), etc., and find their density distribution.

The experiments show that after one division, there is essentially none of the original double-stranded $^{15}N–^{15}N$ DNA left. Instead a new species of density intermediate between that of light and heavy DNA appears which must have the composition $^{14}N–^{15}N$. After a second division, about equal amounts of $^{14}N–^{14}N$ (light) DNA and $^{14}N–^{15}N$ DNA were found, showing that the original chains are still intact. This experiment clearly demonstrated the semiconservative mechanism of *in vivo* replication, illustrated in the figure. (A conservative mechanism would have been one in which the original double-stranded $^{15}N–^{15}N$ DNA was retained intact, and a non-conservative mechanism, one in which each strand was fragmented during replication.)

complementary double-stranded structure of DNA. The process is referred to as a *semiconservative* one since both of the original parent strands are kept intact through generation after generation. Even after a large number of divisions in the above experiment the two hybrid molecules ($^{14}N–^{15}N$ DNA) should be detectable if detection methods of sufficient sensitivity were available.

The experiment in Figure 6.3 predicts that there is an enzyme capable of assembling nucleotides into DNA molecules that are exact copies of the parent molecule by a mechanism involving the synthesis of two new polymer chains, each complementary to one of the strands of the parent molecule. Such an enzyme was isolated from *E. coli* some 15 years ago* and, more recently, similar enzymes have also been obtained from other sources (see Table 5.4). Most of the work to date has been done on the *E. coli* enzyme, which will be considered here. This enzyme, a DNA-dependent DNA polymerase, or replicase, catalyzes the following reaction, shows an absolute requirement for a DNA template,† and needs all four deoxynucleoside triphosphates for the reaction to go at a significant rate:

$$\begin{array}{l} n(\text{dATP}) \\ + n(\text{dGTP}) \\ + n(\text{dCTP}) \\ + n(\text{dTTP}) \end{array} \xrightarrow[\text{Replicase}]{\text{DNA, Mg}^{2+}} \text{DNA (product)} + 4n(\text{P—P}_i)$$

In vitro studies with this enzyme have established one very significant feature of the reaction, namely that the nature of the DNA product is determined by the nature of the DNA template and is not affected by altering the ratio of the deoxynucleoside triphosphate reactants (Table 6.3). Before concluding from this observation that this is indeed *the* replicase of *E. coli*, however, one must also take notice of several puzzling features of the in vitro behavior of the enzyme. The rate of DNA synthesis in vivo is far greater than the rates observed in vitro, and the polymerase appears to replicate denatured (single-stranded or fragmented) DNA faster than native (double-stranded) DNA. This problem will be considered further in Chapter 7. The product of the in vitro synthesis shows extensive branching (in the electron microscope) whereas the in vivo DNA is linear, and there is generally no correlation between the length of the template and the length of the product DNA chain. In fact, as shall be seen below this enzyme has been used to synthesize high-molecular-weight polymers using short oligonucleotides as templates. It is also interesting to note that the enzyme is very sensitive to metal ions. In the reaction indicated above Mg^{2+} is required, but if Mn^{2+} is substituted for Mg^{2+} the enzyme also catalyzes the polymerization of ribonucleoside triphosphates, thus losing its specificity for the deoxy series.

One of the most serious problems encountered in research on *E. coli* polymerase was the lack of success in attempts to synthesize biologically active DNA products in vitro. This objection has now been overcome in the case of the recent in vitro synthesis of an infectious phage DNA, the success of which was

* I. R. Lehman, M. J. Bessman, E. S. Simms, and A. Kornberg, *J. Biol. Chem.*, **233**, 163 (1958).

† In many biochemical reactions involving biosynthesis of polymers the reaction will only proceed after addition of preformed polymer. This starter molecule can be considered to have a role either as a *primer* or as a *template*, and there is a good mechanistic distinction between these two roles. A primer acts as the initiator for polymerization and becomes a covalent part of the product, whereas a template acts only as a specific surface upon which the assembly of building blocks can proceed (the template itself does not become involved in covalent bonding).

TABLE 6.3 TEMPLATE–PRODUCT RELATIONSHIP IN THE ENZYMATIC SYNTHESIS OF DNA (REPLICATION)

DNA Source	Base Composition				
	A	T	G	C	$\frac{A+T}{G+C}$
Micrococcus lysodeikticus					
Template	—	—	—	—	0.39
Product	0.15	0.15	0.35	0.35	0.43
Mycobacterium phlei					
Template	0.162	0.165	0.338	0.335	0.49
Product	0.165	0.162	0.335	0.342	0.48
Aerobacter aerogenes					
Template	—	—	—	—	0.82
Product	0.22	0.22	0.28	0.28	0.79
Escherichia coli					
Template	0.25	0.243	0.245	0.262	0.98
Product	0.26	0.25	0.243	0.245	1.03
Calf thymus					
Template	0.29	0.267	0.228	0.215	1.27
Product	0.288	0.277	0.218	0.218	1.29
Bacteriophage T2					
Template	0.327	0.33	0.167	0.175[b]	1.92
Product	0.332	0.322	0.172	0.175	1.89
AT copolymer					
Template	0.50	0.50	—	—	—
Product	0.498	0.483	0.01	0.01	49

[a] The (A + T)/(G + C) ratio is a convenient method for assessing the relatedness of nucleic acids. Since A and T and G and C are always paired in complementary sequences, the ratio must be the same for any comparison of related strands (single-stranded product–double-stranded template, double-stranded product–double-stranded template, etc.).

[b] 5-hydroxymethylcytosine.

in part due to the isolation of a second *E. coli* enzyme that catalyzes the formation of a diester linkage between the two ends of a DNA strand. This enzyme—ligase, or joining enzyme (see Table 5.4)—is responsible for the formation of circular DNA molecules. It should be emphasized that this enzyme does not simply join the ends of any linear DNA strand. There appears to be excellent evidence that the joining of two DNA ends only takes place when they are brought together in the right juxtaposition on a template (complementary DNA) that is continuous in the region where the new link is formed.

Bacteriophage ϕX174 is a nucleoprotein particle containing DNA having a molecular weight of 2×10^6 as well as five different proteins. The DNA is circular and its base ratio—$A = 1.00$, $G = 0.98$, $C = 0.75$, $T = 1.33$ ($A \neq T$, $G \neq C$)—shows that it is also single stranded. It was used as template for DNA synthesis with *E. coli* polymerase in an experiment where 5-bromodeoxyuridine triphosphate was substituted for TTP to obtain a product containing the heavier bromouracil in place of T and thus allow separation of the template and product on the basis of density after nuclease treatment and denaturation. The experiment is illustrated in Figure 6.4. The bases of the newly synthesized

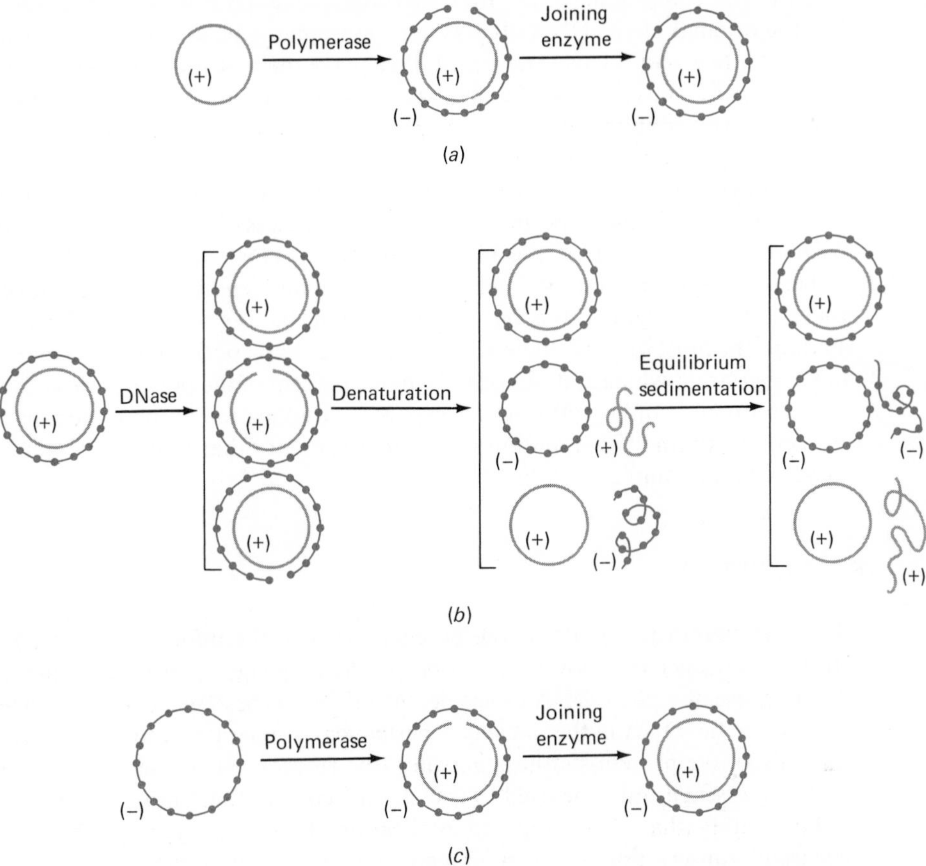

Figure 6.4 The *in vitro* synthesis of biologically active (infective) virus DNA. The single-stranded DNA from coli phage ϕX174 (+ ring) was used for template in the step (*a*), and the complementary − ring was synthesized with deoxy 5-bromo-uridine triphosphate (BU) substituted for deoxythymidine triphosphate to facilitate subsequent separation. The ends of the linear polymer produced were joined to form closed − rings with the "joining enzyme," ligase. In step (*b*) limited nuclease treatment and heat denaturation allowed separation of double-stranded rings, single-stranded rings, and denatured linear DNA. The BU-containing rings (− rings) could be separated from the original + rings because of the former's higher density, and in step (*c*), the synthetic process in (*a*) was repeated using the newly synthesized − rings as template for the polymerase. In this way a completely synthetic DNA, identical to original virus DNA, was produced in the test tube for the first time. [Reproduced from Goulian, M., A. Kornberg, and R. L. Sinsheimer, *Proc. Nat. Acad. Sci.*, **58**, 2321 (1967), with permission.]

and isolated minus-rings were found in the proportions of A = 1.33, G = 0.75, C = 0.98, and BrU = 1.00, which again shows single-stranded DNA, and also precise complementarity to the original "plus-rings." When the new minus-rings were used as templates in a second synthetic cycle using ATP, GTP,

CTP, and TTP, a product (plus-rings) was obtained that was in every way exactly like the original phage DNA. It was infectious, giving phage production when injected into *E. coli.* This experiment, then, gave the first test-tube synthesis of a biologically active DNA. At the same time it confirmed the very accurate copying involved in the replicative process catalyzed by the *E. coli* polymerase, which had previously been demonstrated through nearest-neighbor analyses of many other template-product pairs. It is true that only single-stranded templates were used in this experiment and that the success of the experiment could be based on the fact that the biologically active DNA was single stranded.

There are thus many experimental facts that would challenge the conclusion that the DNA polymerase is *the* enzyme responsible for replication in *E. coli.* Perhaps the most serious challenge is the finding that a polymerase-less mutant (an *E. coli* strain which contains less than one percent of the polymerase activity of the normal strain) is viable and is capable of duplication at the same rates as the normal strain.* This finding alone suggests that other replicating systems must be functioning in *E. coli.*

6.4 TRANSCRIPTION

The next step in the model is the process in which the information stored in the base sequence of DNA is *transcribed* to the language of an RNA sequence. The purpose of such an RNA molecule, the mRNA, is best seen in the nucleated cells where the DNA in the nucleus is permanently separated and distinct from the bulk of the protein-synthesizing sites (the ribosomes). Under these conditions a messenger must be used to transfer the code to the synthetic site.

Postulating that in analogy to replication, base pairing is the basis for accurate transcription, we can immediately predict that there must be an enzyme that assembles polyribonucleotides on a DNA template. Since the RNA product should be a complement of one of the DNA chains, we also can predict that it should be possible to demonstrate formation of double-stranded complexes (hybrids) from the DNA template and the RNA product. Both of these predictions have been experimentally confirmed and provide the main support for the occurrence of this step in the model.

The enzyme responsible for the process of transcription, DNA-dependent RNA polymerase (sometimes referred to as transcriptase) has been isolated from several different sources (originally from liver) and catalyzes the following general reaction:

$$\begin{array}{l} \text{n(ATP)} \\ +\ \text{n(GTP)} \\ +\ \text{n(CTP)} \\ +\ \text{n(UTP)} \end{array} \xrightarrow[\text{Enzyme}]{\text{DNA, } Mg^{2+}} \text{RNA} + 4\text{n}(\text{P—P}_i)$$

* R. DeLucia and J. Cairns, *Nature*, **224**, 1164 (1969).

As in the case involving the DNA polymerase, the composition of the product is independent of the proportions of nucleoside triphosphates and dependent only on the DNA provided as template (Table 6.4). Different enzymes show differences in their template requirement, however. In vitro, the enzyme will use either single-stranded or double-stranded DNA template, and in the latter case both chains are transcribed; in vivo, it appears that only double-stranded DNA is used and that only one of the chains is copied. This statement is again based on observations with bacteriophage ϕX174 infection of *E. coli.* When the single-stranded bacteriophage ϕX174–DNA plus circle is injected into the host cell, the first event that can be observed is the synthesis of the minus circle to form the DNA duplex (double-stranded DNA). When the duplex is completed, and not before, RNA synthesis starts, and the finished mRNA product is a complement of the newly synthesized minus DNA and a copy of the original plus DNA of the virus.

The complementarity of the RNA and DNA template has been demonstrated in a number of ways, the classical demonstrations being illustrated in Figure 6.5. The basis for all these experiments demonstrating formation of the DNA-RNA hybrid is the reversible "melting" of the hydrogen-bonded structure at elevated temperature.

Specifically this experiment deals with the question of whether it is possible to demonstrate that infection of *E. coli* with DNA from bacteriophage T_2 leads to the synthesis of a new RNA molecule not related to normal *E. coli* RNA but complementary to phage DNA. The RNA synthesized after infection is labeled with uridylic acid having a high specific radioactivity, and the new RNA is then investigated by two separate techniques [Figure 6.5(*a*) and (*b*)].

It is interesting to note that the technique illustrated in Figure 6.5(*b*) could

TABLE 6.4 TEMPLATE–PRODUCT RELATIONSHIP IN THE ENZYMATIC SYNTHESIS OF RNA (TRANSCRIPTION)[a]

Template DNA	*Template* $\frac{A+T}{G+C}$	*Product* $\frac{A+U}{G+C}$
T2 phage	1.86[b]	1.85
ϕX-174 phage (single-stranded)	1.38	1.33[c]
Calf thymus	1.35	1.52
Escherichia coli	0.98	0.93
Micrococcus lysodeikticus	0.40	0.48

[a] The data, obtained with *E. coli* RNA polymerase, were taken from J. Hurwitz, *et al., Cold Spring Harbor Symp. Quant. Biol.*, **26**, 91 (1961), and S. Spiegelman and M. Hayashi, *Cold Spring Harbor Symp. Quant. Biol.*, **28**, 161 (1963).

[b] Contains 5-hydroxymethylcytosine instead of cytosine.

[c] In vitro, using the single-stranded phage DNA as template, the RNA produced is complementary to the DNA. In vivo, however, the mRNA produced after infection with phage is a copy of the original phage DNA (a complement of the complement), showing that a double-stranded DNA is synthesized prior to the transcription process. (In either case the same (A + T)/(G + C) ratio will be obtained; see Table 6.3.)

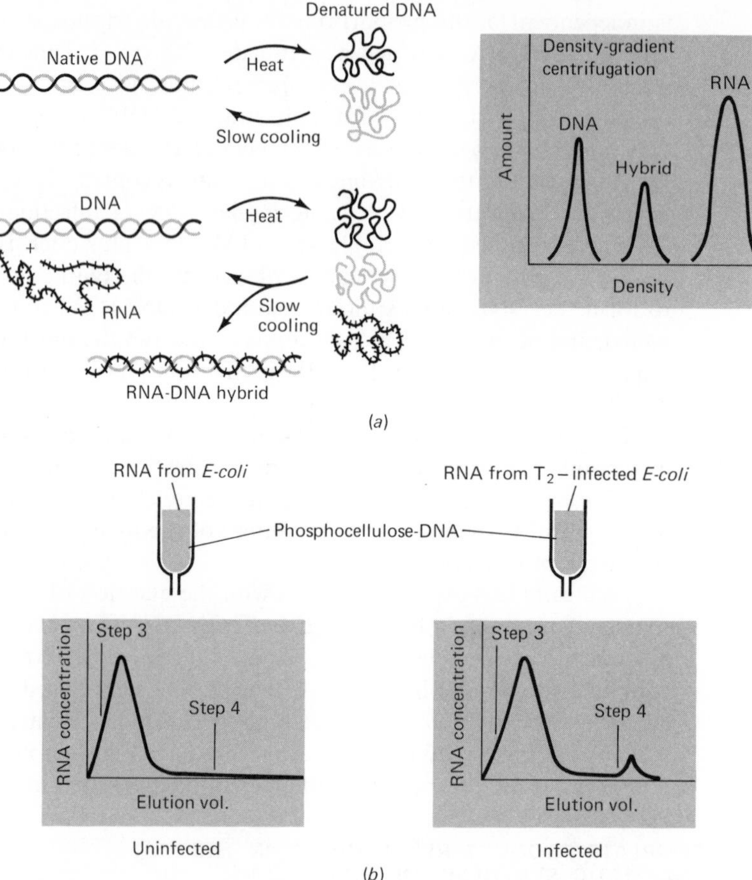

Figure 6.5 (*a*) Demonstration of complementary DNA–RNA hybrids by direct annealing. The mixture of phage DNA and RNA is heated to denature the DNA and then is allowed to cool slowly to favor reformation of double-stranded complementary nucleic acids. When the resulting reaction mixture is subjected to density-gradient centrifugation, a very small radioactive peak is formed with a density intermediate between the larger DNA and RNA peaks. (RNA is synthesized from ^{14}C precursors.) This peak is not apparent in uninfected cells. When tritium-labeled phage DNA is used the small intermediate peak is found to contain both tritium and ^{14}C; this shows that it is a hybrid containing both RNA and DNA.

(*b*) Demonstration of complementary DNA–RNA hybrids by binding on a bacteriophage T2-DNA column: phosphocellulose-DNA (see Figure 5.2). Bacteriophage T2 contains 5-hydroxymethylcytosine bases and several of the hydroxymethyl groups are glucosylated. Treating a mixture of phosphocellulose and DNA-(glucose)$_n$ with a condensing reagent, dicyclohexylcarbodiimide (DCC), the following reaction should take place:

$$\text{Cellulose–O–P(=O)(O^-)_2} + \text{HO–CH}_2\text{–(glucose ring: OH, HO, OH)–O–DNA} \xrightarrow{\text{DCC}} \text{Cellulose–O–P(=O)(O^-)–O–CH}_2\text{–(glucose ring: OH, HO, OH)–O–DNA}$$

[As it turns out (see below), it is not necessary to form the covalent bond, as DNA (and other high-molecular-weight nucleic acid complexes) form very strong noncovalent "bonds"

be used to consider another important question. If the total *E. coli* DNA were immobilized and hybridization with the total *E. coli* RNA (rRNA, tRNA, and mRNA) were allowed to take place, one should be able to find how much of the total RNA is complementary to the cell's own DNA. The very important answer is that *all* the RNA of the cell has complementary DNA sequences and binds to the DNA matrix. This must mean that all the RNA—rRNA, tRNA, and mRNA—is synthesized on a DNA template. It may also be noteworthy that the part of the DNA genome that is complementary to the rRNA and tRNA (98 percent of the total RNA) constitutes less than 1 percent of the total DNA sequence.

6.5 TRANSLATION

In the fourth process in the model, *translation*, the RNA transcript (mRNA) of the genetic code (DNA) is translated into the language of amino acid sequence (protein) in the assembly of the polypeptide-chain product.

It is now well established that the protein synthesis apparatus consists of the following components: mRNA and ribosomes (polysome), tRNA, amino acids, and a number of enzymes and cofactors. The fundamental problem in reaching an understanding of the translation process was the difficulty in applying the simple template hypothesis, which worked so well in explaining replication and transcription, to the assembly of amino acids on a polynucleotide template. Amino acids simply do not have any special affinity for RNA. The postulation of an adaptor (the adaptor hypothesis) presented a very attractive mechanism for the solution of this dilemma (see Figure 6.6), and it rapidly became apparent that tRNA, together with the proper specific enzymes, would fulfill the role of adaptor very adequately.

As illustrated in Figure 6.6, the translation process should have two distinct chemical steps. The first is the specific matching of an amino acid and a tRNA molecule to form the amino acyl–tRNA complex (charged tRNA or adaptor), and the second is the specific matching of the charged tRNA (containing an anticodon triplet) with its "site" (the codon) on the mRNA template. The third step leading to protein synthesis, then, is the catalysis of peptide-bond formation between the amino acyl groups as they are brought onto the template in the proper order according to the genetic code transcribed on the mRNA molecule. Many of the arguments for this mechanism have direct bearing on the elucidation of the genetic code, and will be discussed below. At this point

---

with cellulose and cellulose derivatives.] (1) Two small identical columns are now prepared with this cellulose-DNA complex, and radioactive RNA from uninfected and from T2-infected cells is allowed to saturate the columns. (2) Again, annealing (heating and slow cooling) is performed and (3) the two columns are eluted exhaustively with dilute buffer. In both cases RNA is eluted, and this RNA is identical to *E. coli* RNA. (4) Elution is next continued with high-ionic-strength buffer, strong enough to disrupt hydrogen-bonded double-stranded nucleic acids. This second buffer gives no more RNA from the column treated with uninfected *E. coli* RNA, but a small but significant amount of RNA is obtained from the column treated with RNA from infected cells. The base composition of this RNA is quite different from *E. coli* RNA, and it can be concluded that since it is bound to phage DNA, it must be mRNA synthesized in response to phage infection.

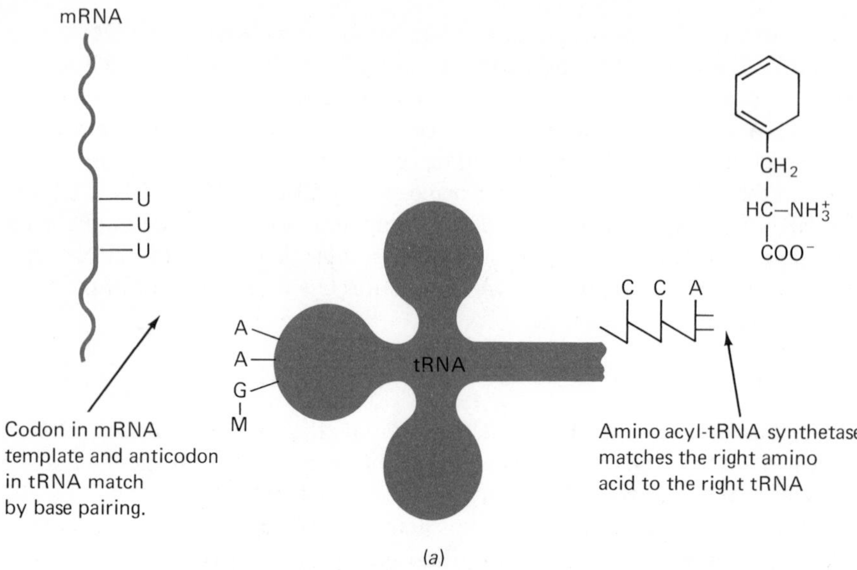

Figure 6.6(*a*) An illustration of tRNA as the "adaptor" in matching the nucleotide language code to the specific amino acid "translation." In the first step, the specific amino acyl-tRNA synthetase matches the proper amino acid to the proper tRNA, and in the next step the anticodon of the tRNA matches the codon of the mRNA to put the amino acid in the proper position in the polypeptide. In the example illustrated the codon U—U—U matches the amino acid phenylalanine through the mediation of the $tRNA_{phe}$ adaptor.

```
A–G–U
     C–C–U–G–C–A–G–U–U–G–U–C–G–C–C–A_OH
G    :  :  :  :  :  :  :  :  :  :
     G–G–A–C–G  U–C–A–A–C–G–G_P
C–Ψ–T
              C–M   U                  H2
U          G           G         G–A–G–U
   G–C–C–C           G–C–C–Ac
U  :  :  :  :        :  :  :              G–M
   C–G–G–G           C–G–G       H2 H2
U          U–M    G–M2         A–A–U–U–G
              U··A
              U··A
              U··A
              C··G
              Ψ··A
           A·········Ψ
           A-isop    U
           A         I
               G
              (b)
```

tRNA$_{ser\ 2}$

Figure 6.6(*b*)
The "cloverleaf" representation of the primary nucleotide sequence in $tRNA_{ser\ 2}$.

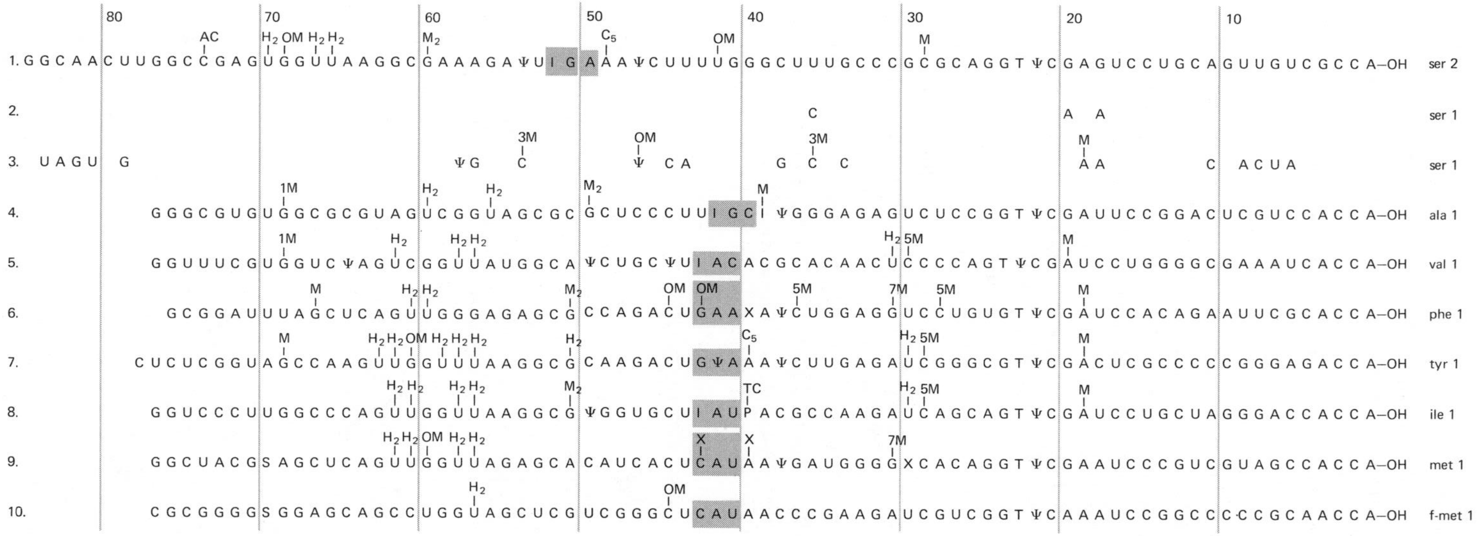

```
      80        70           60          50          40          30          20          10
            AC    H2 OM H2 H2  M2          C5          OM          M
1. GGCAA CUUGGCCGAG UGGUUAAGGC GAAAGAΨUIG AAΨCUUUUG GGCUUUGCCC GCGCAGGTΨC GAGUCCUGCA GUUGUCGCCA-OH  ser 2

2.                                                     C                      A  A                       ser 1
                                  3M        OM         3M                     M
3. UAGU G                      ΨG  C        Ψ  CA   G  C  C                  AA        C  ACUA          ser 1
              1M        H2    H2         M2       M
4.     GGGCGUG UGGCGCGUAG UCGGUAGCGC GCUCCCUUIGCI ΨGGAGAG UCUCCGGTΨC GAUUCCGGAC UCGUCCACCA-OH          ala 1
              1M       H2   H2H2                              H2 5M     M
5.     GGUUUCG UGGUCΨAGUC GGUUAUGGCA ΨCUGCΨUIAC ACGCACAAC UCCCCAGTΨCG AUCCUGGGGC GAAAUCACCA-OH         val 1
             M       H2H2          M2     OM OM       5M      7M 5M      M
6.     GCGGAUU UAGCUCAGUU GGGAGAGCG CCAGACUGAA XAΨCUGGAGG UCCUGUGTΨC GAUCCACAGA AUUCGCACCA-OH         phe 1
             M      H2H2OM H2H2H2     H2         C5       H2 5M       M
7.   CUCUCGGU AGCCAAGUUG GUUUAAGGCG CAAGACUGΨA AAΨCUUGAGA UCGGGCGTΨC GACUCGCCCC CGGGAGACCA-OH        tyr 1
               H2H2  H2H2        M2          TC           H2 5M       M
8.     GGUCCCU UGGCCCAGUU GGUUAAGGCG ΨGGUGCUIAU PACGCCAAGA UCAGCAGTΨC GAUCCUGCUA GGGACCACCA-OH         ile 1
               H2H2OM H2H2                  X  X          7M
9.     GGCUACG SAGCUCAGUU GGUUAGAGCA CAUCACUCAU AAΨGAUGGGG XCACAGGTΨC GAAUCCCGUC GUAGCCACCA-OH         met 1
                     H2               OM
10.    CGCGGGG SGGAGCAGCC UGGUAGCUCG UCGGGCUCAU AACCCGAAGA UCGUCGGTΨC AAAUCCGGCC CCCGCAACCA-OH         f-met 1
```

(c)

Figure 6.6(c) The linear sequences of several tRNAs. Since the 3′-terminal end is identical for all tRNAs (the amino acid attachment site), the sequences have been aligned and numbered from the 3′ end. The structures are (1) $tRNA_{ser\,2}$ (baker's yeast); (2) $tRNA_{ser\,1}$ (baker's yeast) [from Zachau, H. G. *et al.*, *Z. Physiol. Chem.*, **347**, 212 (1966)]; (3) $tRNA_{ser\,1}$ (rat) [from Staehelin, M. *et al.*, *Nature*, **219**, 1363 (1968)]; (4) $tRNA_{ala\,1}$ (baker's yeast) [from Holley, R. W. *et al.*, *Science*, **147**, 1462 (1965)—this was the first tRNA sequence to be determined]; (5) $tRNA_{val\,1}$ (baker's yeast) [from Venkstern, T. V. *et al.*, *Molek. Biol.*, **2**, 597 (1968)]; (6) $tRNA_{phe\,1}$ (baker's yeast) [from RajBhandary, U. L. *et al.*, *Proc. Nat. Acad. Sci.*, **57**, 751 (1967)]; (7) $tRNA_{tyr\,1}$ (baker's yeast) [from Madison, J. T. *et al.*, *Science*, **153**, 531 (1966)]; (8) $tRNA_{ile\,1}$ (torula yeast) [from Takemura, S. *et al.*, *J. Biochem.*, **65**, 553 (1969)]; (9) $tRNA_{met\,1}$ (*E. coli*) [from Cory, S. *et al.*, *Nature*, **220**, 1039 (1968)]; (10) $tRNA_{N\text{-}formyl\,met\,1}$ (*E. coli*) [from Dube, S. K. *et al.*, *Nature*, **218**, 232 (1968)]. Both differences and similarities should be conveyed by these structures. Thus the three $tRNA_{ser}$ should give a picture of the similarities of two distinct "iso-tRNA" from a single organism (1 and 2) and also between $tRNA_{ser}$ from two very different species (1, 2 vs 3). Only the nucleotides that differ from $tRNA_{ser\,2}$ are given in 2 and 3. Notice that the replacements in nucleotides 6, 7, 8, 9, 11 in 3 are matched by complementary replacements 79, 81, 82, 83, 84, indicating that the cloverleaf representation which attempts to maximize base-pairing may be a realistic model of the actual structure. Although the differences both in chain length and in sequence are noticeable, there are also some remarkable similarities in the different tRNA's (the CΨTG sequence 21–24 for example). The anticodon is shown in the box for each tRNA.

we will only briefly list some experiments that were extremely important in establishing the translation process and confirming the adaptor hypothesis.

In early experiments on protein synthesis it was found that radioactive amino acids were concentrated on the ribosomes prior to any release of newly synthesized protein, thus implicating the ribosomes as part of the protein-synthesis apparatus. This implication has since been confirmed both in vivo and in vitro, and it is now unequivocally established that the ribosomes represent the site of protein synthesis. The association of ribosomes and mRNA is equally well established by visual observation of polysomes (the association of several ribosomes with mRNA) in the electron microscope and by the chemical demonstration of association between ribosomes and polynucleotides in vitro, which will be discussed below. The evidence for binding of charged tRNA to the mRNA-ribosome complex will also be discussed later and is a well-established phenomenon. Thus all the prerequisites exist for the formation of the ribosome-messenger-adaptor-amino acid complex. If the adaptor representation in Figure 6.6 is correct, one should be able to predict that the mRNA does not recognize the amino acid on the adaptor (tRNA) and that the matching of a given tRNA with a given amino acid must be a specific process distinct from the codon-anticodon recognition. These two predictions have been tested and found to be correct. Starting with the second one, an enzyme, amino acyl–tRNA synthetase, has been isolated (or rather several such enzymes have been isolated), each one specific for a given amino acid and for a corresponding tRNA.* The amino acyl–tRNA synthetases are found in the soluble fraction of a cell homogenate, have been isolated from a number of sources, and all have been found to catalyze the same general reaction:

$$\mathrm{R{-}CH(\overset{+}{N}H_3){-}COO^-} + {}^-\mathrm{O{-}P(=O)(O^-){-}O{-}P(=O)(O^-){-}O{-}P(=O)(O^-){-}OCH_2{-}ribose(OH)(OH){-}A} \underset{}{\overset{E}{\rightleftharpoons}} \left[\mathrm{R{-}CH(\overset{+}{N}H_3){-}C(=O){-}O{-}P(=O)(O^-){-}O{-}CH_2{-}ribose(OH)(OH){-}A}\cdots E\right] + {}^-\mathrm{O{-}P(=O)(O^-){-}O{-}P(=O)(O^-){-}O^-}$$

$$\left[\cdots\right] + \mathrm{tRNA{-}pC{-}pC{-}A(OH(2'))(OH(3'))} \longrightarrow \mathrm{tRNA{-}pC{-}pC{-}A(OH){-}O{-}C(=O){-}C(H)(R){-}\overset{+}{N}H_3} + E + \mathrm{AMP}$$

* For a review, see G. D. Novelli, *Ann. Revs. Biochem.*, **36**, 449 (1967).

The amino acid is activated in the formation of the enzyme-bound amino acyl–AMP mixed anhydride, and the amino acyl group is next transferred to tRNA (recall that all tRNA molecules end with the 3′-terminal CpCpA) to form the amino acyl ester with one of the free hydroxyl groups (2′ or 3′) of the terminal adenosine. The structure above is represented as the 3′-ester but this is actually not firmly established. There is a rapid acyl migration between the 2′ and 3′ sites and no one has been able to cleanly isolate one specific ester derivative.

Firm conclusions as to the specificity of each new synthetase (or "amino acid-activating enzyme" according to early nomenclature) are based on the fact that the synthetases can be tested with both pure amino acids and tRNA molecules, as recent developments in chromatographic techniques have made tRNA purification possible. Each new pure enzyme shows an absolute specificity for only one amino acid and one, two, three or four different tRNA molecules (see Section 7.4). If a pure tRNA molecule has been found to accept a given amino acid, however, it will not function with any other enzyme or amino acid.

The postulate that mRNA recognizes only the adaptor tRNA and not the attached amino acyl group, has also been tested.* Cysteinyl-$tRNA_{cys}$ was prepared with radioactive (^{14}C) cysteine, and the incorporation of radioactive cysteine was demonstrated in a rabbit reticulocyte system capable of synthesizing hemoglobin (or at least something similar to hemoglobin) in vitro. Next the cyteinyl-$tRNA_{cys}$ was chemically modified as follows:

$$tRNA_{cys}{-}O{-}{}^{14}C(=O){-}\underset{\overset{+}{N}H_3}{\overset{H}{C}}{-}CH_2SH \xrightarrow{H_2,\ Ni} tRNA_{cys}{-}O{}^{14}C(=O){-}\underset{\overset{+}{N}H_3}{\overset{H}{C}}{-}CH_3 + H_2S$$

Thus the adaptor specific for cysteine has been made to carry alanine instead, and when this alanyl-$tRNA_{cys}$ was supplied to the reticulocyte system, alanine was incorporated into the protein in place of cysteine at the same rate as the unmodified cysteine had been incorporated.

These experiments establish the translation process in terms of two highly specific recognition events on each "end" of the adaptor molecule (the synthetase-catalyzed formation of specific amino acyl-tRNA, and the mRNA-tRNA alignment) but, as already indicated, the bulk of the evidence is best derived from the experiments in which the nature of the code was established.

6.6 THE NATURE OF THE CODE

All the preceding discussion has been based on the concept of a nucleotide language or code, and the unraveling of this code was obviously of fundamental importance to the whole idea that nucleotide sequences in DNA determine amino acid sequences in the proteins.

* F. Chapeville, *et al.*, *Proc. Nat. Acad. Sci.*, **48**, 1086 (1962), and G. Von Ehrenstein, *et al.*, *Proc. Nat. Acad. Sci.*, **49**, 669 (1963).

A numerical study of the equivalence of the two chemical languages involved is straightforward. The protein language has 20 words (20 amino acids), and so the nucleic acid language must have at least 20 words to give an unequivocal translation. The nucleic acid alphabet contains only four letters, however, and so it seems reasonable to propose that each word must be three letters long (there are only $4^2 = 16$ ways of arranging the four letters in two-letter words, which is not sufficient, but $4^3 = 64$ different three-letter words can be made from the four-letter alphabet; this is more than enough to designate the 20 amino acids). The possibility of having both two- and three-letter words could not be disregarded but, in setting up an hypothesis for experimental test, it is common practice to focus on the simplest model—in this case the uniform "triplet" or three-letter code. (Elegant genetic experiments were carried out to confirm the triplet-code hypothesis, but we will focus here on the direct chemical evidence.

Let us assume a triplet code and then, in more or less chronological order, follow the experimental approach to deciphering the nucleotide language. The genetic equivalent of the Rosetta stone was unearthed, figuratively speaking, when it was found that synthetic messages will work in the protein-synthesizing system. This made possible the first experiment designed to test the ingeniously simple postulate that if the mRNA is a simple homopolymer of a single nucleotide (in the actual case, polyuridylic acid), the protein product should be a polymer of a single amino acid. The now classical experiment* was conducted as follows: "poly-U" (polyuridylic acid)† was added to purified *E. coli* ribosomes and incubated with a mixture of tRNA's, a crude mixture of amino acyl–tRNA synthetases (designated at that time "pH-5 soluble enzymes"), and a mixture of amino acids, where a different amino acid was radioactively labeled in each incubation mixture. The test for a polypeptide product was simply based on finding acid-insoluble radioactivity (free amino acids and short peptides are soluble in 5- to 10-percent trichloroacetic acid, while longer peptides—from about 10 amino acids on—are insoluble). The results of the experiment showed that when poly-U was the message, polyphenylalanine was synthesized; thus the first word known in the "dictionary" was U—U—U = phenylalanine. This was a very important beginning, the first direct support for the language concept, the definition of an experimental approach, and, more importantly, a clear demonstration of the feasibility of eventually decoding the chemical language of the genetic message.

The next step was the synthesis of copolymers and the assignment of code-letters, based on the following statistical argument. Let us say that a polymer is synthesized again with the phosphorylase from UDP and GDP in a ratio of 5:1:

$$5n\mathrm{UDP} + n\mathrm{GDP} \xrightarrow{\text{enzyme}} (\mathrm{U_5G})_n + 6n\mathrm{P}_i$$

* The specific references to each step are given in Table 6.5.

† Poly-U was prepared enzymatically by the action of polynucleotide phosphorylase (see Table 5.4) on UDP (UDP $\xrightarrow{\text{enzyme}}$ p(Up)$_n$ + nP$_i$).

The relative probability of finding a triplet of three U in this polymer is $5 \times 5 \times 5$ while the relative probability of finding a triplet containing one U and two G is $5 \times 1 \times 1$. If U_3 corresponds to phenylalanine and UG_2 to amino acid X, it can be predicted that the frequency of incorporation of X relative to the frequency of incorporation of phenylalanine should have the same value as the ratio of the frequency of the respective triplets:

$$\frac{UG_2}{U_3} = \frac{\text{frequency of incorporation of } X}{\text{frequency of incorporation of phe}} = \frac{5}{125} = 4\%$$

The relative abundance of X should thus be 4 (that of phenylalanine is 100), and when the polypeptide product was analyzed, it was found that the abundance of glycine was 4 per cent and that of tryptophan 5 per cent, suggesting that triplets of UG_2 correspond to these two amino acids. Since this method does not give

TABLE 6.5 THE MAJOR EXPERIMENTAL EVENTS LEADING TO THE ELUCIDATION OF THE GENETIC CODE[a]

Synthetic Polyribonucleotides as mRNA

General reaction: aminoacyl-tRNA + synthetic "mRNA" $\xrightarrow{\text{ribosomes}}$ polypeptide (coded by "mRNA")

1. Homopolymers[b]	*Synthetic "mRNA"*	*Polypeptide Produced*	*Coding Information*
	Polyuridylic acid	Polyphenylalanine	U-U-U = phe
	Polyadenylic acid	Polylysine	A-A-A = lys
	Polycytidylic acid	Polyproline	C-C-C = pro

2. Random coplymers[c]	*Synthetic "mRNA"*	*Possible Triplets*	*Probability*	*Polypeptide Produced* Amino acid	*Polypeptide Produced* Incorporation frequency, %	*Coding Information*
	Poly-UG (5:1)	U_3 (5 × 5 × 5)	125 (100)	phe:	100	U_3 = phe
		U_2G (5 × 5 × 1)	25 (20)	cys, val:	20, 20	U_2G = cys, val
		UG_2 (5 × 1 × 1)	5 (4)	gly, try:	4, 5	UG_2 = gly, try
	Poly-AU(5:1)	A_3	125 (100)	lys:	100	A_3 = lys
		A_2U	25 (20)	ile, asn:	20, 28	A_2U = ile, asn
		AU_2	5 (4)	leu, tyr:	3, 3	AU_2 = leu, tyr
	Poly-CG (5:1)	C_3	125 (100)	pro:	100	C_3 = pro
		C_2G	25 (20)	ala, arg:	22, 19	C_2G = ala, arg
		CG_2	5 (4)	gly:	5	CG_2 = gly

3. Polymers of known sequence[d]	*Synthetic "mRNA"*		*Polypeptide Produced*	*Coding Information*
	Poly(U-C):	$(U\text{-}C)_n$	$(\text{ser-leu})_n$ [or $(\text{leu-ser})_n$]	UCU, CUC = ser, leu
	Poly(A-G):	$(A\text{-}G)_n$	$(\text{arg-glu})_n$ [or $(\text{glu-arg})_n$]	AGA, GAG = glu, arg
	Poly(A-C):	$(A\text{-}C)_n$	$(\text{thr-his})_n$ [or $(\text{his-thr})_n$]	ACA, CAC = thr, his
	Poly(U-U-C):	$(U\text{-}C\text{-}C)_n$	phe_n or ser_n or leu_n	UUC, UCU, CUU = phe, ser, leu
	Poly(C-A-A):	$(C\text{-}A\text{-}A)_n$	gln_n or thr_n or asn_n	CAA, AAC, ACA = gln, thr, asn
	Poly(G-A-A):	$(G\text{-}A\text{-}A)_n$	lys_n or glu_n or arg_n	GAA, AAG, AGA = lys, glu, arg
	Poly(U-A-U-C):	$(U\text{-}A\text{-}U\text{-}C)_n$	$(\text{tyr-leu-ser-ile})_n$	UAU, CUA, UGU, AUC = tyr, leu, ser, ile

TABLE 6.5 THE MAJOR EXPERIMENTAL EVENTS LEADING TO THE ELUCIDATION OF THE GENETIC CODE[a] *(Cont.)*

Direct Binding Experiment[e]

General reaction: Ribosome + aminoacyl-tRNA ⟶ no complex

Ribosome + aminoacyl$_x$-tRNA$_x$ + trinucleotide codon$_x$ ⟶ complex

Di- and trinucleotides Added	*p Mole of ^{14}C-amino Acid Bound*		*Coding Information*
	cys-tRNA$_{cys}$	leu-tRNA$_{leu}$	
None	0.29	0.76	—
UGU	1.46	0.78	UGU = cys
UUG	0.32	1.74	UUG = leu
GUU	0.34	0.92	—
UG	0.21	0.92	—
GU	0.34	0.86	—

The Dictionary

First Letter	*Second Letter* U	C	A	G	*Third Letter*
U	phe	ser	tyr	cys	U
	phe	ser	tyr	cys	C
	leu	ser	(t.c.)[f]	(t.c.)[f]	A
	leu	ser	(t.c.)[f]	try	G
C	leu	pro	his	arg	U
	leu	pro	his	arg	C
	leu	pro	gln	arg	A
	leu	pro	gln	arg	G
A	ile	thr	asn	ser	U
	ile	thr	asn	ser	C
	ile	thr	lys	arg	A
	met (i.c.)[f]	thr	lys	arg	G
G	val	ala	asp	gly	U
	val	ala	asp	gly	C
	val	ala	gln	gly	A
	val (i.c.)[f]	ala	gln	gly	G

[a] All the experiments in this system were carried out under conditions which do not select for a specific starting point in the reading of the message (the polymer U—U—C—U—U—C—U—U—C— can thus be read either as a repeating sequence of UUC, UCU, or CUU, depending on whether one reads the message from the first, second, or third nucleotide).

[b] M. W. Nirenberg and J. H. Mathaei, *Proc. Nat. Acad. Sci.*, **47**, 1558 (1961).

[c] Data taken from M. W. Nirenberg, *Sci. Amer.*, **208**, 80 (1963); P. Lengyel, J. F. Speyer, and S. Ochoa, *Proc. Nat. Acad. Sci.*, **47**, 1936 (1961).

[d] Data taken from H. G. Khorana, *The Harvey Lectures*, **62**, 79 (1966–67); H. G. Khorana, et al., *Cold Spring Harbor Symp. Quant. Biol.*, **31**, 39 (1966). The details of the synthesis of the polymers can be found in these references and are also discussed in Barker's book, in this series.

[e] Data taken from P. Leder and M. W. Nirenberg, *Proc. Nat. Acad. Sci.*, **52**, 1521 (1964); see also D. Söll, et al., *Cold Spring Harbor Symp. Quant. Biol.*, **31**, 51 (1966).

[f] i.c. = initiation codon; t.c. = termination codon. These will be discussed in Chapter 7.

any information about the nucleotide sequence in the triplet, the results (see Table 6.5) were ambiguous. It was established unambiguously, however, that A_3 corresponds to lysine and C_3 to proline. (Poly-G is quite insoluble in water and cannot readily be made with phosphorylase; thus it was not tested.)

The next step in the development was based on a specific binding experiment that actually provided information relevant to both the dictionary and the mechanism of protein synthesis. The postulate was that charged tRNA will not bind to free ribosomes, but the proper nucleotide triplet will mediate the binding of amino acyl-tRNA to the ribosomes. Again the experimental design was remarkably simple. Ribosomes are retained by millipore (cellulose nitrate) filters, whereas amino acyl-tRNA and oligonucleotides are not, and it is thus possible to measure binding to ribosomes simply in terms of retention on the filter after a standard procedure of washing. Synthetic dinucleotides and trinucleotides of *known sequence* and amino acyl-tRNA with high specific radioactivity in the amino acid were prepared and the effect of the nucleotides on the binding of the adaptor to the ribosome was tested. Typical results are given in Table 6.5, and they demonstrate that dinucleotides do not affect binding but that trinucleotides do. The idea of the triplet code thus is the correct one (a "coding ratio" of three nucleotides per amino acid is effective), and it is also clear that the sequence of each triplet is important.

The most recent step in research on the code is based entirely on work with synthetic polyribonucleotides of known sequence. It combines the use of chemical synthesis and enzymatic synthesis in a sophisticated series of experiments that very nicely duplicates and illustrates the whole sequence of events from DNA to protein.

First di- and trideoxyribonucleotides are chemically synthesized using specific blocking groups and condensing agents (see Barker in this series). Next these nucleotides are chemically polymerized with dicyclohexyl carbodiimide to give oligomers of the original starting material. This product is then used as template for DNA-specific DNA polymerase (replicase). It is interesting to note that for this step to give high-molecular-weight product, it is essential that the complement of the template also be provided. Under these conditions high-molecular-weight double-stranded products are obtained with a degree of polymerization much higher than that of the template. This product is now used as template for DNA-dependent RNA polymerase (transcriptase), forming specific polyribonucleotides that can be separated into two mRNA strands of completely defined sequence. When tested in the protein-synthesizing system, the polypeptide sequence and composition should be completely and rigidly defined if the triplet code is correct. The sequence of steps and the products obtained are included in Table 6.5.

The combined results from all of these experiments (chronologically outlined in Table 6.5) have led to the unequivocal elucidation of the genetic dictionary given in Table 6.5. Several features of this dictionary as well as its elegant experimental foundations have not been elaborated on in this discussion but will be considered in more detail in the next chapter.

6.7 VIRUS INFECTION

As already indicated, if our general DNA-RNA-protein model is correct, one can best visualize virus infection in terms of the virus forcing on to the host cell a new genetic message, either at the level of DNA (the DNA viruses) or at the level of mRNA (the RNA viruses).

In the case of DNA viruses, the replication and transcription of the virus genome probably does not require any virus-specific replicase or transcriptase, but could use that of the host cell. This does not mean, however, that the virus genome does not code for specific replicases or transcriptases in some cases. In some virus-infected cells, the rate of DNA synthesis is increased to 4 or 5 times the normal rates, indicating, perhaps, the presence of an increased number of replicase enzymes. If the virus-DNA replication only uses the host-cell–DNA polymerase, one may ask how the virus manages to compete so effectively with normal host-cell–DNA synthesis. There is no single answer to this question. In some infected cells, host-DNA synthesis proceeds in essentially normal fashion; in others, it appears to stop; in yet others, normal host DNA is actually degraded after infection. Each individual case is obviously determined by the total makeup of each individual virus, involving both protein (enzyme) components and genetic information. The example of host-DNA degradation mentioned applies to the bacteriophage T2. The DNA of T2 contains no cytosine; instead it has the derivative 5-hydroxymethylcytosine, which forms hydrogen bonds with guanosine just as does cytosine. Upon infection with T2 DNA, the virus genome is transcribed and several new enzymes are synthesized. One of these hydrolyzes host DNA, another hydrolyzes CTP to CMP and $P{-}P_i$, a third hydroxymethylates CMP to 5-hydroxymethyl-CMP, a fourth catalyzes the reaction 5-CH_2OH—CMP + ATP $\longrightarrow$ 5-CH_2OH—CTP + AMP, and the polymerase next synthesizes new DNA on the virus-DNA template. Since there is no pool of CTP (only 5-CH_2OH—CTP is available), no host DNA can be synthesized. Furthermore, since host polymerase does not use 5-CH_2OH—CTP as substrate, the polymerase must be a unique one coded for by the virus genome. In this case then, there are clear mechanistic reasons for the domination by the virus DNA.

It should be mentioned here that the DNA of bacteriophage T2 also contains glycosyl residues attached to the 5-CH_2OHC bases. The glycosyl units are incorporated into the finished polymer by yet another unique virus-DNA–coded enzyme, which catalyzes the transfer of glucose from uridine diphosphate glucose (UDPG) to the DNA.

In the few known instances in which the virus DNA is single stranded, we have already seen evidence that the first step after infection is the formation of DNA duplex (replicating form = *RF*), which is subsequently replicated and transcribed.

The effective domination of virus DNA over host DNA in transcription is perhaps even a greater puzzle. Unless the host DNA is indeed altered, it is difficult to explain why the virus genome is preferentially transcribed and translated. It appears that specific nucleases may furnish the answer. Normal host mRNA appears to be degraded very quickly after infection. If the virus could cause synthesis of nucleases specific for host RNA, the complete dominance of the virus over the host would be ensured.

With RNA viruses we have essentially the same problems and perhaps the same type of answers. In addition, however, there is a unique problem to be considered in the case of RNA viruses, namely the fact that the RNA is simultaneously the message and the genome. For new virus to be made, the RNA must be replicated; however, there is no enzyme available in the normal host cell that will catalyze RNA replication. Thus the virus genome *must* code for an RNA-specific RNA polymerase (RNA replicase), which is synthesized by the host's synthetic apparatus after infection with RNA virus. Such an enzyme has been obtained in pure form from *E. coli* cells infected with either of two RNA phages, HS2 and Qβ (RNA phages are relatively rare; most phages are DNA viruses). The knowledge that such an enzyme is formed immediately after infection confronts us with another dilemma similar to that of DNA replication. The host has a normal complement of a large number of RNA molecules—tRNA, mRNA, and rRNA. Why, then, is the *single* molecule of virus RNA specifically replicated in preference to all the host RNA? In this case the answer appears unequivocal: The virus-coded replicase is template specific and will only replicate the virus RNA. In fact, if the virus RNA is fragmented, it no longer is accepted as template for replication; thus the specificity appears to be absolute for intact homologous virus RNA (see Table 6.6). It is interesting to note that Mg^{2+} again is an essential ion for the replicase and the high template specificity is observed specifically with this metal ion. If Mn^{2+} is also added, the specificity is decreased and the enzyme will replicate other RNA's as well (Table 6.6). This is reminiscent of DNA polymerase, which loses specificity for deoxyribonucleoside triphosphates in the presence of Mn^{2+}.

The RNA-dependent RNA polymerase has been studied extensively, and this enzyme was indeed used for the first de novo test-tube synthesis of a biologically active (infectious) nucleic acid.* The sequence of events of the RNA replication appears to be quite analogous to single-stranded DNA replication. First a complement of opposite polarity is synthesized on the original single-stranded (plus) RNA template. The resulting duplex is now the replicating form of the RNA, and it appears that the complement, the minus strand, acts as template for the synthesis of a large number of copies of the original plus-strand.

An extremely important discovery has recently been made with animal RNA viruses; namely that after infection with the virus RNA, the host responds by synthesizing a new DNA molecule.† This process clearly involves information transfer from RNA to DNA in the reverse of the normal direction, and requires the virus-induced synthesis of yet another specific enzyme, an RNA-dependent DNA polymerase. In any host cell which does not go through a lysogenic

* I. Haruna and S. Spiegelman, *Science*, **150**, 886 (1965).

† D. Baltimore, H. M. Temin, and S. Mizutani, *Nature*, **226**, 1209 (1970).

TABLE 6.6 SPECIFICITY OF RNA-DEPENDENT RNA POLYMERASE (RNA REPLICASE) FROM Qβ-INFECTED *E. coli* CELLS

RNA Template	*^{32}P Incorporation, Counts per Minute*
None	30
$Q\beta$-RNA (intact)	4930
$Q\beta$-RNA (fragment of $\frac{1}{2}$ size)	109
MS-2 virus RNA	35
Turnip yellow mosaic virus (TYMV) RNA	146
E. coli rRNA	45
E. coli rRNA	15
E. coli total (bulk) RNA (from infected cells)	146
Satellite tobacco necrosis virus RNA	61

RNA Template	*^{32}P Incorporation (In Counts per Minute)*		
	Mg^{++} Only (4 *μmoles*)	*Mn^{++} Only* (0.12 *μmoles*)	*Mg^{++} + Mn^{++}* (3 *μmoles*) (0.2 *μmoles*)
None	1	103	103
$Q\beta$ RNA (intact)	4655	474	2804
$Q\beta$ RNA (fragment)	109	222	240
TYMV	218	766	1037

The reaction mixture (0.25 ml) contained in addition to the enzyme Tris-buffer, pH 7.4, Mg (and/or Mn), 1 μg of template, 0.2 μmoles each of ATP, CTP, GTP, and ^{32}P-UTP. The incorporation of ^{32}P was determined after 20 minutes incubation at 35° (similar template specificity was also found with phage MS-2 RNA replicase). From the data of Spiegelman and co-workers. [I. Haruna and S. Spiegelman, *Proc. Nat. Acad. Sci.*, **54**, 579, 1189 (1965.]

cycle (does not lyse), the production of the virus-induced DNA leads to a stable host-cell containing a foreign DNA molecule. This could well represent the basis for the cell transformations involved in virus-induced cancer. In the case of cell lysis, the new DNA is presumably simply released as a non-infective "side product," and only the released RNA viruses are capable of infecting new host cells.

Before leaving the viruses, it may be interesting to review the small satellite tobacco-necrosis virus (STNV; see Table 1.2). This virus is unable to multiply in the host (tobacco plant) cell in the absence of tobacco-necrosis virus (TNV). When TNV and STNV infect together, however, STNV dominates; very few new TNV are produced, but a large number of satellite virus result.

The physical and chemical properties of STNV are summarized in Table 6.7. The RNA, with a molecular weight of 400,000, contains 1,200 nucleotides, and the protein appears to be a polymer of a single molecule that has a molecular weight of 39,000 and contains 375 amino acids. This means that the RNA is just large enough to code for the protein of the STNV coat according to the triplet code; the STNV particle thus represents a natural particle made up of a single message and a single corresponding protein product.

One aspect of virus replication that will not be treated here must at least be mentioned—namely, the assembly of the complete virus particle from the

TABLE 6.7 SOME PROPERTIES OF SATELLITE TOBACCO NECROSIS VIRUS[a]

Properties of nucleocapsid:	particle weight = 2×10^6 daltons, made up of 80 percent protein and 20 percent RNA
Properties of RNA:	Molecular weight = 395,000 daltons, with base ratio 24.9 percent = U, 28 percent = A, 24 percent = C, 22.1 percent = G single stranded, corresponds to 1,200 nucleotides or 400 triplets
Properties of protein:	Molecular weight = 1.6×10^6 daltons
	Minimum molecular weight from amino acid analyses (based on 1 cyS per molecule) = 13,000
	Amino acid composition in moles per 13,000 g of protein (only relevant amino acids given) = lys, 5.0; arg, 8.2; met, 3.1. Total residues 124 molecules
	Amino-terminal analysis = 0.3 moles of alanine per 13,000 g Carboxy-terminal analysis = 0.31 moles of alanine per 13,000 g
	Number of tryptic peptides = 39 (lys + arg per mole = 38) Number of methionine peptides (after treatment with cyanogen bromide) = 10 (met per mole = 9)
	Molecular weight of smallest unique protein sequence = 39,000 ($3 \times$ minimum molecular weight above) = 372 residues

The protein capsid thus consists of 42 identical subunits of molecular weight 39,000, and the RNA is almost exactly the right size to code for only the coat protein subunit. The icosahedral structure of the virus observed in the electron microscope is consistent with this 42-subunit structure.

[a] Summarized from M. E. Reichman, *Proc. Nat. Acad. Sci.*, **52**, 1009 (1964).

individual macromolecules synthesized by the host cell.* Clearly the viruses represent excellent systems in which to study specific molecular assembly, and results obtained from such studies will eventually be essential in understanding the types of interaction sites and forces that determine the formation of a particular type of molecular aggregate (the complete virus particle) from the infinite number that would be possible with random interactions.

6.8 INHIBITORS OF PROTEIN SYNTHESIS

What is the experimental evidence that the sequence of steps really is DNA ⟶ RNA ⟶ protein? This has really been answered several times in the preceding discussion, since no other sequence would be consistent with all the data. Other experiments could have been cited that would present direct evidence for the flow of events, and it may be appropriate to briefly mention here the kind of experiments by which such evidence is obtained. Keeping in mind that since the measurement of incorporation of specific radioactively labeled precursors into macromolecules makes it possible to study the patterns of DNA synthesis, RNA synthesis, and protein synthesis simultaneously, one can always in principle determine the relation between these three processes.

* T. F. Anderson, "The Molecular Organization of Virus Particles," in *Molecular Organization and Biological Function*, J. A. Allen (Ed.), Harper and Row, New York (1967).

Thus, if one can specifically inhibit any one of the steps in a sequence of reactions, then intermediates and products in front of the site of inhibition will accumulate while those behind will not be formed. (This has been an essential tool in the elucidation of metabolic pathways and is discussed in detail by Larner in this series). This principle can be used to study the events of information transfer and protein synthesis both in vivo and in vitro, since a very large number of specific inhibitors of DNA synthesis, RNA synthesis, and protein synthesis are known (see Table 7.5). Very briefly and considerably oversimplified, the findings using such inhibitors can be summarized as follows: Inhibitors of any part of the translation process block protein synthesis without immediately affecting DNA and RNA synthesis, and inhibitors of RNA synthesis block both RNA and protein synthesis while leaving DNA synthesis to proceed normally. In addition to establishing the sequence of events, inhibitors have also been invaluable in the study of the mechanisms of individual reactions (see Chapter 7).

SEVEN | MOLECULAR MECHANISMS

In this chapter the general model presented in the previous chapter will be reexamined at the next higher level of mechanistic detail. Having set the stage by introducing the main components involved in the process and establishing the sequence of steps in the flow of information from DNA to specific protein sequences, we wish to examine the question of *how* the reactions proceed. In this chapter we shall therefore reinvestigate the model in more chemical terms and then examine the current mechanistic status of (1) replication (including transformation, transduction, and conjugation); (2) transcription; (3) translation and protein synthesis; (4) specific inhibitors as tools in the study of molecular genetics; (5) mutation; and finally (6) the universality of the model in the synthesis of all proteins.

7.1 MOLECULAR POLARITY IN INFORMATION TRANSFER

The model is restated in Figure 7.1. It is important to note that the conventions in representation can be ambiguous if they are not used properly. Thus one is faced with the problem of representing polymer chains in a consistent manner while illustrating the complementarity between pairs of chains. The model thus becomes a series of antiparallel chains (consistent with the actual structure of double-stranded DNA), and it is important to always indicate which is the

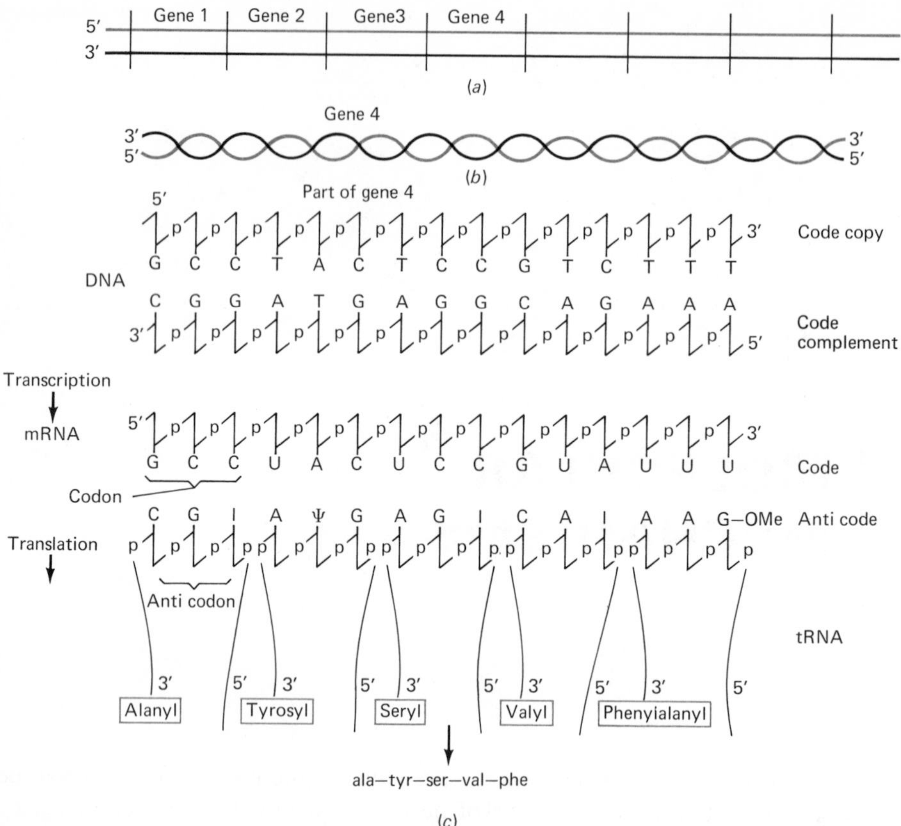

Figure 7.1 A "molecular" model of the DNA ⟶ RNA ⟶ protein flow of events, emphasizing the polarity of the polymer chains and some of the common conventions. If the 5′ ⟶ 3′ nucleotide sequence in mRNA represents the actual code, then the DNA strand having the complement of that sequence is the direct template for its synthesis, and the other DNA strand is the copy of the code (with T substituted for U). (*a*) is a long double-stranded DNA segment containing several individual genes; (*b*) a shorter segment in an actual helix; and (*c*) a short 15-nucleotide sequence (about $1\frac{1}{2}$ turn of the helix) in shorthand notation. The complete nucleotide sequence of each of the five tRNA's used for illustration above is given in Figure 6.6.

5′- and 3′-terminal end in such representations. If this is done there is no ambiguity. If only letter codes are used, however, there can be confusion. Thus a representation of the first codon in Figure 7.1 through the complete transfer, writing the triplet always from the 5′- to the 3′-terminal end, would give: GCC (DNA code), GGC (DNA complement), GCC (mRNA), and IGC (anticodon). On the other hand, writing the matching codons in the more conventional way in terms of base pairing—GCC, CGG, GCC, and CGI, respectively—is highly ambiguous unless the polarity of the chain is also given. This is readily done with an arrow indicating the 5′ ⟶ 3′ direction: $\overrightarrow{\text{GCC}}$, $\overleftarrow{\text{CGG}}$,

$\overrightarrow{GCC}$, $\overleftarrow{CGI}$. The convention to indicate the $5' \longrightarrow 3'$ direction is based on the direction of biosynthesis of nucleic acids. All the evidence to date shows that new DNA and RNA chains grow from the 5′-terminal end, and also that the polypeptide chain grows from the amino-terminal end. Typical experiments leading to these conclusions are outlined in Table 7.1. Let us then turn to the molecular aspects of the various reactions of this model.

TABLE 7.1 EXPERIMENTAL BASES FOR DETERMINING DIRECTION OF CHAIN GROWTH IN NUCLEIC ACID BIOSYNTHESIS

(A) End-group analysis in incomplete chains

In four parallel experiments, ^{32}P-labeled nucleoside (deoxynucleoside) triphosphate is used:

(1)	(2)	(3)	(4)
A-P-P-^{32}P	A-P-P-P	A-P-P-P	A-P-P-P
G-P-P-P	G-P-P-^{32}P	G-P-P-P	G-P-P-P
C-P-P-P	C-P-P-P	C-P-P-^{32}P	C-P-P-P
U(T)-P-P-P	U(T)-P-P-P	U(T)-P-P-P	U(T)-P-P-^{32}P

Samples are removed at different times and checked for ^{32}P in polymer.

Prediction: Growth from the 5′ end would result in all new molecules, finished or unfinished, having only one kind of 5′ terminal; hence ^{32}P incorporation would be observed in only one of the four tubes.
Growth from 3′ end would result in the 5′ terminal (containing the triphosphate) varying from molecule to molecule according to state of completion; hence ^{32}P incorporation would be observed in all four tubes.

Result: Only one kind of terminal triphosphate is found.

(B) Pulse labeling.

Synthesis is initiated with radioactive nucleoside triphosphates (base is radioactive). After a short time of growth, the radioactivity is diluted out with "cold" precursors, and further synthesis is allowed. The product is analyzed for the location of radioactivity—either toward the 5′ end or 3′ end.

Result: Radioactive bases are found at the 5′ end.

7.2 REPLICATION

If the semiconservative replication mechanism is correct, then the process of synthesizing new chains must consist of two steps: (1) the unraveling of the double-stranded DNA template and (2) the (perhaps) simultaneous replication of each of the individual chains. Some possible models for the replication process are indicated in Figure 7.2. If we accept the evidence that no $3' \longrightarrow 5'$ synthesis takes place [Figure 7.2(*a*)], however, we are left with the dilemma of the simultaneous $5' \longrightarrow 3'$ replication of two chains of opposite polarity as illustrated in Figure 7.2(*b*): At the halfway point of replication, 50 percent of the DNA would be single stranded, but there is strong evidence that this never happens. This dilemma has not been fully resolved, but an interesting possible mechanism was suggested, that utilizes the polynucleotide ligase as indicated in Figure 7.2(*c*). In this proposal, one chain grows continuously $5' \longrightarrow 3'$, whereas the other chain also synthesized $5' \longrightarrow 3'$, but discontinuously in small segments on the single-stranded segment that results from the continuously copied template. These short segments are subsequently linked together by the ligase to give the completed DNA strand. There is one piece of evidence in favor of this hypothesis,

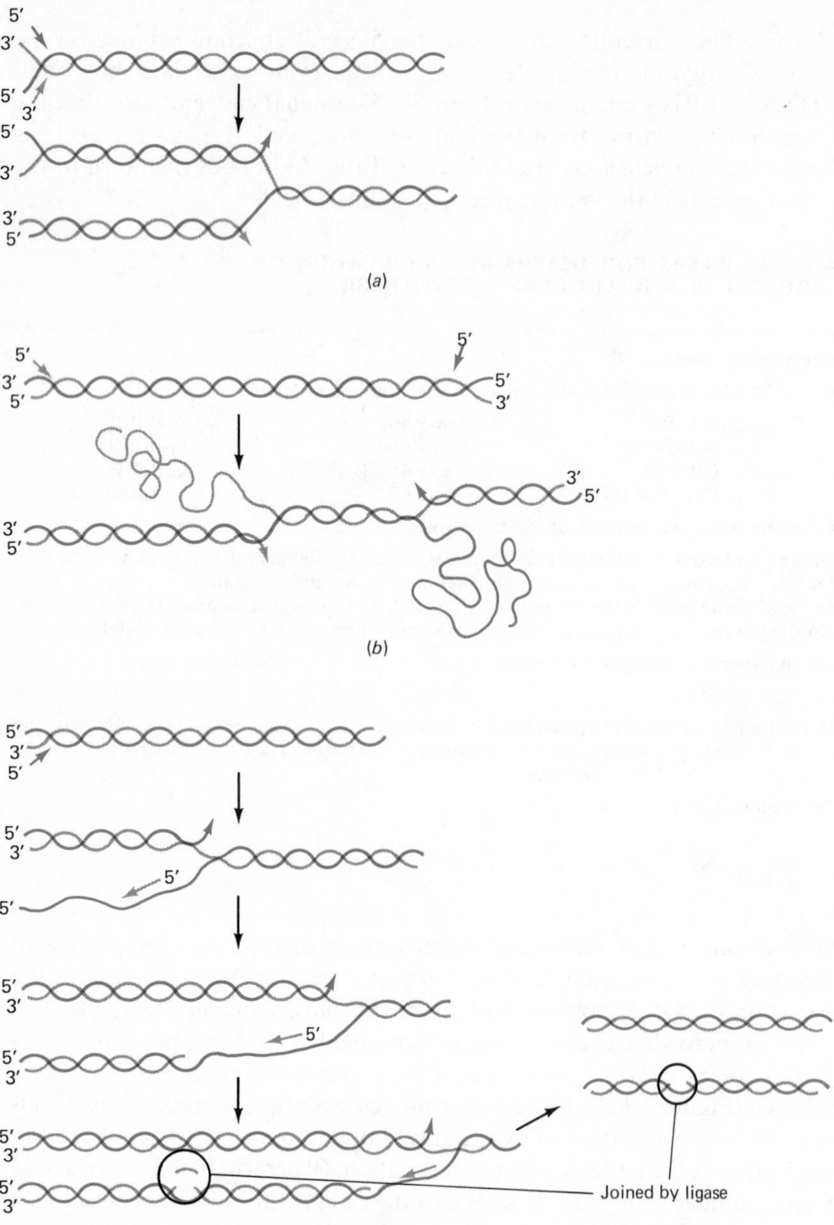

Figure 7.2 Possible replication mechanisms (see also Figure 7.4). (*a*) The simultaneous replication of the two antiparallel chains in the same linear direction would require two enzymes: one, the normal polymerase, assembling the new chain from the 5′ end, and another, assembling the other chain from the 3′ end. No enzyme of the second type has yet been found. (*b*) The simultaneous replication of the two antiparallel chains in opposite directions requires only the normal polymerase but leaves large segments (half of each chain) of single-

namely the recent discovery of newly replicated DNA fragments with a chain length of about 1,000 nucleotides, found both in normal *E. coli* cells and in phage-infected cells.* By using bacteriophage T4 mutants with a temperature-sensitive ligase, the replication experiments could be done at a temperature in which the polymerase worked normally but the ligase was inactive; under these conditions, large quantities of DNA fragments accumulated (a further refinement of this mechanism will be considered below). Another interesting feature of the latter mechanism is that it also explains the insertion of different genetic material into the existing genome. Figure 7.3(*a*) schematically illustrates the three different types of DNA transfer already mentioned in Chapter 6. In all three cases (transformation, conjugation, and transduction) transfer of DNA from a capable (+) mutant (donor) to a deficient (−) mutant (recipient) leads to "repair" of the deficient gene, and the acquired character becomes a permanent part of the transformed cell, and is passed on to its progeny. In all of these cases it has been unequivocally demonstrated that new genetic material (a DNA segment) has actually become an integral part of the genome in the recipient. The representations in Figure 7.3 are somewhat misleading in showing only the transformed recipients. The frequency of "repair" by any of these processes is very low, but the experiments are designed in such a way that only the transformed cells are observed (growth on medium where *a* is omitted, so that only $A+$ cells, which can synthesize *a*, will grow, for example).

The mechanism proposed as a general explanation of all these cases of *recombination* through a process called *crossover* can be illustrated by two possible molecular models, "breaking and reunion" and "copy choice" in Figure 7.4(*a*). Breaking and reunion is schematically illustrated for two double-stranded parent molecules. In the first step a single stranded "nick" is introduced in both molecules and some unraveling of the original double-stranded structure and perhaps a base pairing of the two loose strands follows. The intact strand is next used as template for the synthesis of a short new polymer segment from the free 3′ end liberated by the initial nick. The new segments next form a double-stranded segment, while the corresponding segments of the original intact strands are digested by nucleases. Finally a ligase joins the proper ends together (circles) to give a new pair of double-stranded "recombinant" DNA molecules, each containing a segment from each parent molecule. It would clearly require two separate breaking and reunion steps to insert a DNA segment into the interior of the recipient DNA. This mechanism should be recognized by manifestation of crossover with very little synthesis of new DNA (only a very small segment of the total DNA had to be synthesized).

* R. Okazaki, *et al.*, *Proc. Nat. Acad. Sci.*, **59**, 598 (1968).

stranded template. This has never been observed. (*c*) The continuous replication of one chain from the 5′ end with the normal polymerase, with a second polymerase molecule synthesizing short 5′ ⟶ 3′ sequences in the opposite linear direction on the single-stranded template of the complementary chain left behind. The short segments are assembled into the finished chain through the action of ligase (see also mechanism in Figure 7.4).

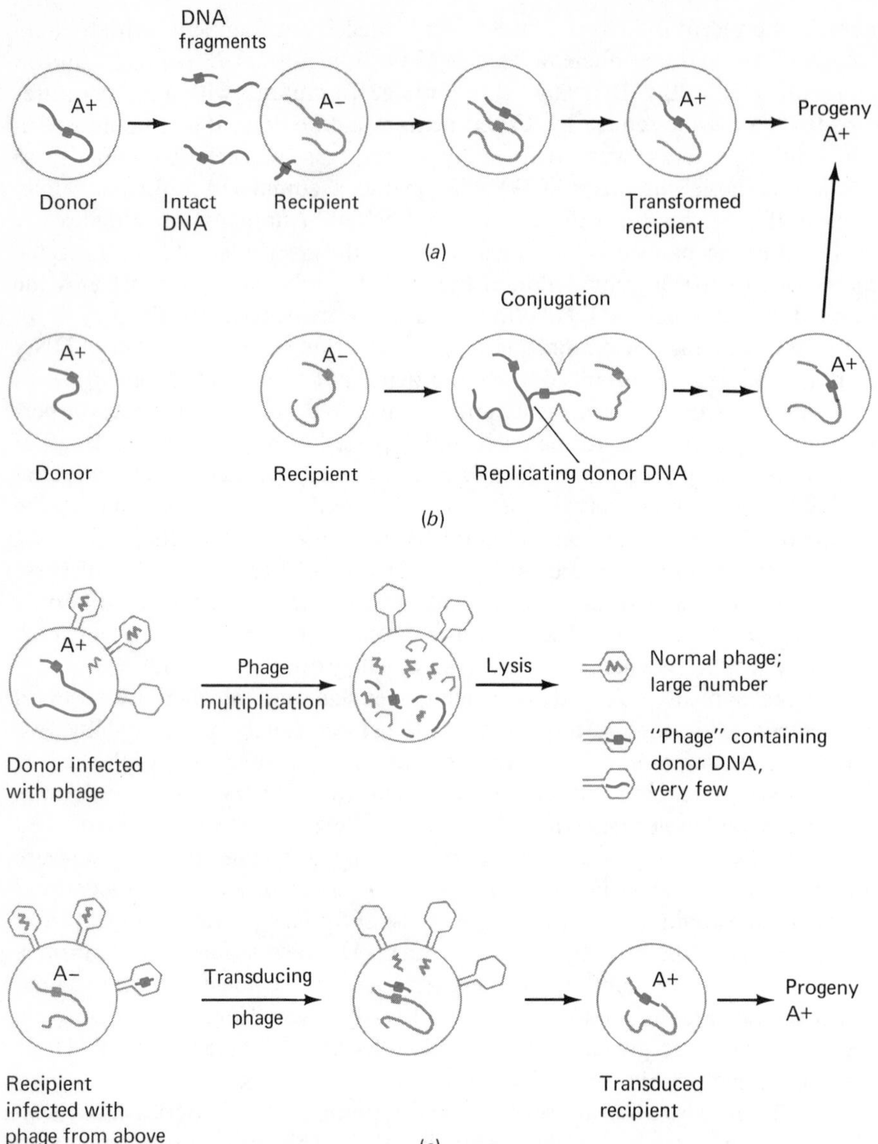

Figure 7.3 A schematic representation of the different types of DNA transfer (see Chapter 6). (*a*) Transformation: the transfer of free DNA or DNA fragments. (Transformation has only been successful with a very limited number of bacterial species.) (*b*) Conjugation: the direct transfer of DNA from donor to recipient through cell attachment. It is suggested that this type of transfer requires simultaneous replication of the donor DNA. (*c*) Transduction: transfer of DNA from donor to recipient by phage. As indicated in Chapter 6, the success of this process depends on the recipient being lysogenic ("immune") to the infection by phage. The introduction of the normal phage DNA into lysogenic cells does not lead to phage production and lysis, and the donor DNA fragment thus can be incorporated into the recipient's chromosomes.

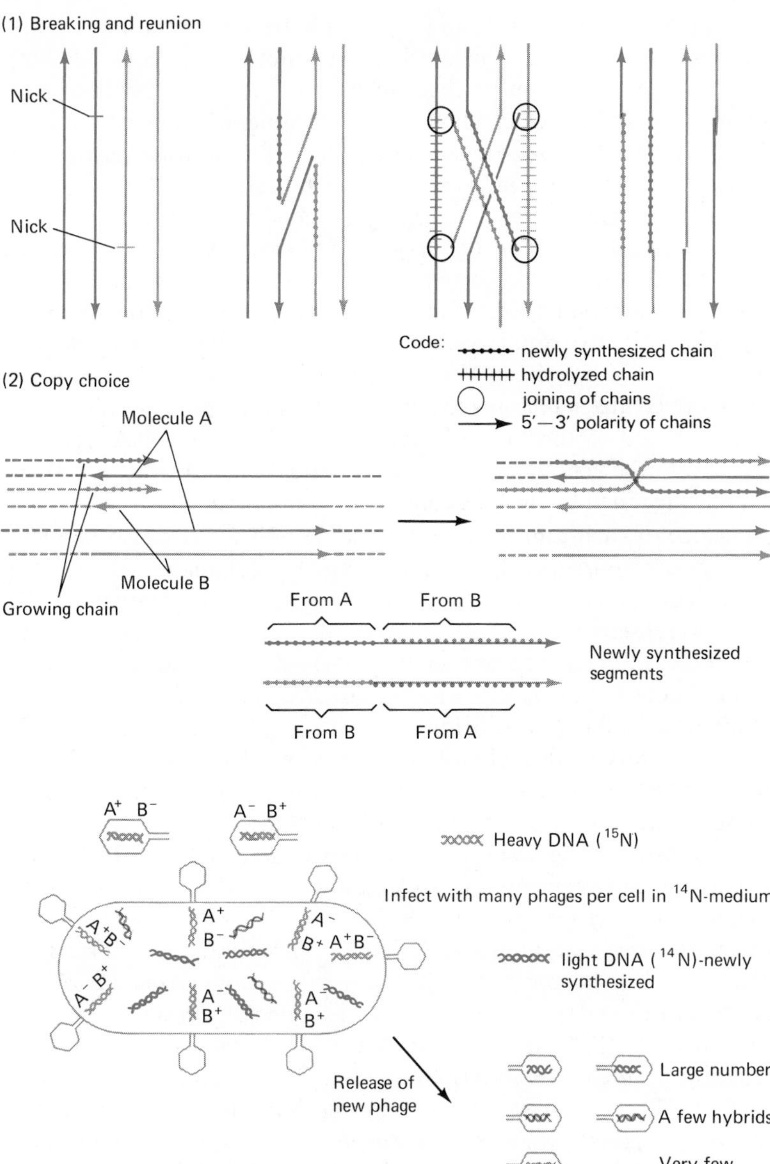

Figure 7.4 (*a*) Molecular mechanisms for DNA transfer (recombination). See text. (*b*) Experimental evidence in favor of the breaking and reunion model. The two phage mutants are obtained from infection of cells grown in ^{15}N-containing medium. The phages (genotype A^+B^- and A^-B^+, respectively) contain "heavy" DNA. Reinfecting with these phages in medium containing ^{14}N makes it possible to ask whether any phages can be obtained which are of the recombinant A^+B^+ genotype and still contain only "heavy" DNA. The fact that such phage could be found shows that crossover had taken place without synthesis of new ("light") DNA, and that the breaking and reunion mechanism is operative.

Copy choice is based on DNA synthesis using both the parent DNA molecules alternatingly as template. This mechanism is not easy to reconcile with the semiconservative replication mechanism but, nevertheless, has been considered as a real possibility. Copy choice can be distinguished experimentally from the breaking and reunion mechanism by the fact that crossover by copy choice can only be observed during active DNA synthesis. (Only one strand of each of the parent strands is shown as being replicated in the figure.)

In order to decide between these two models, an experiment was designed to show whether the crossover could be demonstrated in the absence of newly synthesized DNA. Positive results would favor model 1, the "breaking and reunion" mechanism, while a negative result would support model 2, the "copying choice" mechanism. The type of experiment designed to obtain the answer was based on recombination of two phage mutants to give back the wild type and is illustrated in Figure 7.4(*b*). Based on this kind of evidence, it is clear that DNA segments can be inserted into the genome in the absence of active DNA synthesis along the general path suggested by the breaking and reunion model.

The significance of such a process to "repair" of DNA must also be mentioned. If it is desirable to remove a small segment of "damaged" DNA, the breaking and reunion coupled with a nuclease-catalyzed removal of the damaged segment represents an attractively possible mechanism. The similarity between the process of breaking and reunion and the discontinuous replication mechanism in Figure 7.2(*c*) should be obvious. We shall return to a more detailed formulation of these ideas below.

Perhaps the best way of formulating a coherent mechanism of replication is to review the most recent data on the various enzymes that appear to be responsible for the process. As indicated in Chapter 5, there are several different types of DNases, both of the endo- and exo- type; there are also the ligases, which can join together segments of DNA; and finally, there are the DNA-dependent DNA polymerases, responsible for the chain-elongation process. This latter type of enzyme has now been isolated from several different kinds of cells, and the *E. coli* polymerase has been thoroughly characterized during the last few years.* The pure enzyme appears to be a single polypeptide chain with a molecular weight of 110,000 daltons (the conclusion that it is a single chain is based on finding a single amino-terminal residue (methionine) per 110,000 daltons and the failure to dissociate the enzyme into smaller units by dissociating agents). This pure enzyme, in addition to catalyzing polymerization of deoxynucleoside triphosphates (when in the presence of the proper template and aided by oligodeoxynucleotide primers), has also 3′-exonuclease, 5′-exonuclease, and 3′-pyrophosphorylase activity. This important enzyme thus turns out to have an extremely complex functional makeup; the multiple activities are all associated with a single polypeptide chain. For the discussion here, we will only concentrate on one other property of the enzyme, the template specificity. By labeling the enzyme with a radioactive isotope, its binding to

* See reference in Figure 7.6. See also A. Kornberg, and co-workers, *J. Biol. Chem.*, **244**, 2996–3052 (1969) and **245**, 39 (1970).

different DNA templates can be measured directly in a density-gradient sedimentation system by measuring the amount of enzyme associated with the characteristic DNA band. The results of such experiments are given in Figure 7.5. While the enzyme can bind extensively to both linear and circular single-stranded DNA, its binding to double-stranded DNA is much more limited. In fact, there is an exact correspondence between the number of enzyme molecules bound per DNA template and the number of free ends plus the number of "nicks" in the double-stranded template. A "nick" is a hydrolytic cleavage on only one strand of the double-stranded structure; it can be produced by a number of endonucleases, such as pancreatic DNase (this type of endonuclease is sometimes referred to as "nickase").

Based on these observations, a general replication mechanism has been proposed (2) that is consistent with all the facts. It should be emphasized, however, that the proposal is still only hypothetical and is based on observations made mostly on *E. coli* cells. This mechanism is illustrated in Figure 7.6(*a*).

If the chromosomes are intact, double-stranded circles, they are completely resistant to replication (no polymerase binds). If the chromosomes are linear double helices, polymerase is bound at the ends. However, in this case there is no primer, so replication is slow. The first step in replication, therefore, must be the introduction of a nick, catalyzed by an endonuclease, perhaps a highly specific one. This nick represents a binding site for the polymerase, and replication can start immediately at this point. However, it is also possible that, prior to replication, the exonuclease action of polymerase is used to digest a short section of the nicked chain from the 5′-terminal end or from the 3′-terminal end, causing some unraveling of the double-stranded structure. Using the 3′ end of the nicked strand as primer and the intact strand as template, chain elongation can now proceed in the $5' \longrightarrow 3'$ direction. To explain the simultaneous unidirectional replication of both strands, it is suggested that perhaps the replication switches from the original template strand to the complementary strand to form a fork (Figure 7.4). This fork must then in turn be cleaved by an endonuclease to free a new 3′-terminal, from which replication can proceed to the next fork. The net result is an apparently continuous synthesis of one strand and discontinuous synthesis of the other (see Figure 7.2), with the ligase furnishing the final catalytic step in linking the pieces together. Ligases have been isolated from different sources and some information is now available on the details of the ligase reaction. For example, it is well established that cofactors are required, either NAD (in *E. coli*) or ATP (in phage-induced and mammalian ligase). With this scheme, the early dilemma (that double-stranded DNA was such a poor in vitro template for the polymerase whereas denatured DNA was such a good one) has been resolved. The combination of exonuclease and polymerase activity suggests an excellent safeguard against and repair of errors in replication. Thus any wrong base incorporated by mistake would lead to a weak point in the double-stranded product, and since such a weak point would be susceptible to nuclease action, the error would be eliminated and corrected before the chain growth could continue [Fig. 7.6(*b*)].

	Number of nicks/mole	Number of ends/mole	Moles of polymerase bound/mole of template	Relative rate of replication (DNA synthesis) with template
1. ϕX–174 DNA (1.7×10^6 mw) (single-stranded closed circle)	 0	0	20	Slow[a]
2. ϕX–174-RF				
(a) double-stranded closed circle	 0	0	0	None
(b) Denatured	 0	0	21	Slow[a]
(c) Nicked with pancreatic DNAase (–OP(5′), –OH(3′))	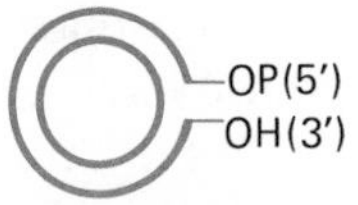 1	0	1	Fast
(d) Nicked with micrococcal DNAase ((3′)PO, OH(5′), (5′)HO, OP(3′), OH(5′), (3′)OP, OP(3′), (5′)OH, OP(3′), (5′)HO)	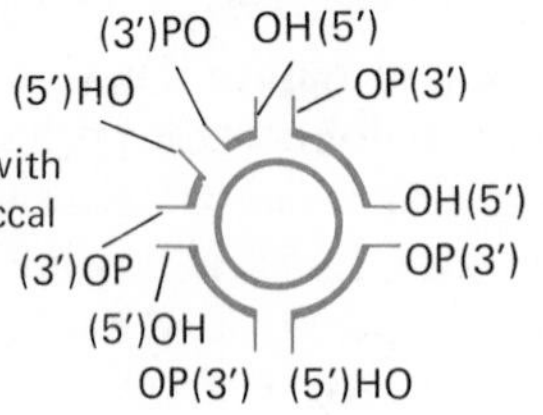 5	0	6	None
3. T7 DNA (26×10^6 mw)				
(a) Linear, double stranded.	 0	2	–	Very slow
(b) Denatured	 0	2	240	Slow[a]
(c) Nicked ((3′)OH OP(5′))	 0.5	2	2.6	Fast
4. Synthetic $d(A\text{-}T)_{12}$ (8000mw) A–T–A–T–A–T–A–T–A–T–A–T / T–A–T–A–T–A–T–A–T–A–T–A (hairpin)	0	1	1	Very slow

[a] If any fragmentation takes place so that short polynucleotide segments are present to act as primers, fast DNA synthesis is observed in these systems. This must be the explanation for the early observations that denatured DNA gave much faster in vitro DNA synthesis than did native DNA.

Figure 7.5 A summary of the results of *in vitro* studies on the binding of DNA polymerase to well-characterized template DNA molecules and of the relative effectiveness of the same template DNA's in giving fast DNA synthesis. The results

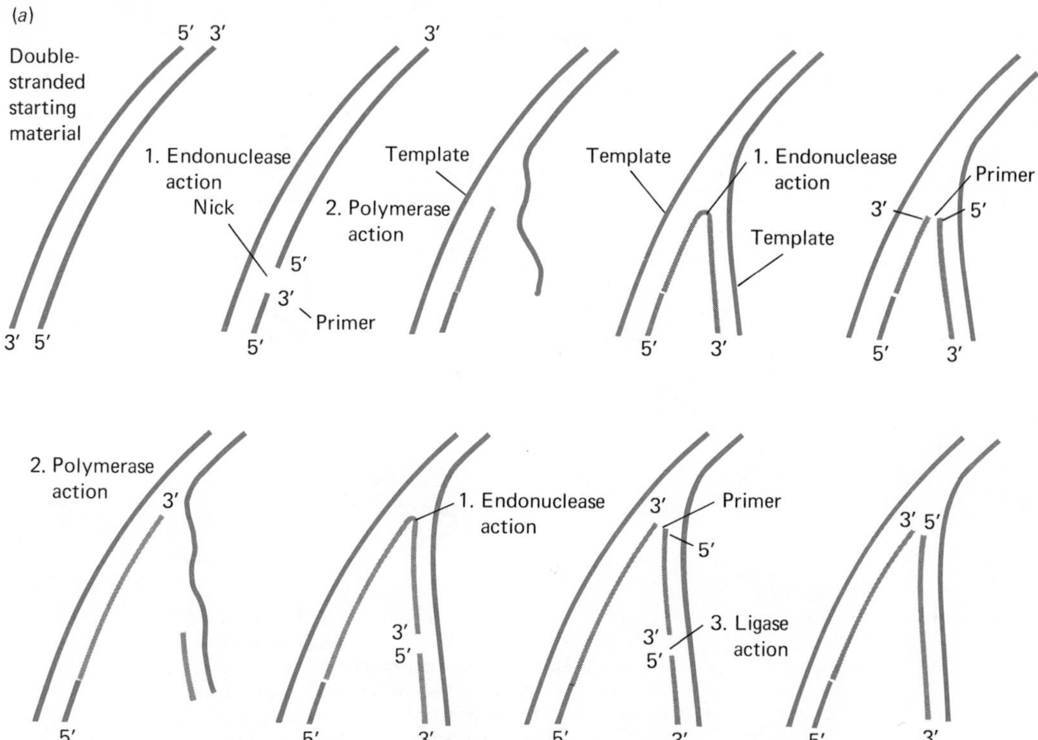

Figure 7.6(a) A hypothetical scheme for the replication process, based on the proposed continuous replication of one template chain and discontinuous replication of the other, both in the $5' \longrightarrow 3'$ direction (the "knife and fork" model). The process as presented involves at least three separate enzyme activities: (1) an endonuclease (two separate endonucleases may be required, one for the introduction of the nick and another for the cleavage of the fork), (2) the polymerase, for the chain elongation, and (3) the ligase for the fusion of the discontinuous segments. The unique features of this model as compared with the model in Figure 7.2(*c*) are the incorporation of the nicked template as the starting point for replication and the proposal that the chain elongation is a continuous process, first along one template chain and then forming a fork along the other template chain, and that the nuclease cleavage of the fork is a prerequisite for the second step of the chain elongation (full green colored line in figure).

---

show that polymerase can bind to the ends or to "nicks" in double-stranded DNA or to single-stranded DNA. The first prerequisite for replication must clearly be the binding of the polymerase to the template. In addition, however, it is also clear from the data that rapid replication requires a free 3′-OH group as primer for the initiation of DNA synthesis (the free 5′-OH groups in 2(d) do not serve as primers).

(b)

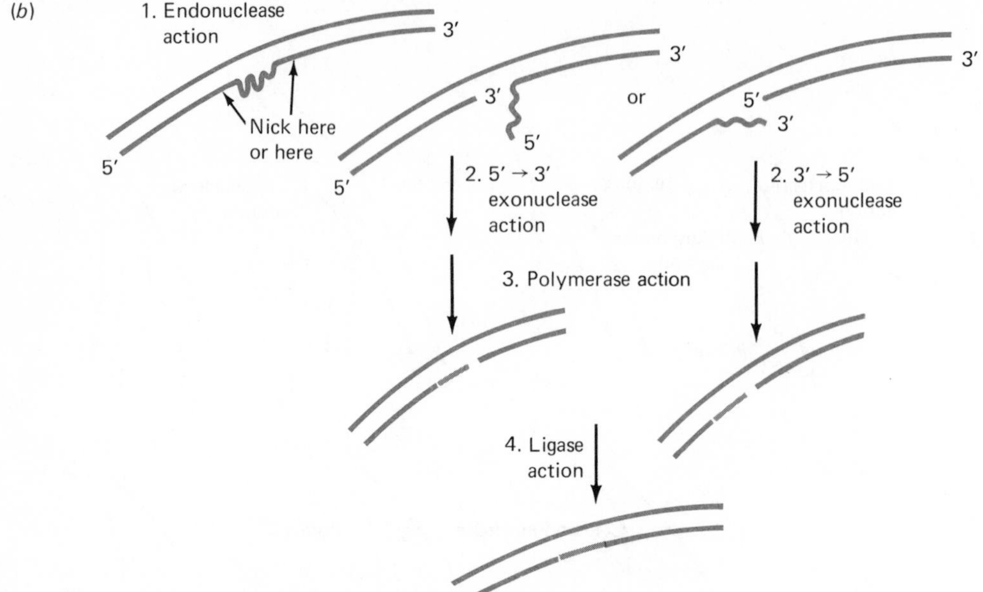

Figure 7.6(*b*) A possible repair mechanism for DNA. A weak spot (mutation for example) in the double-stranded DNA may be excised by the use of the same three enzymes used in (*a*). In addition this repair mechanism uses the exonuclease (either the 5′ ⟶ 3′ or the 3′ ⟶ 5′ exonuclease may be involved, as indicated in the figure) activity associated with the polymerase to remove the "faulty" segment. [It is quite possible that this exonuclease activity may also be an important part of replication in widening the "nick" before replication starts, but this step has not been incorporated in (*a*).] [The data and most of the mechanistic proposals in this figure are taken from Kornberg, A., *Science*, **163**, 1410 (1969).]

It must be emphasized here that the polymerase discussed above in fact appears so well suited for DNA repair that it has been suggested that repair is its only function, and that a different enzyme system thus must be responsible for DNA replication. There is now some evidence that such a second DNA polymerase system exists. In accordance with this suggestion and for the sake of easy distinction, the polymerase discussed above will be referred to as "repair polymerase," and the proposed new enzyme as "replication polymerase." By the use of different mutants, different inhibitors and cofactors, and the special pretreatment of cells with toluene to render them permeable to nucleoside triphosphates, it was possible to demonstrate two types of polymerase activity in the intact, albeit nonviable, *E. coli* cells. The first of these, the repair polymerase, required nuclease activity for DNA synthesis and was found only in Pol+ mutants (strains with normal polymerase levels). It did not require ATP, nor was it inhibited by N-ethylmaleimide. The second type of polymerase, the replication polymerase, required ATP, was inhibited by N-ethylmaleimide, was independent of nuclease activity, and was operative in Pol− mutants (the polymerase-less strains referred to in Section 6.3). Both systems catalyzed semiconservative DNA-synthesis.* Thus it appears that there is a replicating process which is distinct from the one assigned to the well-characterized "repair polymerase," and the mechanism of DNA replication must still be considered an open question.

7.3 TRANSCRIPTION

The transcription process is not well understood at the molecular level. The salient features that remain to be elucidated are the chemical processes that trigger the transcription, the chemical marker on the chromosome that designates the initiation and termination of a given gene, and the mechanism by which transcription of a given gene is stopped. Some aspects of these questions will be discussed in Chapter 8, in connection with regulation of gene expression.

A brief review of the enzyme responsible for transcription, RNA polymerase, may again be the best way to survey the status of this process. RNA polymerases have been isolated from many sources, and again, most information is available from the *E. coli* enzyme. *Escherichia coli* RNA polymerase is a complex multisubunit enzyme, with a molecular weight of about 400,000. There are at least four different subunits, α, β, β', and σ, and the enzyme appears to have the composition $\alpha_2\beta\beta'\sigma$. The specific function of each subunit is not clear, except perhaps for the growing body of evidence that σ is responsible for initiation of mRNA synthesis. The current proposal is that σ, which is only loosely bound to the complex, is responsible for the initial binding to the DNA template and therefore probably also for the selection of the proper transcribing DNA strand. Once RNA synthesis has started, σ dissociates from the enzyme-template complex and initiates synthesis at a new site. Chain elongation then proceeds in the absence of σ. The template requirement of the polymerase appears to be double-stranded DNA. Although single-stranded DNA will serve in vitro, the product in this case is of too low a molecular weight to be true mRNA. The polymerization reaction does not require a primer, as is clearly demonstrated by the fact that the 5′-terminal nucleoside retains its 5′-triphosphate in the finished product. The termination mechanism appears to be associated with ionic strength and, especially, Mg^{2+} concentration. In vitro, at low ionic strength, termination becomes faulty and very large RNA molecules result, whereas higher ionic strength and Mg^{2+} concentrations appear to result in normal product. The transcription process must have some features in common with replication, but the requirement for specific initiation and termination signals make it necessary to look for new and unique chemical mechanisms to fully understand replication.†

* R. E. Moses and C. C. Richardson, *Proc. Nat. Acad. Sci.*, **67**, 674 (1970).

† E. P. Geiduschek and R. Haselkorn, *Ann. Revs. Biochem.*, **38**, 647 (1969).

7.4 TRANSLATION: PROTEIN SYNTHESIS

One of the most complex systems studied in biochemistry is the protein synthesizing apparatus. Our current knowledge of the steps by which free amino acids are activated and condensed into polypeptide chains in a sequence specified by an mRNA sequence is summarized below.

Amino acid activation:

1(*a*) Amino acid$_x$ + ATP + $E_{1(x)}$ $\rightleftharpoons$ [$E_{1(x)}$—amino acyl$_x$—AMP] + P − P$_i$ (P − P$_i$ $\xrightarrow{H_2O}$ 2P$_i$)

(*b*) [$E_{1(x)}$—amino acyl$_x$—AMP] + tRNA$_x$ $\rightleftharpoons$ amino acyl$_x$—tRNA$_x$ + $E_{1(x)}$ + AMP

2(*a*) Amino acid$_y$ + ATP + $E_{1(y)}$ $\rightleftharpoons$ [$E_{1(y)}$—amino acyl$_y$—AMP] + P − P$_i$

(*b*) [$E_{1(y)}$—amino acyl$_y$—AMP] + tRNA$_y$ $\rightleftharpoons$ amino acyl$_y$—tRNA$_y$ + $E_{1(y)}$ and so forth

Directed peptide-bond synthesis:

3(*a*) Amino acyl$_x$—tRNA$_x$ + ribosome—mRNA(n site) [with (amino acids)$_n$—tRNA$_z$ attached] $\xrightarrow{E_2}$ ribosome—mRNA(n site) [with tRNA$_z$ and (amino acids)$_{n+1}$—tRNA$_x$ attached]

(*b*) Ribosome—mRNA(n site) [with tRNA$_z$ and (amino acids)$_{n+1}$—tRNA$_x$ attached] + GTP $\xrightarrow{E_3}$ ribosome—mRNA(n + 1 site) [with (amino acids)$_{n+1}$—tRNA$_x$ attached] + tRNA$_z$ + GDP + P$_i$

4(*a*) Amino acyl$_y$—tRNA$_y$ + ribosome—mRNA(n + 1 site) [with (amino acids)$_{n+1}$—tRNA$_x$ attached] $\xrightarrow{E_2}$ ribosome—mRNA(n + 1 site) [with tRNA$_x$ and (amino acids)$_{n+2}$—tRNA$_y$ attached]

(*b*) Ribosome—mRNA(n + 1 site) [with tRNA$_x$ and (amino acids)$_{n+2}$—tRNA$_y$ attached] + GTP $\xrightarrow{E_3}$ ribosome—mRNA(n + 2 site) [with (amino acids)$_{n+2}$—tRNA$_y$ attached] + tRNA$_x$ + GDP + P$_i$

The complete synthetic apparatus is even more complex than indicated in this scheme. In addition to amino acids, tRNA, ATP, mRNA, ribosomes, GTP, and the enzymes E_1, E_2, and E_3, several other "factors" are also required. To formulate a mechanistic model of protein synthesis, it is necessary to keep in mind that we again can visualize the total process as consisting of three distinct steps,

(1) chain initiation, (2) chain elongation, and (3) chain termination, and that all these steps must be accounted for in the model (the scheme only shows chain elongation).

In the following paragraphs the individual components and steps of protein synthesis will be briefly reviewed as the background for the current detailed mechanistic model. It should perhaps be stated here that the amino acid activation steps above are well understood mechanistically. The role of ATP in the acyl activation step, with the hydrolysis of $P - P_i$ to P_i as the driving force, and the formation of the amino acyl ester of tRNA were mentioned in Chapter 6 and are also discussed by Larner in this series. Our main concern here is the process of protein synthesis from amino acyl-tRNA.

Amino Acyl–tRNA Synthetase

Amino acyl–tRNA synthetase (E_1) is an important component of the translating mechanism, catalyzing reactions 1 and 2 above.* Several amino acyl–tRNA synthetases have been obtained quite highly purified from different sources. The molecular weights fall in the range of 90,000 to 115,000, except for two phenylalanyl-tRNA synthetases (from yeast and *E. coli*), which both have a molecular weight of 180,000. All the synthetases appear to contain a free —SH group, which is essential for activity. The K_m for the respective tRNA is about 10^{-7} M in the cases of all the enzymes studied, but the K_m for ATP and amino acid varies considerably and is sensitive to Mg^{2+} concentration. In one instance it has been clearly established that the two-substrate reaction (1 and 2 above) is distinctly ordered: The ATP-E complex forms first, and then reacts with the amino acid.

Since amino acyl–tRNA synthetase represents the key to the translation process in matching a given tRNA (from the nucleotide language) with a given amino acid (from the protein language), it is clear that the *specificity* of the enzyme is highly relevant to the model. With several pure enzymes available for study, it is now certain that the synthetases indeed have very stringent requirements for a specific amino acid (generally for the natural L-isomer, although some will also activate the D-isomer), as well as for a specific tRNA. It has also been demonstrated in one case that a "wrong" AA-AMP-enzyme complex (produced by saturation with a wrong amino acid) is hydrolyzed *when the correct tRNA* is added. The overall reaction in matching the correct amino acid to the correct tRNA thus is safeguarded first by substrate specificity and secondly by specificity of the transfer reaction itself (the wrong amino acyl group is transferred from AMP to water instead of to tRNA).

Since the genetic code is universal, it might be reasonable to expect a fairly uniform structure in the different specific tRNA's in all living cells, and therefore also in the synthetases; however, this does not appear to be the case. Some synthetases will charge heterologous tRNA's (tRNA from different organisms) whereas others will only use homologous tRNA (tRNA from the same organism)

* G. D. Novelli, *Ann. Revs. Biochem.*, **36**, 449 (1967).

as substrates, indicating that tRNA's differ from organism to organism.* It has also been shown that although tyrosyl-tRNA synthetase from *E. coli* and from *Bacillus subtilis* have identical kinetic properties and use tRNA from either organism, these enzymes differ in molecular weight and show no immunochemical cross-reactivity (antibodies to the *E. coli* enzyme do not react with the *B. subtilis* enzyme).

Transfer Ribonucleic Acids

In Chapter 5 brief mention was made of the new chromatographic techniques developed for tRNA fractionation, and tRNA's for all 20 amino acids have now been purified to some extent, some of them to a high degree. The complete covalent structure has been determined for several of them (see Figure 6.5). More importantly, it has recently been possible to crystallize tRNA's and, with well-defined crystals, it should be possible to carry out analyses by X-ray diffraction and thus obtain three-dimensional structural models. With 64 codons available in the dictionary, it has generally been assumed that there will turn out to be a corresponding number (or very nearly so) of different tRNA molecules, and in that case perhaps an equal number of amino acyl synthetases as well. The experiments pertaining to this question are ambiguous at present. There is no doubt that there are several chromatographically distinct tRNA's for a given amino acid; in *E. coli* there are five $tRNA_{leu}$, three each of $tRNA_{ser}$ and $tRNA_{val}$, and two each of $tRNA_{tyr}$ and $tRNA_{phe}$. However, just as in the case of establishing unequivocally the existence of isozymes, very stringent experimental tests are required to eliminate the possibility that chromatographically distinct forms of tRNA are artifacts of isolation. In most cases such tests have not been run. Some recent experiments have, however, given good reason to be concerned about this point. Thus one of two yeast $tRNA_{ser}$ peaks, active when tested with crude synthetase preparations, was not active when tested with a pure enzyme; that particular fraction was later found to have lost the 3′-terminal adenosyl residue, very likely as a result of purification procedures. *Escherichia coli* $tRNA_{try}$ has also been shown to exist in two chromatographically distinct forms, one of which can be converted to the other at low pH or at low ionic strength (under conditions where hydrogen bonds are weakened). These two forms are thus most likely to be two conformers of the same tRNA molecule, produced during isolation and stabilized by noncovalent forces.

It is thus possible that the number of different tRNA molecules may be considerably lower than the theoretical 64; there has indeed been a proposal, the "wobble" hypothesis, to rationalize that possibility. The proposal is that while the first and second nucleotide in each codon triplet will pair with the first and

* This matching of amino acyl-tRNA synthetases with heterologous tRNA's represents yet another basis for studying species relatedness (compare to Figures 3.18, 3.19, and 6.2). The more closely related two species are, the better their synthetases and tRNA's interact. [T. Yamane, and N. Sueoka, *Proc. Nat. Acad. Sci.*, **50**, 1093 (1963).]

second nucleotide in the anticodon strictly by normal hydrogen-bonded base pairing, the third anticodon nucleotide has more flexibility in selecting its partner. Thus a third anticodon nucleotide, G, can pair with either U or C; U can pair with either A or G, and I with either A, U, or C (all of these can form two hydrogen bonds if they are not restricted by any regular helical structure, such as found in DNA). This wobble in the third anticodon nucleotide is consistent with the lack of specificity in the third codon nucleotide observed in the dictionary (one $tRNA_{ser}$ with the anticodon AGI should thus be able to handle at least three of the six serine codons—UCU, UCC, UCA—whereas the other three—UCG, AGU, and AGC—may require separate tRNA molecules). With complete sequence analyses of several "iso-tRNA's," the question whether each codon in the dictionary has its own tRNA may soon be settled, especially if X-ray crystallography will now lead to the determination of their three-dimensional structure.

Ribosomes

Ribosomes have also been isolated from a variety of sources. Their general chemical composition—65 percent RNA and 35 percent protein—and the base composition of their rRNA—high G and low C contents—appear to be quite uniform for different organisms. Again, for the purpose of studying ribosomes in protein synthesis it is necessary to obtain pure ribosomes. Bacterial ribosomes have been by far the simplest to prepare pure. In higher organisms the ribosomes are persistently associated with lipoprotein components in the microsomal fraction and can only be purified by relatively severe treatments with enzymes and detergents. Nucleases represent another group of very troublesome contaminants encountered in the preparation of ribosomes from higher organisms. For these reasons the bacterial ribosomes have been most thoroughly studied and are best characterized.

The role of the ribosome in protein synthesis is not fully understood, nor is there any unequivocal evidence for specific functional roles of either the ribosomal proteins or the rRNA. Recent success in dissociating both 30-S* and 50-S† ribosomes into several protein fractions and RNA and subsequently reconstituting fully active ribosomes by adding all the components back together promises that such evidence is forthcoming. By systematically studying the structural and functional deficiencies produced by omitting each component in the reconstitution experiment, the role of each component can hopefully be established. It is clear, however, that the ribosome is the site of protein synthesis. In addition, it is generally accepted that the ribosome, utilizing the energy released from GTP hydrolysis, moves along the mRNA as the polypeptide chain is synthesized, and that each ribosome in a polysome

* C. G. Kurland, *et al.*, *Cold Spring Harbor Symp. Quant. Biol.*, **34**, 17 (1969), and N. Nomura, *et al.*, *Cold Spring Harbor Symp. Quant. Biol.*, **34**, 49, 63, 69 (1969).

† R. R. Traut, *et al.*, *Cold Spring Harbor Symp. Quant. Biol.*, **34**, 25 (1969), and T. Staehelin, *et al.*, *Cold Spring Harbor Symp. Quant. Biol.*, **34**, 39 (1969).

thus is the site for separate, growing (nascent) polypeptides on a single mRNA molecule. (It was originally thought that a message was translated only once and was then destroyed. On the basis of the observed polysomes it appears, however, that a message is translated several times.)

Peptidyl Transferase

Peptidyl transferase, E_2, is associated with the 50-S ribosome and catalyzes the transfer of the peptidyl group from $tRNA_x$ to the amino group in the next amino acyl–$tRNA_y$ molecule:

$$\sim NH{-}CH(R_{(x)}){-}C(=O){-}O{-}tRNA_x + H{-}\ddot{N}H{-}CH(R_{(y)}){-}C(=O){-}O{-}tRNA_y \longrightarrow \sim NH{-}CH(R_{(x)}){-}C(=O){-}NH{-}CH(R_{(y)}){-}C(=O){-}O{-}tRNA_y + tRNA_x$$

At least two different enzymes appear to be involved, but they have not been extensively purified and characterized. They are unstable and require a free —SH group in the active site. These enzymes are also species specific (enzymes obtained from *E. coli* will not catalyze transfer with mammalian ribosomes and vice versa).

Guanosine Triphosphatase or G-Factor

Although a supernatant enzyme with no known biological activity by itself, *G*-factor (E_3), when combined with ribosomes, catalyzes the hydrolysis of GTP to GDP and P_i. In the complete protein-synthesis apparatus, the *G*-factor–catalyzed GTP hydrolysis is most likely involved in the translocation reaction [reactions 3(*b*) and 4(*b*) above], and it is now clear that the release of tRNA from the synthetic site is also associated with this step.*

Other Factors

Initiation factors, F_1, F_2, and F_3, are supernatant components that are required for chain initiation. It may be that F_3 is involved in the formation of the initial mRNA-ribosome complex (only the 30-S ribosome is involved in this step), while F_1 and F_2, together with GTP, are involved in the binding of the first amino acyl tRNA, which must have a blocked amino terminal (see below), to the active ribosome.

* F. Lipman, *Science*, **164**, 1024 (1969), and J. Lucas-Lenard, and A.–L. Haenni, *Proc. Nat. Acad. Sci.*, **63,** 93 (1969).

Another supernatant component, T factor, is involved in chain elongation. In analogy to the F factors, it is proposed to be a requirement for the binding of the incoming amino acyl tRNA to the ribosome.

The chain termination factors, R_1 and R_2, are also supernatant components. Their role is to cause the release of the completed chain from the ribosome-bound tRNA (still attached to the carboxy-terminal amino acid in the newly synthesized polypeptide chain). Although their mechanism of action is not understood, it is important to note that the two factors respond to specific termination codons in the mRNA. Specifically, R_1 is active in the presence of UAA or UAG, and R_2, in the presence of UAA or UGA (see dictionary in Table 6.5).* All the above-mentioned factors appear to be proteins.

Initiation and Termination Processes

In Table 6.5 it was seen that if a synthetic polymer of repeating triplets were used as messenger in protein synthesis in vitro, more than one amino acid could be converted into a homopolymer. Thus, with the polymer $(GAA)_n$, either polylysine, polyglutamic acid or polyarginine could be obtained. The explanation of this fact must be that the starting point on the message is random under the experimental conditions used. The polymer GAAGAAGAAGA has three possible *reading frames*; it will spell glu-glu-glu (GAA = glutamic acid) when read from the first nucleotide, lys-lys-lys (AAG = lysine) when read from the second nucleotide, and arg-arg-arg (AGA = arginine) when read from the third nucleotide. Once the starting point has been fixed, a repeating single-word code is always obtained. Clearly such a randomly selected initiation point would be disastrous in vivo, and it is necessary to look for the answer to the question of how the chain is initiated to ensure that the reading frame is always the same. Since it is also clear that a given mRNA may carry the code for several different polypeptides (viral RNA, for example, is with few exceptions a single RNA molecule carrying several transcribed genes), we also need to know how the message for one polypeptide chain terminates. Connected with these two problems is the question of the direction of peptide synthesis and the direction of message reading. If we know the mechanism of chain initiation and chain termination, the question of direction will automatically be answered.

The clue to the initiation came as a rather surprising result of an experiment demonstrating the following reaction catalyzed by an *E. coli*–soluble fraction (the same fraction that contains the amino acyl–tRNA synthetases):†

$$1\ \text{Methionine} + \text{ATP} + \text{tRNA}_{\text{met}} \xrightleftharpoons{E_{1(\text{met})}} \text{methionyl-tRNA}_{\text{met}} + \text{P} - \text{P}_i + \text{AMP}$$

$$2\ \text{Methionyl-tRNA}_{\text{met}}\ (\text{I}) \xrightarrow[\text{folate + enzyme}]{\text{N}^{10}\text{-formyl tetrahydro-}} \text{N-formylmethionyl-tRNA}_{\text{met}}\ (\text{II})$$

* G. Milman, *et al.*, *Proc. Nat. Acad. Sci.*, **63**, 183 (1969).

† K. Marcker and F. Sanger, *J. Mol. Biol.*, **8**, 835 (1964).

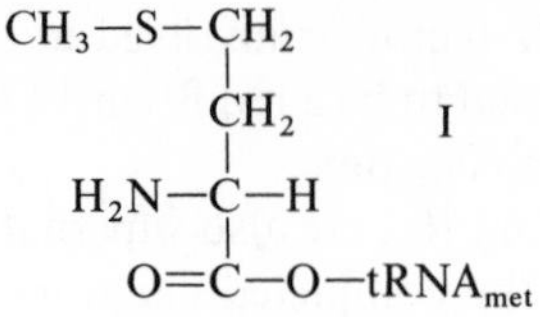

$$\begin{array}{c} CH_3-S-CH_2 \\ | \\ CH_2 \\ | \\ H_2N-C-H \\ | \\ O=C-O-tRNA_{met} \end{array} \quad \text{I} \qquad\qquad \begin{array}{c} CH_3-S-CH_2 \\ | \\ O \;\; H \;\; CH_2 \\ \| \;\;\; | \;\;\; | \\ H_2C-N-C-H \\ | \\ O=C-O-tRNA_{met} \end{array} \quad \text{II}$$

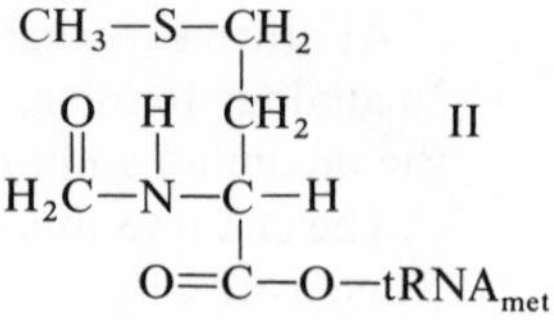

When it was also found that the N-formylmethionine was subsequently incorporated into proteins, this appeared to be the ideal chain initiator since its α-amino group was blocked, thus preventing it from forming peptide bonds. This predicts, however, that all *E. coli* proteins should have an N-formylmethionine as the amino terminal amino acid. When this prediction was tested, it was found that although the amino-terminal residues in *E. coli* are definitely not a random population of amino acids, they are not uniformly N-formylmethionine either. Free, unblocked amino terminals of alanine, serine, and methionine made up the bulk of the amino-terminal amino acids. To resolve this problem it was proposed that the proteins are indeed all synthesized with an amino terminal N-formylmethionine but that subsequent enzymatic attack releases either the formyl group or N-formylmethionine, along with other amino acids. The demonstrated presence of such hydrolytic enzymes in *E. coli* and the relatively narrow spectrum of amino-terminal amino acids in that organism both lend support to this proposal. It was therefore proposed that the only known methionine codon, AUG, was the *initiator codon* in *E. coli.*

Chain termination codons were also established in a fairly direct way. The arguments could be better discussed later, but one of them will be briefly mentioned here. A phage mutant (amber mutant) with a single base changed—a *point mutation*—was found to produce an incomplete polypeptide chain. The wild-type codon CAG (= glutamine) had in the mutant become UAG, and instead of synthesizing a protein with glutamine in a given position, the synthesis stopped at this position and the incomplete chain was released from the ribosomes. This was the first indication that some codons (in this case UAG) specify chain termination. Three codons, UAA, UAG, and UGA, do not code for any amino acid; it now appears clear that all these codons in some way trigger the release of the growing polypeptide chain from its ester link to the last tRNA molecule.

These two postulates have both been tested by a more direct experimental approach. The experiment in Table 7.2 illustrates such a test using synthetic messages, and provides direct evidence for AUG as an initiation codon and UAA as a termination codon. It is now clearly of interest to establish whether the initiation and termination codons and mechanisms are universal. Host specificity of viruses could, for example, be explained on the basis of initiation codon matching, if different organisms have different initiators. N-formylmethionine and AUG have been found to initiate protein synthesis in yeast as well as in *E. coli*, but the information on this point is yet insufficient for any firm conclusions.

If AUG is accepted as initiator and UAA as terminator, the experiment reported in Table 7.2 clearly establishes the direction of message reading (from

TABLE 7.2 DEMONSTRATION OF INITIATION AND TERMINATION CODONS[a]

Synthetic Polymer	*Acid insoluble peptide product*
AUG(A)$_n$	N-formylmet–(lys)$_n$
AUGUUA(A)$_n$	N-formylmet–leu–(lys)$_n$
AUGUUU(A)$_n$	N-formylmet–phe–(lys)$_n$
AUGUAA(A)$_n$	No product
AUGUUUUAA(A)$_n$	No product[b]

[a] System: *E. coli* ribosomes, amino acyl-tRNA, synthetic polynucleotide, and the soluble factors (F, T, G, R) required for proper initiation, synthesis, and termination. (From W. M. Stanley, Jr., *7th Intern. Congress of Biochem.*, Tokyo, 1967, Abstract I, p. 45.)

[b] No acid insoluble product was obtained, but the dipeptide N-formylmethionyl–phenylalanine was demonstrated to be present in the reaction mixture by paper chromatography.

the 5′-terminal end) and of polypeptide growth (from the amino-terminal end), confirming earlier findings based on more complicated experiments.

The Model

All the data reviewed above are presented schematically in Figure 7.7. The participants are represented on the top of the figure: ribosomes, mRNA, amino acyl-tRNA, and the different soluble "factors." The scheme is based on current knowledge of the specific site of action of the different factors and the characteristic properties of the synthetic apparatus. The first step is the combination of mRNA with the 30-S ribosome requiring factor F_3; the subsequent combination with 50-S ribosome gives the active mRNA-ribosome complex with two amino acyl-tRNA binding sites, *D* and *A*, located in such a way that the anticodon of the tRNA can be aligned with the proper codon in mRNA. In the next step an amino acyl-tRNA with a blocked amino group (initiation) is brought into site *D* under the influence of factors F_1 and F_2 and GTP. (It is possible that a translocation step is involved in the initiation, with site *A* being occupied first with codon 1 at the site-*A* position, but there is no evidence available on this point.) Site *A* is next occupied by the amino acyl-tRNA specified by codon 2 (*T*-factor is required) and the 50-S ribosomal enzyme, peptidyl transferase catalyzes the transfer of the amino acyl group from tRNA (1) to the free amino group of the acceptor (2) to form the first peptide bond. The next step is translocation, requiring *G*-factor and utilizing the energy released in the hydrolysis of GTP. In the translocation the uncharged tRNA (1) is released from site *D* and the peptidyl tRNA (2) is transferred from the site *A* to the site *D* with a concomitant movement of the ribosome relative to mRNA to position codon 3 under site *A*. Site *A* is next occupied by the amino acyl-tRNA specified by codon 3 and the whole cycle is repeated, building up the polypeptide chain

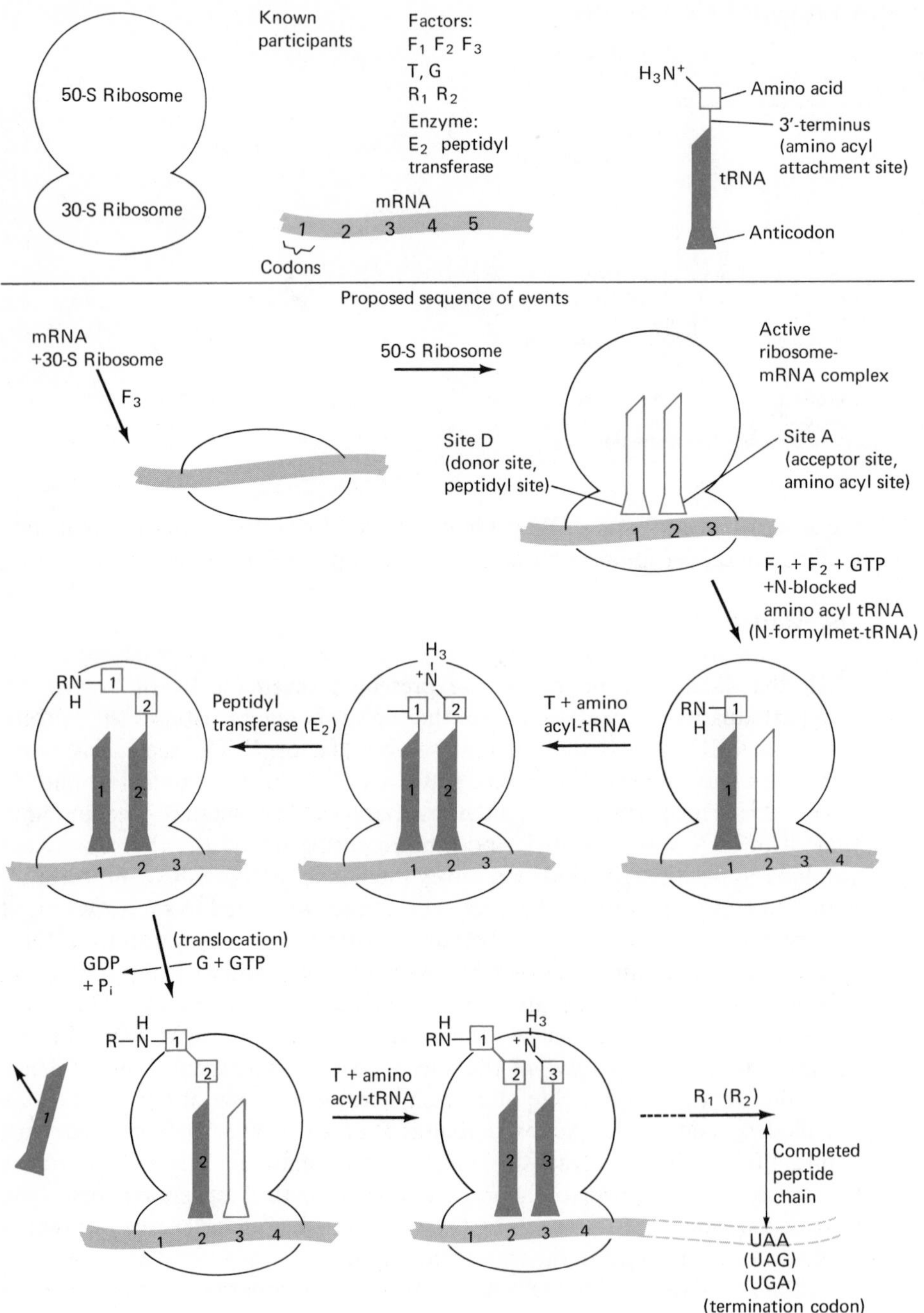

Figure 7.7 A hypothetical scheme illustrating the steps in protein synthesis. [Based on the models and data from Lipman, F., *Science*, **164**, 1024 (1969).]

step by step according to the information stored in the mRNA. Finally, when one of the termination codons is encountered (as in the *n*th codon), and in the presence of the proper *R*-factor, the completed polypeptide is released from the $(n - 1)$th tRNA molecule.

In order to fully appreciate all aspects of this model and understand apparent contradictions in the data reported in Chapters 6 and 7, two aspects of the experimental conditions must be given the strongest possible emphasis. One of these is the purity of the components used and the other is the Mg^{2+} concentration. The first one is quite obvious when one keeps in mind that the various factors and enzymes tend to associate with the ribosomes. Whether the addition of a given factor will or will not stimulate a given function must thus depend on the extent to which that factor was originally removed from the system. Standardized methods of preparing and washing ribosomes and purifying the supernatant fractions have been of great importance in establishing the role of the individual components and assuring uniform and consistent results. The effects of Mg^{2+} on the protein-synthesis system should perhaps also be obvious, since the divalent metal ions bind so strongly to the polyanionic nucleic acids. Two cases of Mg^{2+}-affected specificity change were mentioned earlier in connection with DNA replication (see Section 6.3) and RNA replication (see Table 6.5), and one more case must be mentioned in order to explain the data on protein synthesis. The important point is that the various interactions that determine the specificity of the coding appear to become less specific as the Mg^{2+} concentration increases. This effect is perhaps significant in the amino acyl-tRNA–trinucleotide–ribosome binding studies listed in Table 6.5, since these were carried out at high Mg^{2+} concentration (0.035 *M*) and show some ambiguities in the specificity of binding.

One of the most dramatic effects of Mg^{2+}, however, appears to be involved in the initiation of the polypeptide chain. In addition to initiation factors, a low Mg^{2+} concentration is also important in effecting proper initiation and consistent reading frames. The following proposal is made to incorporate this effect into the model in Figure 7.7; it may provide a useful and simple (and very hypothetical) picture of the protein-synthesizing sites until more conclusive data are available. In the presence of low Mg^{2+} concentration (0.001 *M*), the specificity of binding in the two ribosomal sites is such that site *D* (the donor site) will only accept amino acyl-tRNA (or peptidyl-tRNA) after its amino-terminal α-amino group has been blocked, while site *A* (the acceptor site) binds amino acyl-tRNA with a free, charged α-amino group. At increasing Mg^{2+} concentrations the repulsion of site *D* for the positive charge of a free amino group decreases until, at sufficiently high Mg^{2+} concentrations, unblocked amino acyl-tRNA can occupy both site *A* and site *D*, thus eliminating the initiation fidelity and permitting random reading frames.

Allowing for such experimental variables in the results reported over the last 8 to 10 yr, it appears that all the available data now consistently fit the proposed model.

One more problem must be mentioned in the closing of this section—the conversion of the linear polymer of amino acids into the folded, biologically active form. Based on the information available on reversible denaturations of proteins (see sections on ribonuclease in Chapter 3 and insulin in Chapter 4), it is clear that, in some instances at least, the folded, native three-dimensional structure of a polypeptide chain represents the stable conformation of that particular primary structure in dilute aqueous solution. Does this mean that the growing polypeptide chain attached to the polysome is always a random coil and that the characteristic, folded structure only forms after the completed polypeptide chain has been released from the synthetic site? Of course no such conclusion can be drawn from the denaturation data; in fact, evidence has been obtained in recent years that polysome-bound protein will recognize and bind substrate as well as other proteins that are natural partners in multisubunit structures. Thus it must be safe to conclude that some folding of the nascent polypeptide chain takes place before it is released from the polysome, so that at least some features of the substrate site or other "recognition" sites, characteristic of the finished product, are present in the growing polypeptide chain. The regularity of the disulfide bonds in γ-globulin (Figure 4.4) would be consistent with stepwise oxidation of —SH pairs as the chain is synthesized.

7.5 EFFECT OF INHIBITORS

As already mentioned in Chapter 6, one of the fundamental biochemical tools in determining the proper sequence of a set of linked reactions is the use of specific inhibitors. This approach has been very fruitful in formulating both the sequence of steps and the mechanism of some of the steps in information transfer and in protein synthesis. In addition, inhibitors of protein synthesis are of great interest as drugs and regulatory substances because they affect the cell's most vital chemical reactions. The partial list in Table 7.3 of various inhibitors indeed is a catalogue of antibiotics and antitumor agents. The sites of action of the inhibitors are well established, and the specific mechanisms of action of some of them are also quite well understood. The inhibitors in group 1 are structural analogues of different substrates and cofactors and thus act by competitively inhibiting the respective enzymes. In group 4, 5-methyltryptophan also fits into this category.

Puromycin is another group 4 inhibitor that can be described quite accurately in terms of both structure and function. Structurally, it is similar to a tyrosyl tRNA with most of the RNA missing, suggesting that its mode of action might be at the level of amino acid assembly. This hypothesis has now been well documented. When treated with puromycin, short polypeptides containing the puromycin residue as the "carboxyl" terminal end are released from the ribosomes, and no finished polypeptide chain is observed. There is no obvious reason why puromycin should be accepted by the synthetic site; it has no anticodon, but perhaps the small molecule has a nonspecific "fit" to the acceptor site on the ribosome.

TABLE 7.3 INHIBITORS OF PROTEIN SYNTHESIS

	Inhibitor	Action
Group 1. Inhibitors of nucleoside (deoxynucleoside) triphosphate synthesis:		
	(a) Azaserine	Inhibits reactions that use NH_2 from glutamine
	(b) Aminopterin and amethopterin sulfonamides	Inhibit synthesis of tetrahydrofolic acid (coenzyme in the synthesis of nucleotides)
	(c) 5-fluorodeoxyuridine	Inhibits (competitively) thymidine biosynthesis (specific inhibitor of DNA synthesis)
Group 2. Inhibitors of DNA-synthesis (replication)		
	(a) Mitomycin C Phleomycin	Cause cross-link formation in DNA (?)
	(b) Hydroxyurea	Polymerase inhibition
Group 3. Inhibitors of RNA-synthesis		
	(a) Actinomycin D Chromomycin A3 Acridine orange	Complex with G in DNA to block transcription (not clear why replication is not affected)
	(b) Chloramphenicol 5-fluorouracil	Block synthesis of rRNA and functional ribosomes
Group 4. Inhibitors of protein synthesis		
	(a) Streptomycin	Bind to ribosomes and give faulty (?) reading of mRNA
	(b) 5-methyl tryptophan	Competes with tryptophan for $tRNA_{try}$
	(c) Puromycin Tenuazonic acid Cycloheximide	Blocks growth of polypeptide chain Block release of finished product from ribosomes

The action of the rest of the inhibitors listed in Table 7.3 can only be described empirically as blocking a given step, confounding codon translation, and so forth, and there is no obvious way of explaining their action on the basis of

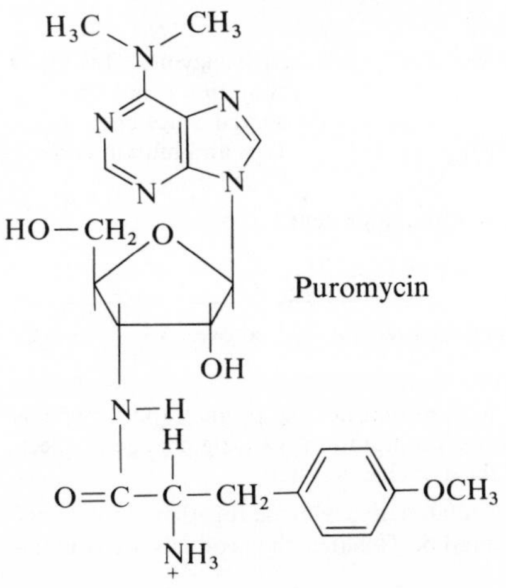

Puromycin

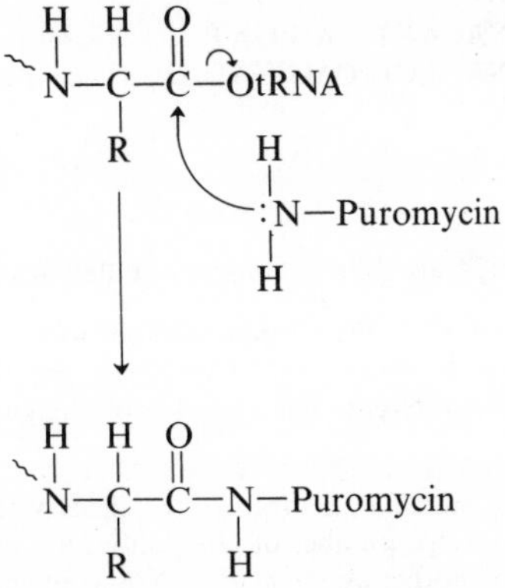

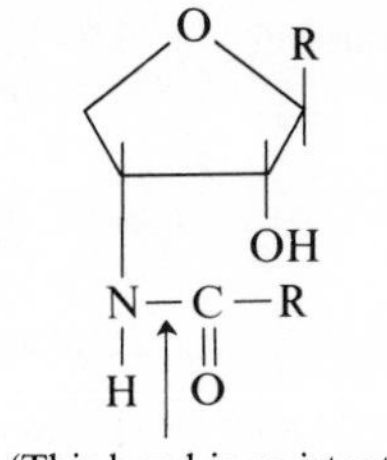

(This bond is resistant to attack by an incoming amino acyl RNA)

their structures. The elucidation of the mode of action of all of these inhibitors is clearly an important challenge, both as a fundamental mechanistic problem and as a question of great relevance to the treatment of viral and bacterial infections and cancer.

7.6 MUTATION

Let us next turn to the problem of explaining mutational changes in terms of molecular events consistent with our models. It is clear that any chemical alterations in DNA leading to an alteration in the base-pairing specificity should lead to a change in a given codon and thus in the final amino acid sequence produced. The initial change could take place at any level of the information transfer. When the change is affected in the genome, it will be transferred to the progeny, leading to a mutant. Different types of mutations can be predicted and have indeed been observed (Table 7.4). The most useful

TABLE 7.4 MUTATIONS, REVERSIONS AND MUTAGENS

Mutations

	Genotype	*Phenotype*: Amino acid sequence	*Phenotype*: Protein product
Wild type (a^+)	DNA: ACT CAG GAC mRNA: ACU CAG GAC	-thr-glu-asp-	Active enzyme
Deletion mutation (a^d)	DNA: ACAGGAC mRNA: ACAGGAC	-thr-gly? -thr-(termination?)	Incomplete or wrong
Point mutations (a_x^{-1})	DNA: GCT CAG GAC mRNA: GCU CAG GAC	-ala-glu-asp-	Complete protein active or inactive depending on importance of thr
(silent)(a_x^{-2})	DNA: ACT CAG GAT mRNA: ACU CAG GAU	-thr-glu-asp-	Active enzyme. This mutation is not observed, since both wild type and mutant codons code for asp.

(a^+, a^d, a_x^{-1}, a_x^{-2} are different *alleles* = different forms of the same gene.)

Reversions. Mutant changes can "revert" to the normal, wild phenotype in three ways:

1. Back mutation (true genotypic reversion). The reversal of the original change (in a_x^{-1} above, $A \xrightarrow{\text{mutation}} G$, $G \xrightarrow{\text{back-mutation}} A$)
2. Suppression mutation. A second mutation neutralizes the effect of the first, so that normal product (phenotype) is produced. (There are probably a large number of possibilities for suppression mutation: causing change in specificity of tRNA, causing change in another amino acid to "complement" the first change, etc.)
3. Complementation. Two mutants with inactive gene product may produce an active enzyme together. (Active enzyme = *ab*, mutant I makes good *a*, deficient *b*; mutant II, deficient *a*, good *b*. Together the products are complementary and give good *a*—good *b* = active enzyme.

TABLE 7.4 ***(Cont.)***

Different types of mutagens[a]

Treatment	DNA Change	Effect
1. High energy irradiation (UV irradiation most commonly used)	Dimer formation 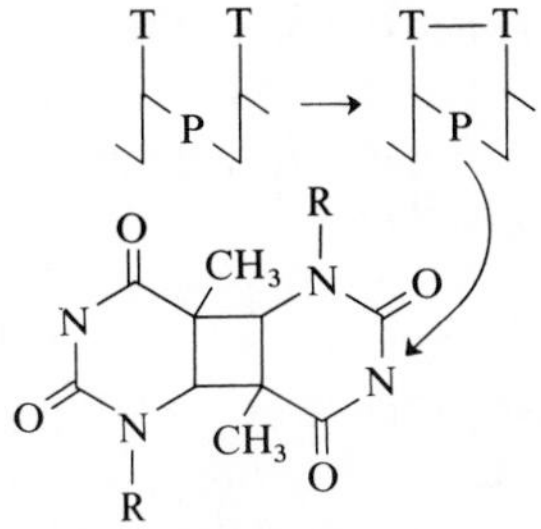	No replication of dimer region. Deletion of dimer?
2. Bifunctional reagents (nitrogen mustards, e.g., $Cl-CH_2-CH_2-N(CH_3)-CH_2-CH_2Cl$)	Cross-linked guanyl residues 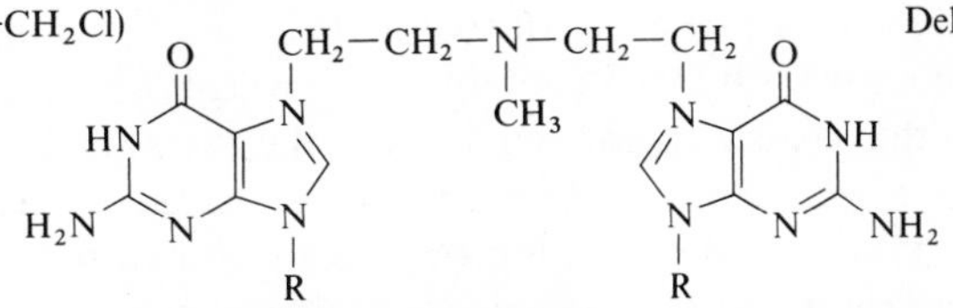	No replication of dimer region. Deletion of dimer?
3. Acridine dyes	Binds to DNA by noncovalent "stacking" or "sandwiching" between bases	Leads to deletions or insertion because of the distortion
4. 2-amino purine (AP) (8-azaguanine and 6-thioguanine are similar)	Is incorporated into DNA, can base-pair with either T or C	Faulty replication and transcription (point mutations)
5. 5-bromo uracil (BU) (5-chloro- and 5-iodo-U are also used)	Incorporated into DNA. Can base-pair with either A or G	Faulty replication and transcription (point mutations)
6. Nitrous acid	1. Deamination: $R-NH_2 \xrightarrow{HONO} ROH + N_2 + H_2O$ $C \longrightarrow U$ $G \longrightarrow$ xanthine [base-pairs with C and T(U)] $A \longrightarrow$ hypoxanthine (base-pairs with C) 2. Diazotization and cross-linking	Faulty replication and transcription (point mutations)
7. Hydroxylamine	Specifically converts C to U	Point mutations

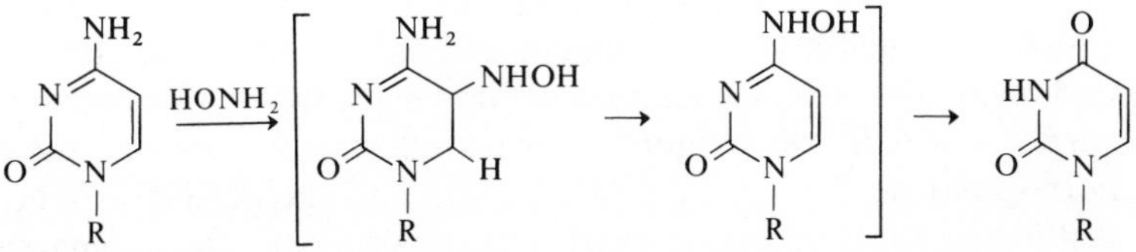

[a] Taken from L. E. Orgel, *Adv. in Enz.* **27**, 289 (1965).

mutations from the biochemist's or molecular biologist's view are the point mutations, in which a single base is altered to change only one particular codon. Mutations probably take place quite frequently in all living cells, and by the standards of natural selection, any given mutation must be considered either inferior, indifferent, or superior to the wild type. (The wild type has obviously also undergone mutations; hence the definition of "normal" is quite arbitrary.) The indifferent and superior ones are retained, while stable inferior ones (in the extreme, lethal ones) are lost in natural selection. In the laboratory mutations can be brought about by chemical reagents of different degrees of specificity, and the resulting mutants can be deliberately selected. The reagents or treatments are called mutagens, some of which are given in Table 7.4. The use of mutants has been essential to the development of the whole field of molecular genetics, and provides most of the critical structure-function tests of the informational molecules (see the discussion of tryptophan synthetase below). Once a mutant allele (see Table 7.4) has been established, there are essentially three ways in which it can revert to the wild type. One way, already mentioned, is by recombination with DNA containing the wild-type allele; a repair mechanism similar to this is also operative by which the thymine dimers formed in UV treatment have been shown to be excised (this process probably involves nucleases, polymerase, and ligase along the lines outlined for replication). Another is back mutation, in which a second mutational event at the same site will give back the original codon. A third is through a *suppressor mutation*, which which a second mutational event in another gene will neutralize or suppress the original one. An example of such a suppressor mutation can be illustrated with the amber mutant mentioned above. This mutant changed from the wild-type CAG (glutamine) to UAG (termination). An amber suppressor mutant has, in addition, an altered $tRNA_{ser}$, which permits the reading of UAG as a codon for serine. In the suppressor mutant, therefore, a complete polypeptide is synthesized with serine substituted for the normal glutamine.*

The study of mutations has been of great importance in clarifying all aspects of the genetic model, and specifically in mapping the gene and thus making available information that is the biological equivalent of actual nucleotide sequencing. The basis of fine genetic mapping is the concept that the closer two mutations are on each of the parental genes, the lower the frequency of mutation-free (wild-type) progeny will be (see Table 7.4). Thus, by examining a very large number of point mutants for frequency of recombination, the relative distance between different mutations in the gene can be determined. If the resulting genetic map now could be compared directly to the amino acid sequence of the protein product, a very meaningful and significant test for the general model and many of its detailed features would be available. Just such a test has been carried out with the tryptophan synthetase system, proving the direct linear correspondence between the DNA chain and the protein product. The main features of the tryptophan synthetase experiments are summarized in Table 7.5.

* N. D. Zinder, *et al.*, *Cold Spring Harbor Symp. Quant. Biol.*, **31**, 251 (1966).

7.7 THE UNIVERSALITY OF THE INFORMATION TRANSFER

Most of the data reviewed in the last two chapters have been obtained from the study of microorganisms, and it is clear that much more information is needed

TABLE 7.5 THE COLINEARITY OF THE GENE AND THE PROTEIN (TRYPTOPHAN SYNTHETASE)[a]

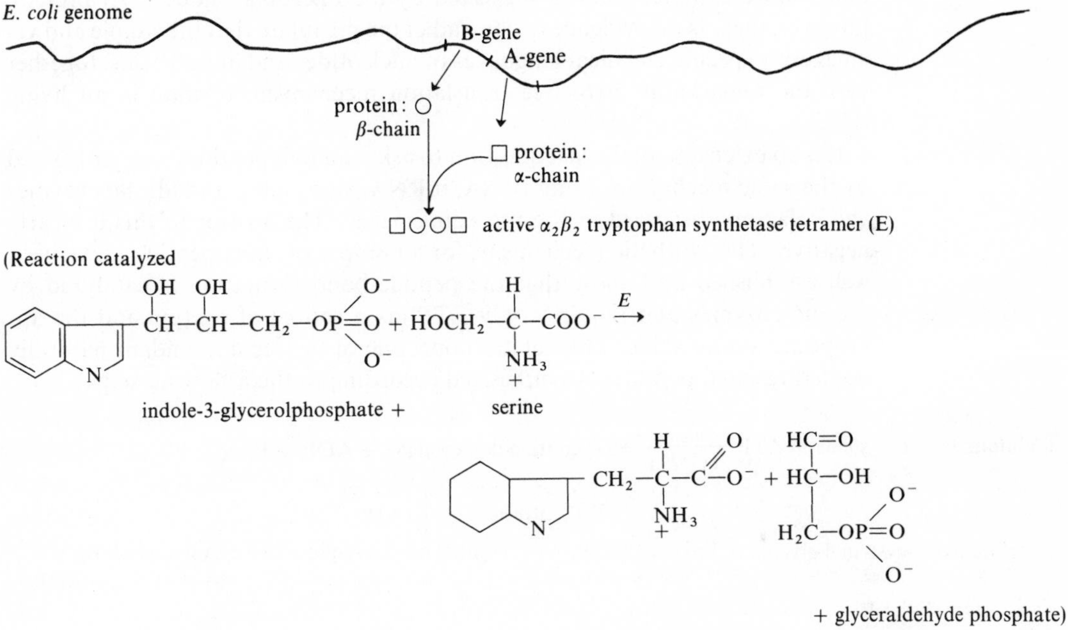

E. coli genome

B-gene

A-gene

protein: ○ β-chain

□ protein: α-chain

□○○□ active $\alpha_2\beta_2$ tryptophan synthetase tetramer (E)

(Reaction catalyzed

indole-3-glycerolphosphate + serine $\xrightarrow{E}$

indole-CH$_2$—CH(NH$_3^+$)—COO$^-$ + HC(=O)—HCOH—H$_2$C—OPO$_3^{2-}$

+ glyceraldehyde phosphate)

Mutant #	*DNA (A-gene)* Distance[b] from starting point in arbitrary units (length of gene = 100)	*Protein Product (α-chain)* Residue position (267 residues for total chain)	Residue change
A 3	18.8	48	glu ⟶ val
A 446	55.8	174	tyr ⟶ cyS
A 487	57	176	leu ⟶ arg
A 223	64.2	182	thr ⟶ ile
A 23	73.4	210	gly ⟶ arg
A 46	73.8	210	gly ⟶ glu
A 187	75.8	212	gly ⟶ val
A 78	87.2	233	gly ⟶ cyS
A 58	87.3	233	gly ⟶ asp
A 169	88.5	234	ser ⟶ leu

[a] From C. Yanofsky, *Scientific American*, **216**, 80 (1967).

[b] Based on fine genetic mapping (frequency of recombination and analysis of deletion mutations).

before a concrete answer can be given to the question of whether or not the genetic code and the mechanism of genetic control of protein synthesis are universal phenomena. The current view is that all the basic features of information storage and transfer and of protein synthesis are common to all living cells. There are aspects of detail that may well differ (the mechanism of chain initiation and termination has already been mentioned as possible variables), and it is quite possible that the complete system in higher animals may well be much more complex than is suggested by the microbial models. At present, however, there is no evidence to contradict the postulate that the simple and yet amazingly specific chemical languages of nucleotides and amino acids, together with the remarkably error-free translation mechanism, operate in all living cells.

It is an extension of the last question to ask if all polypeptides are synthesized by the same mechanism using tRNA, mRNA, ribosomes, and all the enzymes and cofactors discussed in the preceding pages. The answer to this is clearly negative. The synthetic mechanisms for a number of short peptides have been well established and show that the peptide-bond formation is catalyzed by specific enzymes and requires only ATP (as a source of energy) and the appropriate amino acids. Thus glutathione, one of the most abundant naturally occurring small peptides is synthesized according to the following steps:

$$\text{L-glutamate} + \text{L-cysteine} + \text{ATP} \xrightarrow[\text{Mg}^{2+},\ \text{K}^{+}]{\text{enzyme 1}} \text{L-}\gamma\text{-glutamyl–L-cysteine} + \text{ADP} + P_i$$

$$\text{L-}\gamma\text{-glutamyl–L-cysteine} \xrightarrow[\text{ATP}]{\text{enzyme 2}} \text{Enzyme 2–L-}\gamma\text{-glutamyl–L-cysteinyl phosphate} + \text{ADP}$$

$$\text{Enzyme 2–L-}\gamma\text{-glutamyl–L-cysteinyl phosphate} \xrightarrow{\text{glycine}} \text{L-}\gamma\text{-glutamyl–L-cysteinyl-glycine} + \text{Enzyme 2} + P_i$$

More recently a larger polypeptide, the circular decapeptide gramicidin S, an antibiotic produced by microorganisms, has been shown to be assembled by simple activation-condensation steps, again using ATP as energy source.* The enzymes catalyzing the synthesis of gramicidin S were obtained from *Bacillus brevis* in a fraction distinct from amino acyl synthetases. The enzymes were found in the soluble, ribosome-free fraction; addition of tRNA or ribosomes

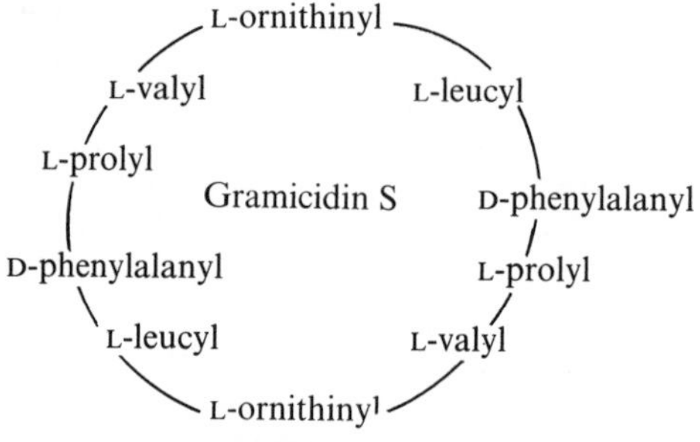

* W. Gevers, *et al.*, *Proc. Nat. Acad. Sci.*, **60**, 269 (1968).

did not affect the rate of synthesis, and the reaction was not inhibited by puromycin. The amino acids are activated by the general reaction

$$\text{amino acid} + \text{ATP} \rightleftharpoons \text{amino acid–AMP} + \text{P} - \text{P}_i$$

and the activated amino acid is passed directly on to an acceptor. The details of the sequence specification, assembly, and cyclization are not known.

REFERENCES

General references in biochemical genetics; research papers; symposia:

Cold Spring Harbor Symp. Quant. Biol., "The Genetic Code," Volume 31, 1966.

Cold Spring Harbor Symp. Quant. Biol., "Replication of DNA in Microorganisms," Volume 33, 1968.

Cold Spring Harbor Symp. Quant. Biol., "The Mechanism of Protein Synthesis," Volume 34, 1969.

San Pietro, A., M. R. Lamberg, and F. T. Kenney (Eds.), *Regulatory Mechanisms for Protein Synthesis in Mammalian Cells*, Academic Press, New York, 1968.

Short and very readable paperbacks:

Hartman, P. E. and S. R. Suskin, *Gene Action*, Prentice-Hall, Inc., Englewood Cliffs, N.J., 1965.

Ingram, V. M., *The Biosynthesis of Macromolecules*, W. A. Benjamin, Inc., New York, 1966.

Stahl, F. W., *The Mechanics of Inheritance*, Prentice-Hall, Inc., Englewood Cliffs, N.J., 1964.

Watson, J. D., *Molecular Biology of the Gene*, 2nd Ed., W. A. Benjamin, Inc., New York, 1971.

Woese, C. R., *The Genetic Code*, Harper and Row Inc., New York, 1967.

Some recent reviews:

Attardi, G. and F. Amaldi, "Structure and Synthesis of Ribosomal RNA," *Ann. Rev. Biochem.*, **39**, 183 (1970).

Howard-Flanders, P., "DNA Repair," *Ann. Rev. Biochem.*, **37**, 175 (1968).

Lark, K. G., "Initiation and Control of DNA Synthesis," *Ann. Rev. Biochem.*, **38**, 569 (1969).

McCarthy, B. J. and R. B. Church, "The Specificity of Molecular Hybridization Reactions," *Ann. Rev. Biochem.*, **39**, 131 (1970).

Osawa, S., "Ribosome Formation and Structure," *Ann. Rev. Biochem.*, **37**, 109 (1968).

General reading:

Clark, B. F. C. and K. A. Marcker, "How Proteins Start," *Scientific American*, **218**, January (1968).

Gorini, L., "Antibiotics and the Genetic Code," *Scientific American*, **214**, April (1966).

Hanawalt, P. C. and R. H. Haynes, "The Repair of DNA," *Scientific American*, **216**, February (1967).

Kornberg, A., "The Synthesis of DNA," *Scientific American*, **219**, October (1968).

Nomura, M., "Ribosomes," *Scientific American*, **221**, October (1969).

EIGHT CONTROL MECHANISMS

It is the aim of this chapter to attempt to show how the regulation (or control) of biological processes can be related to the properties of proteins and nucleic acids discussed in the previous chapters. Since all the chemical reactions that determine each cell's characteristic properties are catalyzed by enzymes, regulation of biological processes becomes a question of how enzymes can be "turned on" and "turned off." Such enzyme regulation is considered to occur at two distinct levels; (1) through regulation of the amount of enzyme (control of enzyme synthesis); and (2) through regulation of the activity of a given quantity of enzyme (control of enzyme activity). The discussion in this chapter will briefly document the existence of biological regulation and then focus on some of the current models and concepts by which control is explained at the different levels. One of the most challenging problems in biological control, cell differentiation, is also introduced.

8.1 MANIFESTATION OF CONTROL

Perhaps the most convincing argument that cell functions are under stringent regulation is the fact that in most normal systems the total mass of all (thousands) of the low-molecular-weight metabolic intermediates makes up less than 1 percent of the dry weight of the cell. This must mean that the cell operates with high efficiency to make its end products and can turn the production off in such a way that none of the intermediates ever accumulate under normal conditions. More specific and direct manifestations of control have been encountered ever since the beginning of biochemical investigation. Thus the Pasteur effect, the observation that anaerobic glycolysis (or fermentation) is inhibited by the presence of oxygen, and its opposite, the Crabtree effect, the inhibition of respiration rate by high glucose levels (and increased glycolysis), are demonstrations of the mutual regulation between two related branches of glucose degradation and represent classical manifestations of control. Hormonal effects in higher animals and plants, although very complex and not well understood, must undoubtedly also reflect regulatory effects on individual enzymes. The demonstration that the addition of a given metabolic intermediate to microorganisms will turn off the entire pathway leading to the synthesis of that intermediate is a more recent accomplishment and led to the concept of *feedback inhibition* (the end product of a pathway inhibits the first enzyme of the pathway). When it was also found that a metabolic intermediate actually could affect the *concentration* of the enzymes involved in the metabolism of that intermediate, it was clear that enzyme synthesis, as well as enzyme activity, had to be considered as a regulatory mechanism.

Perhaps the most dramatic, though indirect, argument for control comes from studies of the abnormal or pathological systems in which control has been lost. Cancer could, for example, be described as cells or tissues in which some of the regulatory elements have ceased to function, causing the natural character of the cells to change and making them continue to grow and multiply in an uncontrolled fashion.

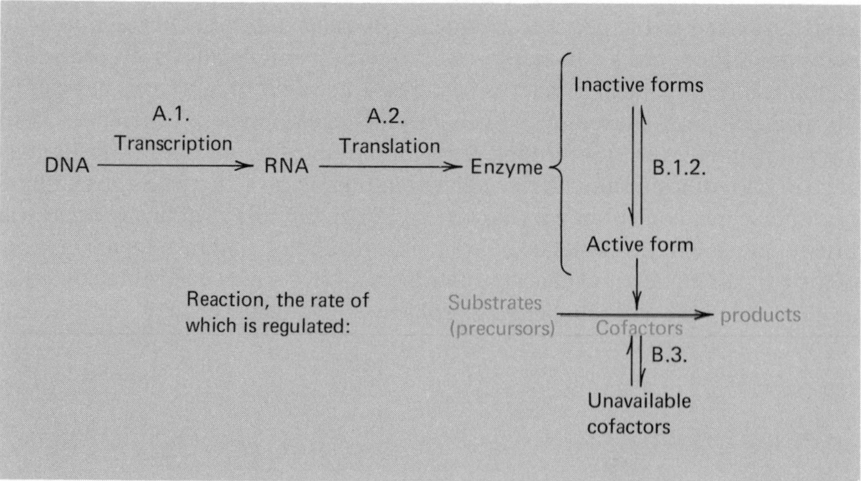

Figure 8.1 A general scheme outlining different points at which the synthesis or the activity of an enzyme can be regulated. It must be emphasized as strongly as possible here that, although only enzymes are discussed in this chapter, everything said about regulation of enzyme synthesis and activity should apply

On the basis of the discussions in the previous chapters, it is easy to arrive at a general postulate for possible regulatory steps; such a postulate is formulated in Figure 8.1 and will serve as a general outline for this chapter. It should be noted that we are concerned only with the direct and specific effect on enzyme quantity and activity. Less specific variables, such as pH, temperature, and ionic strength, could be important in vivo regulatory mechanisms that parallel the well-documented actions on enzymes in vitro. In the absence of much specific information on this point, however, we shall assume that all these factors remain constant in the normal cell. It should also be clear that any fluctuation in substrate concentration should be extremely important in regulating the rate of any one reaction. Since most substrates for one set of enzymes are products of another set of enzymatic reactions, fluctuation in substrate concentrations would tend to link a chain of enzymatic reactions (a metabolic pathway) in such a way that an alteration in one enzyme in the chain could affect the rate of all the reactions. This latter type of regulation, as well as the metabolic significance of all regulation, is discussed in detail by Larner in this series, and will not be stressed here. One final aspect of regulation—the rate of transport of metabolite into and out of cells or different compartments of a single cell—will be briefly mentioned in connection with the barrier properties of membranes, in Chapter 10. It is clear that such transport processes must be extremely important in regulating the levels of substrates, coenzymes, and products in any one cell, and therefore the total biological character of that cell. Since all the information on membrane transport indicates that the molecules responsible for the transport can best be treated as a special kind of enzymes (carrying out translocation instead of catalysis; see Chapter 10), the general outline in Figure 8.1 should in principle apply to transport phenomena as well as to catalysis.

8.2 REGULATION OF ENZYME SYNTHESIS—MANIFESTATION AND MODELS

The concepts of enzyme *induction* (or derepression) and enzyme *repression* are based on the kind of experimental evidence given in Figure 8.2. It has now been clearly established in several cases that a given inducer or repressor will increase or decrease the rate of synthesis of one enzyme or, indeed, of several enzymes involved in the same pathway. It is always necessary to establish that

---

equally well to the other groups of "dynamic proteins" (see Chapter 2) such as regulatory proteins, transport proteins, etc.

A. Regulation of quantity of enzyme (regulation of synthesis):
   1. At the level of transcription
   2. At the level of translation

B. Regulation of quality of enzyme (regulation of activity):
   1. Inhibition and activation by allosteric affectors
   2. Inhibition and activation by another enzyme
   3. Inhibition and activation by removal or addition of cofactors.

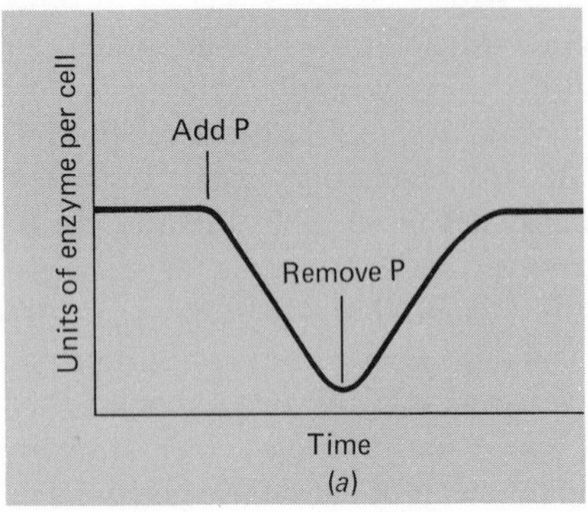

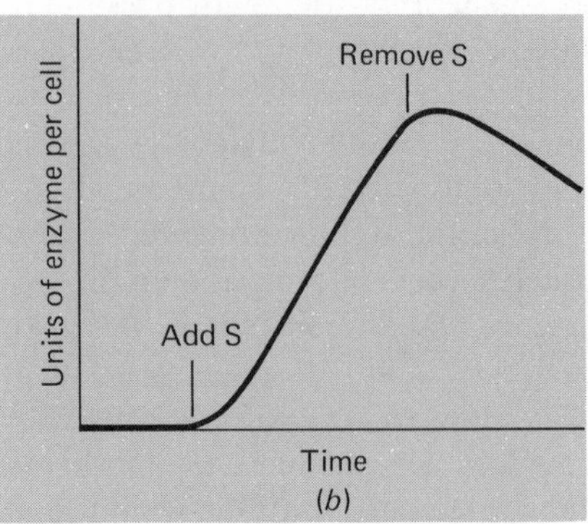

Figure 8.2 The experimental basis for repression and induction (derepression) of the synthesis of enzymes. (*a*) Repression: The effect of an external source of a given metabolite (P) on the cellular concentration of the enzyme responsible for the synthesis of P. The presence of product P *represses* the enzyme, while removal of P restores the original level of the enzyme. (*b*) Induction: The effect of an external source of a metabolite (S) on the cellular concentration of an enzyme which uses S as its substrate. When S is removed, the induced enzyme synthesis stops and the concentration of enzyme returns to the initial low level. In a growing culture this return to low level is simply a dilution of the enzyme due to increased cell mass but, in a resting culture, some degradation of enzyme can also take place. In both the examples above, actual enzyme synthesis can be demonstrated by supplying radioactive amino acids at the point where the enzyme level starts to increase and show that the amino acids were incorporated into the enzyme (isolated) and thus that it was not merely produced by the activation of a preformed inactive polymer precursor. The elimination of the enzyme increases by addition of protein synthesis inhibitors (Table 7.3) is also a valid argument on this point. The general examples in this figure are characteristic of many microbial enzyme systems, but induction of mammalian enzymes has been demonstrated as well.

enzyme synthesis is actually involved in all these cases, since a mere increase in a given activity could also be due to activation of a preformed inactive form of the enzyme. De novo synthesis can be demonstrated in several ways, the most commonly used method being the direct demonstration of incorporation of radioactive amino acid precursors into the product. Supported by such evidence, the induction-repression phenomena are unequivocal demonstrations of regulation at the level of protein synthesis; however, in spite of brilliantly conceived models and ingenious experimentation, the mechanism by which enzyme production is regulated is far from firmly established. A generally accepted hypothesis is the "operator" model illustrated in Figure 8.3. This model very adequately explains the general features of repression and induction of enzymes, either singly or in series, depending on the size of the operon. As originally proposed, and as illustrated in Figure 8.3, the model explicitly defines the regulatory site as part of the chromosome and thus specifies that regulation is affected at the level of transcription. The model predicts the existence of repressors, and a great effort was made to isolate and characterize such repressor substances from repressed cells. This work recently was crowned with success, providing powerful support for the postulated mechanism.*

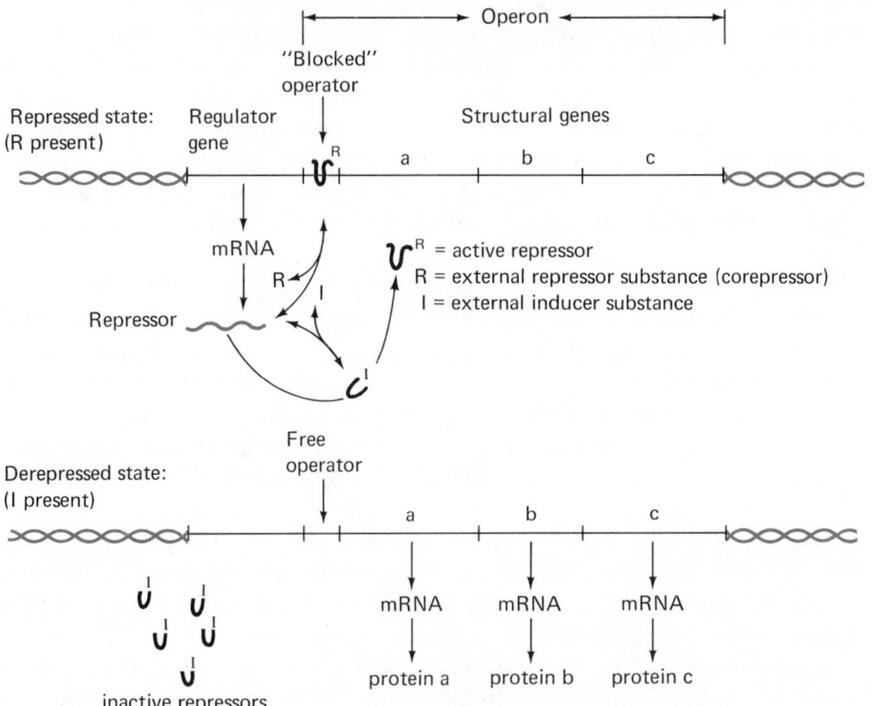

Figure 8.3 The operator model [from Jacob, F. and J. Monod, *J. Mol. Biol.*, **3**, 318 (1961)]. In the simplest interpretation of this model the protein product of the regulator gene is the repressor which is directly involved in the control of the operon. To explain the repression–derepression picture from Figure 8.2, it is easiest to visualize that this repressor recognizes and binds the external effectors of repression and derepression: Thus repressor + external repressor substance, R (P in Figure 8.2), could form the active repressor which blocks the operator, while the repressor + external inducer, I (S in Figure 8.2), could form an inactive form of the repressor, and by mass-action reduce the concentration of active repressor to the point where the operon can be transcribed. The repressor could also be active in the absence of the external repressor substance, and thus be regulated only by inducer.

The experiments demonstrating the existence of a repressor were as follows: The *E. coli* "lac operon" contains three closely linked structural genes, coding for the enzymes thiogalactoside transacetylase, permease, or M-protein (responsible for the transport of galactosides into the cell; see Figure 10.4), and β-galactosidase. The operon is regulated by a regulator gene, designated the *i* gene. The wild-type organism (i^+) characteristically has low levels of the enzymes of the lac operon, but all three enzymes are induced by a number of galactosides, some of which do not act as substrates for the enzymes. One

* W. Gilbert, and B. Muller-Hill, *Proc. Nat. Acad. Sci.*, **56**, 1861 (1966).

mutant (i^-) has an inactive *i* gene and is not inducible; it normally has high levels of all three enzymes, even in the absence of inducer. Another mutant (i^t) has enzyme levels as low as the wild type, but is induced by lower concentrations of inducer (superinducible mutant). After consideration of these three strains, the following postulate was made: The *i* gene regulates the lac operon by producing a repressor that blocks the operon, and the inducer acts by direct combination with and removal of the repressor. Therefore it should be possible to demonstrate the repressor-inducer binding, and it should also follow that if (i^+) has a repressor with a given affinity for inducer, (i^t) should have a repressor with greater affinity for the inducer while (i^-) should have no repressor at all.

Using partially purified cell extracts, the binding of radioactively labeled enzymatically inert inducer, isopropylthiogalactoside, to a component of the extract was sought by equilibrium dialysis. The results were in complete agreement with the predictions. Whereas (i^+) had a component that bound isopropylthiogalactoside with an apparent dissociation constant of 1.3×10^{-6} *M*, (i^t) had a component that bound the inducer with an apparent dissociation constant 6×10^{-7} *M*. No binding component could be found in (i^-). The binding component (repressor) was not affected by treatments with RNase or DNase but was destroyed by proteolytic enzymes. Based on this observation and on the results of sedimentation in sucrose-density gradients, it was concluded that the lac operon repressor in *E. coli* is a protein having a molecular weight of 1 to 2×10^5.

It is interesting to note, however, that the whole concept of the operator model could equally well apply at the level of translation. The only difference would be that the operator would become transcribed together with the structural genes to give an mRNA, which, in addition to the code for the polypeptide chains, would also contain the transcribed operator. When this mRNA operator is associated with repressor, translation of the message would be blocked, and when it is derepressed, translation of the whole message could proceed. In either case the manner in which the operator is blocked and the mechanism by which this blockage prevents either transcription or translation of the information in the structural genes remain obscure.

Several models have been proposed to specifically allow for regulation at the translational level. These proposals arose from a growing volume of information that is difficult to reconcile with the transcriptional operator model and is more in line with regulation at the translational level. Thus histidyl-tRNA is more effective than is histidine in repressing the his operon in Salmonella; regulation of protein synthesis can still be observed in cells whose transcription (mRNA synthesis) is blocked, and genes that are located on distant parts of the chromosome have been found to be under simultaneous control, which, with the operator model, would mean that each would have to have identical operators.

One of the translational regulation models is based on the discovery that partially assembled polypeptide chains on the ribosome have the ability to recognize both their substrates and their future partners of protein subunits.

This makes it reasonable to postulate interaction of the nascent protein with specific regulatory proteins on the polysome, which, in turn, could affect the rate of synthesis of the growing chain.* There is no doubt that this model has some very attractive features. The regulatory protein could be the regulatory subunit that is to become part of the complete multisubunit enzyme; this would neatly tie together the mechanisms of regulation of synthesis and regulation of activity.

Another case that is best handled by translational control is the regulation of synthesis of phage protein after infection with RNA phage. A strict adherence to the transcription model would in this case, require a RNA ⟶ DNA ⟶ RNA cycle, which has no experimental basis in phage-infected bacteria. On the other hand, a recent study has demonstrated one case of translational repression of synthesis of some of the enzymes coded by the bacteriophage f2 RNA genome. The repressor was in this case found to be another protein from the same genome, the coat protein (Figure 8.4). This appears to provide a strong argument for the model of translational control by a protein that is destined to become associated with the other products of the repressed gene (in this case as part of the same virus capsid).

The discussion until now has been limited to microorganisms. It is generally assumed that in principle the same arguments should hold for higher organisms as well. It should be recalled, however, that the procaryotic and eucaryotic chromosomes are distinctly different. One significant difference is that the DNA of the eucaryotic cells is strongly associated with histones. Some recent studies on regulation of protein synthesis in plants have implicated histones as the actual repressor molecules† and have thus lent support to the operator model and the idea of transcriptional control. The experimental results suggest that specific genes are repressed by histones, but only in the presence of a specific form of nuclear RNA, which presumably determines the specificity of the repression.

It seems that at present the regulation of protein synthesis is a well-documented phenomenon, but the mechanisms by which it occurs are not yet elucidated. Most of the models proposed appear restrictive and, implicitly, mutually exclusive. It seems most likely that when more answers are available, all the postulated models will be found to have validity in different cells under different conditions.

We cannot leave this section without mentioning an interesting protein discovered in higher animals in the late 1950's, named *interferon*. This acidic protein, with a molecular weight of about 30,000, has many of the prerequisites of a repressor and represents a naturally occurring regulator of gene expression. Interferon is produced by cells in response to virus infection and can be transferred to neighboring cells; its presence can render all cells resistant to infection by a very broad spectrum of both RNA and DNA viruses. Clearly, this protein is part of an organism's defense mechanism. All the evidence indicates that

* See A. L. Cline, and R. N. Bock, *Cold Spring Harbor Symp. Quant. Biol.*, **31**, 321 (1966).

† J. Bonner, *The Molecular Biology of Development*, Oxford University Press, New York (1965).

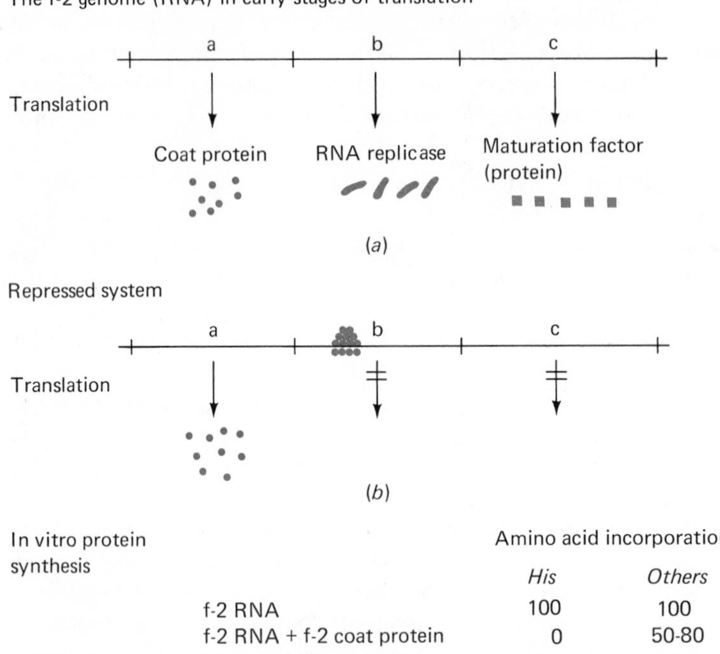

Figure 8.4 A case of repression at the level of translation. The system studied was the RNA virus f2. The RNA genome of this virus codes for at least three different proteins, coat protein, RNA replicase, and maturation factor (m.f.). The coat protein is needed in large quantities to make up the virus capsid, the replicase in small quantities to catalyze the synthesis of viral RNA, and the m.f. in small quantities to produce the completely assembled active virus particle. The coat protein contains no histidine, and incorporation of histidine thus provides a direct method for the assay of replicase and m.f. synthesis. (*a*) A representation of the viral genome with the established sequence of the three genes, and with the three protein products. (*b*) A representation of the repressed genome, proposing that the coat protein combines with the RNA at a point on the first half of the replicase gene, and thus represses all translation beyond that point. (*c*) *In vitro* test to demonstrate that synthesis of replicase and m.f. is inhibited (no histidine incorporation) in the presence of added f2 coat protein while protein synthesis is still going on (other amino acids are incorporated) presumably in the translation of the coat protein gene. (Other virus coat proteins were also tested as repressors at the same concentration; HS2 and f2 coat proteins both gave 100 percent repression, Qβ-coat protein gave 13 percent repression, and tobacco mosaic virus (TMV) coat protein had no effect. Knowing that HS2 and f2 are closely related viruses, while Qβ, another RNA coliphage, is quite different from the other two, as is the plant virus TMV, this finding demonstrates the specificity of the repression.) In an attempt to demonstrate the direct interaction of protein and RNA, the unrepressed genome (a in the figure) and the repressed genome (b in the figure) were digested with ribonuclease. The repressed genome was found to contain a small but consistent fragment of RNA (about 0.5 percent) which was resistant to digestion, presumably because of the tight protein "coat" located at this fragment. The utility of this type of repression is quite obvious. In the early

interferon exerts its effect specifically on the expression of viral genomes without interfering with the expression of the host's genome. Recent experiments in vitro have given evidence that interferon affects the translation step, presumably by inducing the synthesis of a second protein, a *translation-inhibitor protein* which binds to ribosomes and inhibits translation of viral RNA (or of mRNA derived from viral DNA). The effect must be very specific since host mRNA translation can still take place.* Elucidation of the mechanism of interferon action should provide further insight into the regulation of gene expression.

8.3 REGULATION OF ENZYME ACTIVITY

As indicated in Figure 8.1 we will consider the three different types of regulation manifested by (1) the allosteric enzymes; (2) the enzyme-activated (and enzyme-inhibited) enzymes; and (3) the general phenomenon of shared cofactors.

Allosteric Enzymes

Allosteric (Gr. *allos*: another; *stereos*: solid) should mean in chemical terms another structure, but what it actually means biochemically depends on whose paper or book one reads. If for the sake of this discussion we define an allosteric enzyme in a broad sense as an enzyme with two (or more) topologically distinct sites—either interacting catalytic sites or a catalytic site and a regulatory (effector) site—we have a useful concept in general agreement with common usage. The fundamental properties and some of the terminology of allosteric enzymes as thus defined are summarized in Figure 8.5; reference has also been made to such systems in Chapters 2 and 3. Regulation can be affected by like molecules (substrate) binding to interacting catalytic sites (homotropic effects) or by different molecules (effector and substrate) binding to interacting sites. In both cases the binding to the first site can affect the binding to the second site either positively (activation) or negatively (inhibition). With homotropic effectors, activation is referred to as positive cooperativity, inhibition as negative cooperativity.

The characteristic sigmoidal activity-substrate curve† (see Chapters 2 and 3) makes it clear that the inhibition or activation is important only at lower substrate concentrations; at substrate saturation there is no effect (V_{max} is constant

* P. I. Marcus, and J. M. Salls, *Cold Spring Harbor Symp. Quant. Biol.*, **31**, 335 (1966).

† It must be emphasized that sigmoidal rate curves (and binding curves) can be obtained in systems that are not allosteric. The right combination of rate constants and simple isosteric competitive inhibitors can give rise to nonhyperbolic curves.

stages of protein synthesis all three genes are translated. Later the replicase and m.f. genes are turned off by the coat protein while RNA synthesis and coat protein synthesis continues until the quantities needed for the complete virus have been produced. [The data are taken from Ward, R., *Fed. Proc.*, **27**, 645 (1968).]

Overall reaction: $E + S \rightleftharpoons ES$
$ES + S \rightleftharpoons ES_2$

		Preservation of Symmetry	Cooperativity	Reference
A. $K_1 = K_2$	$\xrightleftharpoons{K_1}$ S $\xrightleftharpoons{K_2}$ S S	no	no	1
B. $K_1 > K_2$	$\xrightleftharpoons{K_1}$ S $\xrightleftharpoons{K_2}$ S S	yes	yes: +	2
C. $K_1 > K_2$	$\xrightleftharpoons{K_1}$ S $\xrightleftharpoons{K_2}$ S S	no	yes: +	2
D. $K_1 < K_2$	$\xrightleftharpoons{K_1}$ S $\xrightleftharpoons{K_2}$ S S	no	yes: −	3

(*a*)

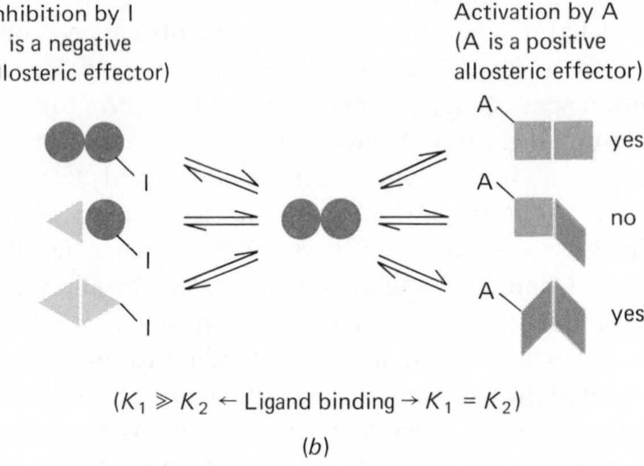

($K_1 \gg K_2$ ← Ligand binding → $K_1 = K_2$)

(*b*)

Figure 8.5 Allosteric protein models. (*a*) Homotropic effect: A model for a simple two subunit protein (E) with two ligand (S) sites. The circles, triangles, squares, and diamonds represent different conformations of the protein *protomers* [identical subunits are referred to as *protomers*, the protein in (*a*) has 2 protomers]. All possible types of protomer interaction (and binding site interaction) are considered in the figure: no interaction in A, two cases of positive cooperativity in B and C, and negative cooperativity in D. In the scheme the triangles are conformers with low affinity for ligand, the circles intermediate affinity, and the squares and the diamonds have high affinity for the ligand. (In the general model squares and diamonds could have the same ligand affinity without having the same conformation, but the affinities could also be different as long as both are greater than that of the circles.) The degree to which the binding of the first ligand affects the binding of the second (the relative values of the *dissociation* constant K_1 and K_2 in the overall reaction) thus reflects whether the protomers have no interaction, give positive or negative cooperativity. (*b*) Heterotropic effects: Using the same simple system from (*a*) and the same conformational symbols, allosteric activators (A) or inhibitors

but K_m is affected by the regulator, or effector, molecule). Since in vivo substrate concentrations normally must be far from saturation levels, this is exactly the kind of regulation required to maintain an efficient metabolic system. The very important aspect of this type of regulation is the fact that, as in induction and repression, the effectors of any one enzyme are intimately connected with the metabolic pathway of which the affected enzyme is a part. Thus the negative effectors are generally end products of the pathway, and the regulated enzyme is often, if not always, one of the first enzymes of the pathway. Similarly the positive effectors are most often the substrate itself or a cofactor. Thus one can visualize a mechanism by which the relative concentration of starting material and end product will turn on and off a whole pathway in feedback regulation.

One interesting aspect of these multisite enzyme systems is related to an often-asked question: Does the effector always simply bind to the effector site, or could the effector site also be a catalytic site carrying out a different catalytic function? There are now some examples in which both sites have been shown to have catalytic activity but with different substrates, and in which each substrate inhibits the other catalytic site. These are cases of multienzyme complexes with mutual regulation. Specific examples are phosphorylase-glutamate-pyruvate transaminase* and homoserine dehydrogenase-aspartokinase.†

The concept of allosteric enzymes as well as of other allosteric proteins is now firmly established, and this concept has given an enormous impact to our thinking in the area of control of enzyme activity. The molecular mechanisms by which the allosteric sites interact are, however, not on equally firm ground. Some very sophisticated models have been proposed both at the structural and mathematical levels, and, because two of these models have become such an important part of biochemical discussion it is necessary to consider them in some detail. The models in question are the symmetry model‡ and the sequential model.§ Both are firmly based on the general concept of flexible protein conformation, and in fact there is only one major point of difference between them.

* G. Bailin, and A. Lukton, *Biochim. Biophys. Acta.*, **128**, 317 (1966).

† P. Truffa-Bachi, *et al.*, *Biochim. Biophys. Acta*, **128**, 440 (1966).

‡ J. Monod, *et al.*, *J. Mol. Biol.*, **12**, 88 (1965).

§ D. E. Koshland, Jr., and K. E. Neet, *Ann. Revs. Biochem.*, **37**, 359 (1968).

(I) can be represented as affecting the affinity for S by inducing or stabilizing the proper high affinity or low affinity conformations. The models are represented both with and without preservation of symmetry. In the case of inhibition the sigmoidicity of the ligand binding curve is accentuated: in the case of activation it is eliminated to give the normal hyperbolic binding curve. The two-subunit model presented here was used for the sake of easy illustration. The actual cases which have been well characterized generally are more complex, and would display the experimental characteristics more clearly than would this simple type of protein. The references refer to some actual cases which could in principle be explained by the individual models: (1) Rabbit muscle enolase: Cardenas, J. M. and F. Wold, *Biochem.* **7**, 2736 (1968). (2) Aspartate transcarbamylase (see Chapter 3): Gerhart, J. C. *et al.*, *Biochem.* **7**, 531, 738 (1968). (3) Triose phosphate dehydrogenase: Conway, A. and D. E. Koshland, Jr., *Biochem.* **7**, 4011 (1968).

To illustrate this difference let us consider the simplest allosteric protein, namely one containing two identical subunits, each with one ligand site. The two sites interact in such a way that when the ligand binds to the first site, the affinity of the second site for the ligand increases (this is the case of positive cooperativity). The symmetry model specifies that molecular symmetry must *always* be retained in the molecule, and therefore that any conformational change in the first subunit associated with the binding of the ligand to the first site is immediately mimicked in the second subunit, thereby inducing the higher-affinity conformation in the second site. Thus the symmetry model requires that one never has two subunits in different (asymmetric) conformations in the same molecule; if one changes, the other changes too. The sequential model, however, considers that different types of site interactions can occur; one that leads to identical changes in both subunits (symmetry preservation) is a special case. But with the sequential model it is also possible to consider a broad spectrum of other interactions. The general case of such a two-subunit protein is presented in Figure 8.5, and includes all possible interactions between sites, from positive cooperativity ($K_1 > K_2$; B and C in Figure 8.5), via no interaction ($K_1 = K_2$; A in Table 8.3) to negative cooperativity ($K_1 < K_2$; D in Figure 8.5). This representation is based on a sequential model, and only in one case, B, are the symmetry requirements satisfied. There is no simple way of distinguishing steps B and C experimentally, so in the case of positive cooperativity (and also in the case of no cooperativity), both models can apply. There is, however, one situation that appears to be explained only by a sequential model, namely the negative cooperativity in D. Since one case of negative cooperativity has now been well documented, the case of NAD binding to triosephosphate dehydrogenase (Figure 8.5), it is clear that the symmetry model cannot have universal validity. The allosteric enzymes with separate effector sites (see aspartate transcarbamylase in Chapter 3) have furnished the most convincing support for the symmetry model, but it appears that these systems can also be handled satisfactorily by a sequential model.

Enzyme-Regulated Enzymes

In this category we are primarily concerned with readily reversible processes and will therefore not include the type of unidirectional activation involved in the many proenzyme-to-enzyme (or zymogen-to-enzyme) conversions. There are, however, enzymes that are reversibly activated and inhibited by the action of specific enzymes in a manner best described as an amplification or cascade effect. The very exciting feature of this type of regulation is the amplification of a minute input concentration of effector to give very large output effects at the critical step. The best-documented cases of this type of regulation are the actions of the enzymes responsible for the interconversion of glucose and glycogen. Figure 8.6 illustrates the phosphorylase system. The growing interest in this type of regulatory mechanism is largely based on its potential for explaining the effects of regulatory substances, both natural and foreign, that

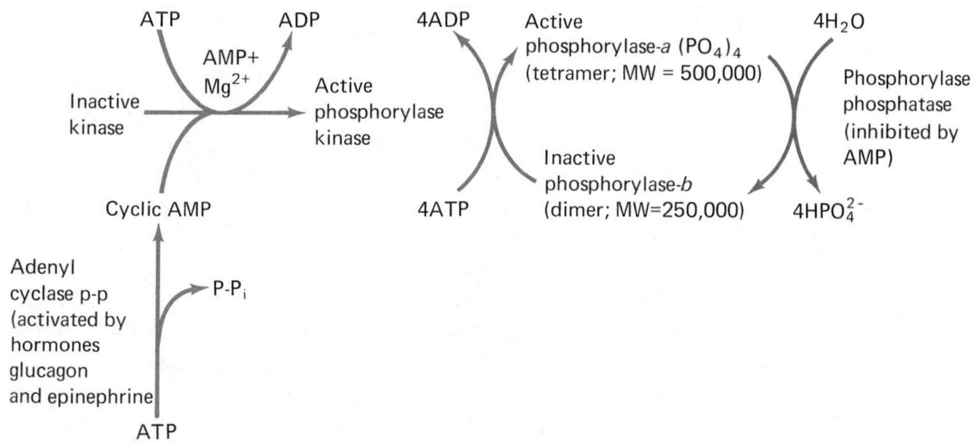

Figure 8.6 Enzyme-regulated enzymes: the phosphorylase system. The activation and deactivation of phosphorylase [catalyzes the phosphorolysis of glycogen: (glucose)$_n$ + $P_i \rightleftharpoons$ (glucose)$_{n-1}$ + glucose 1 − P)] by a kinase and a phosphatase, respectively. The effect of cyclic AMP in activating the kinase is perhaps a characteristic step in the series of events through which hormones can affect enzyme systems (see Figure 4.10). It is of interest to note that a similar phosphorylation–dephosphorylation process for activation and deactivation has been found for several enzyme systems [pyruvate dehydrogenase (Chapter 3) is one of them].

have dramatic physiological effects at extremely low concentrations. Hormones (see Chapter 4) could fit into this category (some insect hormones have been estimated to be detectable and to trigger their characteristic chain of events at the level of one molecule per organism), as could some toxins and drugs.

Shared Cofactors (or Substrates)

The type of regulation involved in shared cofactors is the least specific one and probably is more akin to the general effect of substrates and products on the rate of any enzymatic process. It is of interest to include this type of regulation in the discussion, however, just to emphasize the fact that some of the common coenzymes are obligatory reactants in a very large number of widely different reactions. Any increase in the activity of one reaction could therefore deplete the coenzyme pool and thereby affect the rate of several other reactions. The rate of coenzyme transport from one compartment or cell to another would also furnish strong regulatory effects in this category. Coenzyme regulation is most often accentuated by the mechanisms listed above, in which the coenzyme is an allosteric effector or is involved in cascade-type reactions.

8.4 SPECIAL PROBLEMS OF REGULATION

Having tried in the previous chapters to present a plausible case for a well-defined and well-understood sequence of events in the transfer of information to functional molecules, it would be unfair indeed to omit some problems that are at present completely unsolved.

One of the great challenges of biochemistry today is to explain the molecular mechanisms of the changes that occur during the development of the fertilized egg into the larger number of specialized cells and tissues that make up higher organisms. The process is referred to as differentiation. In the classical biochemical sense, the experimental approach to this problem is severely limited by the fact that the smallest isolatable unit available for study is the developing cell itself. A few facts have been established. It is, for example, clear that the total genetic material (information) does not change during the process of differentiation, and it has also been shown that each step in the series of changes is irreversible. Differentiation is therefore an excellent and very complex example of the phenomenon discussed in the beginning of this chapter, the regulation of gene expression. The original fertilized ovum apparently contains the total information for all the different specialized daughter cells, but only a small fraction of all the genes are actively expressed at any stage of development. During the first several divisions, the daughter cells have essentially all the characters of the parent cell. During this early state, a nucleus can be transferred from any of the daughters to an enucleated, unfertilized ovum, which will then start dividing like a normal fertilized ovum, indicating that no irreversible change in genetic expression has taken place. Suddenly the small cluster of several thousand cells undergoes the first change into early embryonic orders; the present hypothesis is that new genes, which are different for the different cell types are now expressed. If a nucleus is transferred to the unfertilized ovum at this stage, it does not allow normal growth and division, indicating that the original gene expression has been irreversibly lost. Each successive change from each type of cell presumably follows the same pattern of permanent inactivation of previously expressed genes and activation of new ones. The whole process thus looks just like the repression-derepression phenomenon discussed above, with the one added feature that when the derepressed genes of one stage of differentiation are rerepressed in going to the next change, they remain permanently repressed. In spite of much good work in this area, virtually nothing is known about the substances involved or the mechanisms by which the right cell appears at the right place at just the right time for normal embryonic development. It is most likely, however, that with the recent developments in molecular genetics and in experimental techniques applied to these problems, very exciting and extremely important discoveries will be made during the next few years.

The synthesis of antibodies in response to a foreign substance, the antigen (see Chapter 4), can be considered a special case of gene regulation and may well provide an excellent model system for the study of differentiation.

A. Alternate views:

1. On information content:	(a) All antibody-forming cells have *all* the information needed to make any possible antibody	(b) Different antibody-forming cells have different information content (DNA) A given antigen can elicit antibody production only in a certain specific type of cell
2. On the trigger mechanism: (the immune response)	(a) The antigen is the direct trigger (or template)of antibody production and must be present in the antibody-synthesizing cell	(b) The antigen activates a "secondary" trigger (RNA) which is the actual template for antibody synthesis

B. Possible sequence of events in the cell transformation (differentiation) leading to specific antibody production:

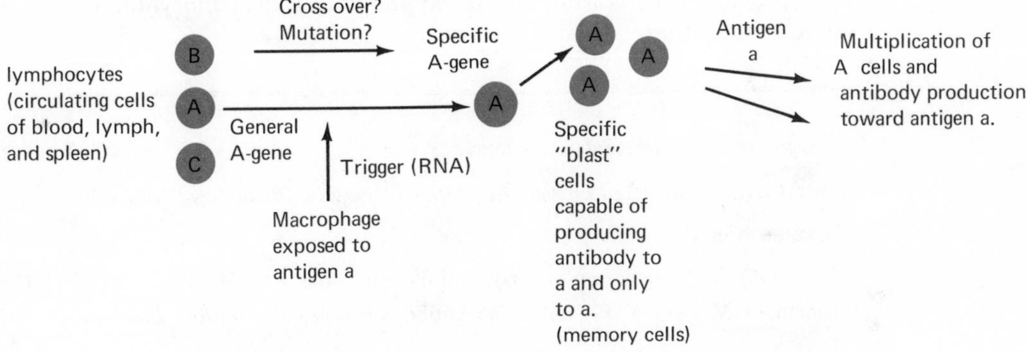

Figure 8.7 Antibody Synthesis: A. Alternate models. The (b) alternatives appear to be the most satisfactory ones according to current views, and the proposal in B is based primarily on those models. [See Lennox, E. L., and M. Cohen, *Ann. Revs. Biochem.*, **36**, 365 (1967).]

The fundamental problem of antibody synthesis is that of determining the mechanism by which the antigen triggers both the "induction" of the specificity of a given antibody and the increased rate of synthesis of this antibody. Certain features of antibody synthesis have been established and are summarized in Figure 8.7. It is, for example, quite clear that the antigen itself does not come into contact with the cell in which the antibody is synthesized. The best antigens are insoluble macromolecules that are first phagocytized by macrophages. In that process the macrophage appears to acquire a new activity that it can transfer to the lymphocyte, thus transforming the lymphocyte to a specific antibody-producing plasmacell. It is highly significant that the inducing activity can be destroyed by treatments with pancreatic ribonuclease. Figure 8.7 also outlines some theories that have been advanced to explain some of the problems and paradoxes encountered in trying to state the basic questions. Is all the genetic information that is needed for the synthesis of antibodies toward all possible natural and artificial antigens present in all the antibody-synthesizing cells, and does the antigen-induced derepressor simply derepress the proper specific gene? On the other extreme, is there only one fundamental antibody gene, which, on response to the antigen inducer, can undergo changes (in the variable region only) by crossover and by selective mutations to become a new specific gene for the synthesis of the desired antibody? The idea of specific mutations is consistent with the fact that most of the amino acid replacements observed in the variable region of γ-globulins can be accounted for by single base changes. However, the high rate of mutation this would require is not easy to reconcile with all the other indications that DNA is very stable. Whatever extreme is correct (and very likely the truth will be found somewhere between the two), the problem of the mechanism for the differentiation of the lymphocyte to give the antibody-producing plasma cell, and the antigen's role in inducing the specific antibody, remain unsolved. However, the tremendous progress in the structural analysis of immune globulins that has been made over the last several years provides a basis from which the problem of their biosynthesis may be solved in the future.

REFERENCES

General (multivolume) reference source:

Weber, G. (Ed.), *Advanced Enzyme Regulation*, Pergamon Press, New York, 1963.

Recent reviews:

Atkinson, D. E., "Regulation of Enzyme Function," *Ann. Rev. Microbiol.*, **23**, 47 (1969).

Edelman, G. M., and W. E. Gall, "The Antibody Problem," *Ann. Rev. Biochem.*, **38**, 415 (1969).

Gross, P. R., "Biochemistry of Differentiation," *Ann. Rev. Biochem.*, **37**, 631 (1968).

Koshland, D. E., Jr., and K. E. Neet, "The Catalytic and Regulatory Properties of Enzymes," *Ann. Rev. Biochem.*, **37**, 359 (1968).

Other general reading:

Changeux, J. P., "The Control of Biochemical Reactions," *Scientific American*, **212**, April (1965).

NINE

MEMBRANES: SYSTEMS AND METHODS

This chapter introduces the third group of ubiquitous molecules, the molecular complexes that make up the membranes of all living cells (and even some viruses). Some of the unique problems associated with the study of membranes are considered. The various types of membranes are introduced and the methods used in the determination of both structure and function are outlined. Some properties of common biological and artificial membranes are tabulated.

The study of any kind of complex of biological macromolecules is fraught with difficulties and uncertainties, some of which have already been alluded to in preceding chapters. The most obvious and immediate problems are encountered in formulating the chemical description of the complex. The first prerequisite to meaningful chemical characterization is some degree of homogeneity in the system to be characterized, and even when a homogeneous complex product is obtained, it must be demonstrated that the product is not an artifact of the purification procedure. This problem is particularly pertinent to the study of membranes, as shall be seen below. If this first problem is overcome and a molecular complex has been obtained in a reproducible manner, the chemical composition (content of amino acids, fatty acids, carbohydrates, phosphate, sulfate, and so forth) can be determined without much difficulty. This corresponds to the first level of structural information and, in the case of membranes, is generally the extent of well-established information available at this time. The further levels of structural determination leading to a complete three-dimensional description of the molecular architecture of the membrane have all the difficulties discussed for proteins and nucleic acids, but these are in the case of the membrane multiplied by some "complexity factor," which undoubtedly is an exponential function of the number of components in the structure studied. The distinction between native and denatured structures also becomes much more difficult and perhaps of more real significance with the multimolecular complexes. This is because it is virtually impossible at present to define the total in vivo function of these complexes and because it generally is difficult to set up in vitro activity assays; any denaturation process therefore becomes a very arbitrary designation of change in either structure or function.

9.1 MEMBRANE TYPES

The group of cellular components referred to as membranes includes readily observable structures found in different forms in all cells, either surrounding the cell (the cytoplasmic or plasma membrane), surrounding the nucleus (the nuclear membrane), or as individual structures in the cytoplasm (the mitochondrial membrane, the endoplasmic reticulum, and so forth). Specialized cells and tissues are also surrounded by membrane structures of more unique character (such as the myelin sheath surrounding nerve cells and the membrane of striated muscle). All cells have at least one kind of membrane—the plasma membrane.

9.2 STRUCTURE

All membranes can broadly be described as lipoprotein complexes containing about 60–75 percent protein and 25–40 percent lipid. The various proteins and lipids making up the intact membrane are held together in the proper architectural arrangement by noncovalent forces. At this descriptive level of definition all membranes are quite uniform throughout the biological world. On closer examination, however, it becomes obvious that very significant and unique differences exist both between membranes within a given cell and between membranes from different cells.

This type of comparison has only become possible quite recently as a result of new advances in analytical procedures that permit (1) the isolation of large quantities of what appear to be real and well-defined membranes and (2) an accurate determination of their chemical constituents. Some typical data are given in Table 9.1. At present it appears that phospholipids constitute the major lipid component in all species, and, in fact, the only lipid-type present in all membranes studied. Other noteworthy features of the lipid composition are the absence of sterols in bacteria, the apparently unique presence of cerebrosides in myelin, and the occurrence of unusual lipids in bacteria (not given in Table 9.1). These unusual lipids include amino acyl–phosphatidyl glycerol, glycosyl lipids, and phosphatides containing unusual fatty acids such as branched chain and cyclopropane acids.

The protein components also appear to vary considerably (Table 9.1), although persistent solubility problems have complicated the complete separation and characterization of all proteins. It seems reasonable to propose that the characteristic functions of each membrane are reflected by its complement of protein components. Many of the different kinds of membrane proteins are enzymes whose catalytic properties are often well characterized. It is significant that a number of the known membrane enzymes require the presence of phospholipids or other typical membrane lipids for activity after isolation. It is also important to bear in mind that specific enzyme components can be used as markers in the isolation of membranes.

There is one protein that appears to be a common component of all membranes, however—the so-called "structural protein." This protein was first obtained from mitochondria a few years ago by extraction with detergents followed by salt fractionation and treatment with organic solvents. The final product accounted for about 50 percent of the total mitochondrial protein. It was found to be surprisingly homogeneous by a number of criteria, quite insoluble in aqueous solutions, and completely inactive when tested for normal mitochondrial-enzyme activities. Based on these facts, it was assigned the role of structural protein. There is clearly an interesting, almost philosophical problem raised here. Treating the membrane with reagents, any one of which could cause a classical type of irreversible denaturation, and obtaining a product with the typical earmarks of a denatured protein (no biological activity and low solubility) are questionable criteria for the postulation of a noncatalytic, structural role for this protein. The absence of enzymatic activity is actually a rather ambiguous concept in itself: Just how many activities must be tested, before such a statement can have any meaning? Much more meaningful to the idea of a structural protein is the fact that this product appeared to be homogeneous and was found in such large quantities. Subsequent work has added further evidence supporting the structural-protein idea in showing that the same protein component (immunochemically cross-reacting) can be isolated in high yield from a large number of different membranes, indicating a ubiquitous role for such a protein. Based on the work to date, it appears safe to conclude that all membranes contain a protein component with the following general characteristics: (1) a subunit molecular weight (solubilized in detergent) of about 20,000; (2) aggregation in the absence of detergents to form insoluble polymers; (3) homogeneity and similarity—if not identity—in different membranes, as indicated by sedimentation behavior, N-terminal analysis, and peptide mapping; and (4) high and specific binding affinity for membrane lipids and other membrane proteins. This last characteristic is perhaps the most significant

TABLE 9.1 CHEMICAL COMPOSITION OF MEMBRANES

	Animal				Bacteria			
Lipids[a] *(Given in percent)*	Myelin	Erythrocyte	Mitochondria	Microsome	*Azotobacter agilis*	*E. coli*	*Agrobacterium tumefaciens*	*Bacillus megaterium*
Cholesterol	25	25	5	6	0	0	0	0
Phosphatides								
Phosphatidyl ethanolamine	14	20	28	17	100	100	90	45
Phosphatidyl serine	7	11	0	0	0	0	0	0
Phosphatidyl choline	11	23	48	64	0	0	10	0
Phosphatidyl inositol	0	2	8	11	0	0	0	0
Phosphatidyl glycerol	0	0	1	2	0	0	0	45
Cardiolipin	0	0	11	0	0	0	0	0
Sphingolipids								
Sphingomyelin	6	18	0	0	0	0	0	0
Ceramide	1	0	0	0	0	0	0	0
Cerebrosides	25	0	0	0	0	0	0	0
Others	11	1	0	0	0	0	0	10

Proteins (enzymes identified with different membranes, transport proteins not included)

Myelin	None
Erythrocyte	Diphosphopyridine nucleotidase, adenosine triphosphatase, glucose-6-phosphatase, acetylcholine esterase, carbonic anhydrase, carnitine palmityl CoA transferase, lecithin synthetase, glycolytic enzymes
Liver-plasma membrane	Adenosine triphosphatase, phosphatase, amidase, phosphodiesterase, 5′-nucleotidase
Mitochondria	Electron-transport chain (cytochromes, quinones, and flavoproteins), tricarboxylic acid–cycle enzymes, fatty acid–oxidation enzymes, transaminases, monoamine oxidase, pyruvate carboxylase, phosphoenolpyruvate carboxylase, nuclease, glutamate dehydrogenase, fatty acyl–CoA synthetase, Mg-activated adenoside triphosphatase, porphyrin-synthesis enzymes and urea-cycle enzymes
Bacteria (gram-positive)	Adenosine triphosphatase, cytochrome oxidase, reduced diphosphopyridine nucleotide oxidase, succinate dehydrogenase, malate dehydrogenase, lactate dehydrogenase, α-ketoglutarate dehydrogenase, malic enzyme, acid phosphatase, α-glycero-phosphatase

TABLE 9.1 ***(Cont.)***

Some Relevant Lipid Structures (Note that each compound listed is a family of compounds depending on the different R groups of the fatty acids)

Phosphatides

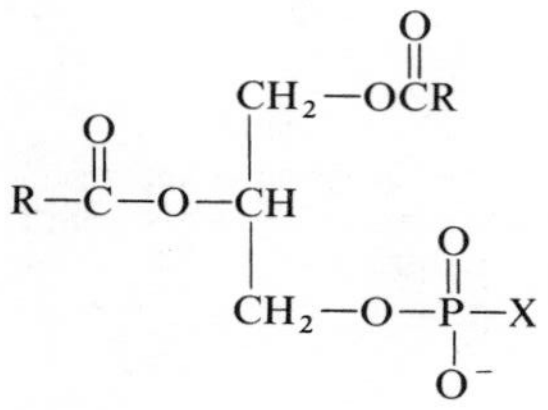

$$R{-}\overset{O}{\overset{\|}{C}}{-}O{-}CH(CH_2{-}O\overset{O}{\overset{\|}{C}}R)(CH_2{-}O{-}\overset{O}{\overset{\|}{P}}(O^-){-}X)$$

$X = -OCH_2CH_2-\overset{+}{N}H_3$	Phosphatidyl ethanolamine
$X = -OCH_2-CH(\overset{+}{N}H_3)-COO^-$	Phosphatidyl serine
$X = -O-CH_2CH_2-\overset{+}{N}(CH_3)_3$	Phosphatidyl choline
$X =$ 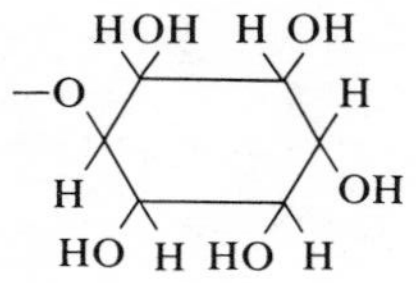 inositol ring, $-O$ attached (ring labels: H OH, H OH, H, OH, H, HO H, HO H)	Phosphatidyl inositol
$X = -OCH_2CHOHCH_2OH$	Phosphatidyl glycerol

Cardiolipins

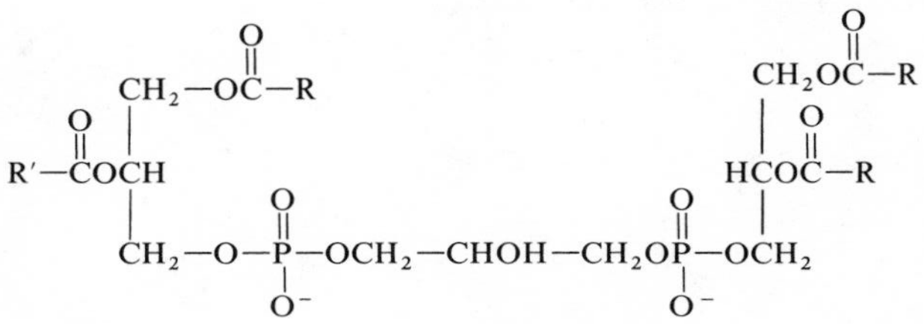

$$R'{-}\overset{O}{\overset{\|}{C}}OCH(CH_2{-}O\overset{O}{\overset{\|}{C}}{-}R){-}CH_2{-}O{-}\overset{O}{\overset{\|}{P}}(O^-){-}OCH_2{-}CHOH{-}CH_2O\overset{O}{\overset{\|}{P}}(O^-){-}OCH_2{-}HCO\overset{O}{\overset{\|}{C}}{-}R(CH_2O\overset{O}{\overset{\|}{C}}{-}R)$$

diphosphatidyl glycerol

TABLE 9.1 CHEMICAL COMPOSITION OF MEMBRANES *(Cont.)*

Sphingolipids

$$CH_3-(CH_2)_{12}-CH{=}CH-\underset{\displaystyle OH}{\underset{|}{CH}}-\overset{\displaystyle NH_2}{\overset{|}{CH}}-CH_2OH$$ sphingosine

(the CH_3 through $CH(OH)$ portion, bracketed, = S)

$$S-\overset{\displaystyle \overset{H}{N}-\overset{\displaystyle O}{\overset{\|}{C}}-R}{\overset{|}{CH}}-CH_2OH$$ ceramide

$$S-\overset{\displaystyle HN-\overset{\displaystyle O}{\overset{\|}{C}}-R}{\overset{|}{CH}}-CH_2O-\underset{\displaystyle O^-}{\underset{|}{\overset{\displaystyle O}{\overset{\|}{P}}}}-CH_2CH_2-\overset{+}{N}(CH_3)_3$$ sphingomyelin

$$S-\overset{\displaystyle HN-\overset{\displaystyle O}{\overset{\|}{C}}-R}{\overset{|}{C}}H-CH_2$$ cerebroside

(CH_2 linked through O to a hexose ring bearing CH_2OH, HO, OH, OH)

[a] The data are the results of many workers; taken in part from a summary by E. D. Korn, *Science*, **153**, 1491 (1966). See also E. D. Korn, *Ann. Rev. Biochem.*, **38**, 263 (1969).

argument against a denatured product. If a structural protein indeed exists, its most important biological function should be specific interaction with the other membrane components.

The data in Table 9.1 suggests that myelin is unique in its lipid content and also in that it contains no enzymes. The myelin is the membrane covering of nerve cells, and its function appears to be to provide the nerve with an electrical insulation.

If we start with this qualitative picture of the lipoprotein nature of membranes, then the next and more fundamental question about membrane structure must be: How are the lipids and proteins organized to form membranes? All the methods discussed in previous chapters in connection with macromolecular structure analysis have been used in attempts to obtain the answer to this question, but so far none of the structural models can be considered even reasonably well established. Figure 9.1 illustrates some architectural models. The model in Figure 9.1(*a*) is not really a molecular model, but a completely hypothetical picture intended to reflect the behavior of membranes during isolation and emphasize the difficulty in defining any one membrane structure. Its purpose is to emphasize the following points: (1) Different groups of proteins show different degrees of interaction with the lipids. (The protein components represented by the different lines, ovals, and triangles each consist of many different proteins and are categorized only on the basis of their relative extractability in a hypothetical series of treatments with different solvents.) (2) The protein components are probably different on the two sides of the membrane. (3) What is "the membrane"? Is the group of proteins represented by the filled ovals an integral part of the membrane or does it simply represent contaminants of cytoplasmic origin? By a similar argument, were some membrane components lost in the initial steps of the preparation? The answers to these questions must often be arrived at quite arbitrarily, but it is clearly essential that the treatments involved in preparing "a membrane" are defined very precisely in order to obtain reproducible results. (4) The membrane must be a dynamic structure undergoing continuous changes. It is hoped that this concept is conveyed by the "water probability diagram." The idea is simply that if an observer could travel back and forth across a membrane at a given point and count water molecules during each crossing, he might observe quite different situations on different trips (the white lines in the graph). The average of all observations over a long period of time (or the average at any given time over the whole membrane) might well be a curve represented by the black line.

Model (*b*) is the classical bilayer model, which was generally accepted as *the* model for all biological membranes, and the foundation of the *unit membrane theory*. The basis for this model was the persistent observation of a dimension of 100 to 120 Å across all membrane structures measured either by X-ray diffraction or with the electron microscope, the suggestion from specific staining (with OsO_4) that the lipid components made up the core of the membrane, and the observation that the quantity of lipid in a membrane closely approximated the amount needed to make a double lipid layer over the area in question.

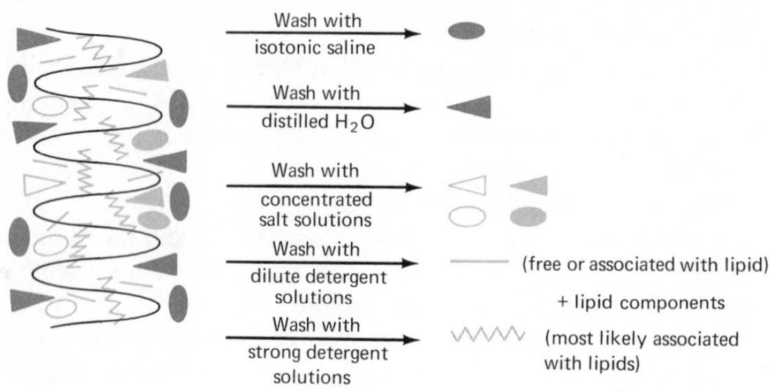

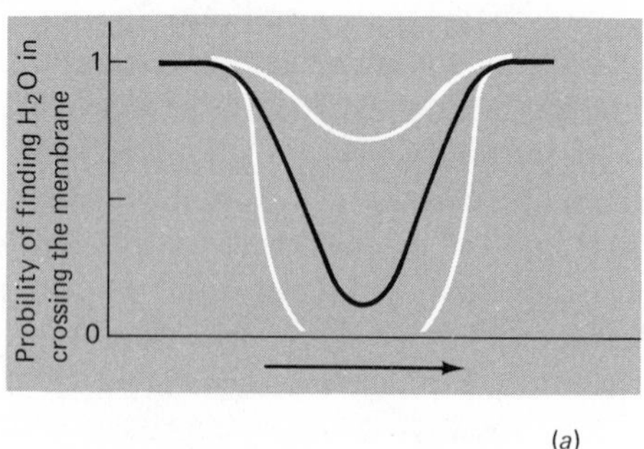

(a)

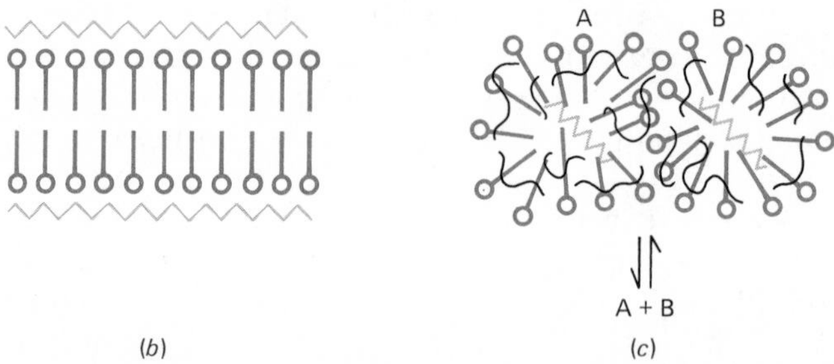

(b) (c)

Figure 9.1 Some models for the structure of membranes. (*a*) A dynamic model in which proteins (illustrated by lines, ovals, and triangles) and water (represented by the probability diagram) interact with the lipid phase to a different degree at different times. (*b*) The lipid bilayer model in which the match sticks represent a regular array of lipid molecules with the nonpolar end in the hydrophobic interior and the polar end interacting with protein (zig-zag lines) in the aqueous phase. [This model was first proposed by J. F. Danielli and H. J. Davson, *J. Cell. Comp. Physiol.*, **5**, 495 (1935).] (*c*) The subunit model with a more irregular protein–lipid water interaction pattern. [First proposed by A. A. Benson, *J. Am. Oil. Chem. Soc.*, **43**, 265 (1966).] [For a more complete discussion of membrane structure, see Robertson, J. D., "The Organization of

In the original models, the protein making up the surface of the membrane was proposed to be an extended β structure. This model has been seriously questioned in recent years. First of all, optical-rotatory dispersion and circular-dichroism measurements have demonstrated that there is practically no protein with β structure in natural membranes; on the contrary, an unusually large amount of α helix was found to exist. The spectral observations further suggested strong interaction between the α-helical proteins and the hydrophobic lipid regions, which is also contrary to the model.

Perhaps the strongest argument against the bilayer-membrane model is a philosophical one. The model implies a certain structural rigidity and uniformity that is very difficult to accept in view of the diverse functional properties of membranes. The bilayer model may be to membranes what Emil Fischer's lock-and-key model was to enzymes, a very important first step in model building and in formulating a direction of thinking, but too restrictive and inflexible to withstand the tests of more and more refined experiments.

Model (*c*) is one example of a "lipoprotein subunit" model. This proposal accounts for the protein as an integral part of the membrane "core" and is consistent with several observations of subunit structures in a number of biological membranes. Several other models have been suggested, and all essentially fit somewhere between these two extremes. It seems that the most serious drawback at this time to any proposed model is the assumption that a single type of structure will apply to all membranes. Especially if one accepts the thesis that membrane function can best be explained on the basis of a dynamic structure in which molecules can move and change their relative position, the best model would seem to be one allowing for all the possible variations between the bilayer and subunit structures.

9.3 FUNCTION

It may be useful to consider membranes as fulfilling two broad types of functional roles: They are the barrier separating two kinds of milieu and regulating the flow of cellular components between them, and they are also the surface or interface upon which functional molecules can be assembled for optimal kinetic and thermodynamic performance, and which in the case of the plasma membrane present to the surroundings a characteristic surface that is the basis for the phenomenon of cellular recognition. The barrier and surface properties are probably closely related, but may represent a practical basis for comparing the function of a plasma membrane (barrier and surface) with that of endoplasmic reticulum (probably surface properties only).

There are four types of functional parameters which can be studied and which have been used to quantitate membrane function: (1) coupled enzyme functions;

Cellular Membranes," in J. M. Allen (ed.), *Mollecular Organization and Biological Function*, Harper and Row, New York, 1967; and Stoeckenius, W. and D. M. Engelman, *J. Cell. Biol.*, **42**, 613 (1969.]

(2) active transport; (3) electrical properties; and (4) cellular interactions. The first three of these will be considered briefly in the following pages. The fourth property is just emerging as a quantitative approach to the study of membrane structure and function and will not be considered here (the reader is referred to references at the end of this chapter).*

Coupled Enzyme Functions

Coupled enzyme function emphasizes the surface or template functions of the membrane. Because of strong associations between membrane components, stable functional complexes can be formed in which individual molecules are linked both structurally and functionally to give maximum efficiency. Electron transport in the mitochondria is a typical example of such a membrane-coupled system, and it is also likely that the enzyme complexes discussed in Chapter 3 represent functional-membrane subunits.

There are two rather speculative aspects of this that may bear some consideration. If a process consists of the chemical steps from *A* to *Z*, for example, and many of the intermediates are unstable, act as ready precursors for side reactions, or are undesirable as products, then it is clear that by putting all the enzymes involved in the total process inside a box, so to speak, all the undesirable possibilities have been eliminated. If only *A* can enter and only *Z* can get out of the box, the compartment of closely linked enzymes represents a closed, highly efficient, single-purpose functional unit. The fatty acid–synthesizing system of yeast could be an example of such a system (see Larner in this series), and it illustrates the utility of this concept in terms of selecting certain specific products in reactions where many similar products are produced. The other point is related to the fact that many reactions associated with membranes may well take place right on a water-lipid interphase. Depending on the solubility of substrates and products, it would seem reasonable that a partitioning of the reactants between the aqueous and nonaqueous phase could readily affect the apparent equilibrium of a reaction. The net effect could be to pull hydrolytic processes in the direction of condensation by pulling the condensation product away from the enzymatic site and water and into the nonaqueous phase (Figure 9.2). This idea is dependent on a vectorial or directional enzyme action which is an inherent property of the models based on enzymes "fixed" on a membrane template.

* The surface of a given cell or group of cells must be as distinctive in relation to other cells and cellular agents as one person's appearance is to other persons. The work to date indicates that the carbohydrate components of membrane glycoproteins play an important part in cellular "individuality" and that immunochemical specificity, for example, is based on antibodies which recognize specific oligosaccharide sequences. Interactions between cells leading to aggregation of like cells or the lack of interaction leading to discrimination between groups of different cells are important consequences of cellular recognition and could well be an essential part of the normal differentiation process.

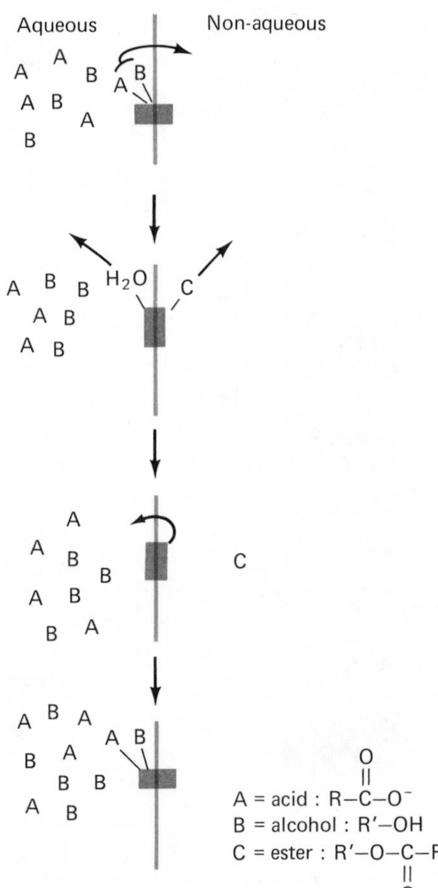

Figure 9.2

A representation of a reaction catalyzed on the interface between aqueous and nonaqueous phases. The reaction could be the interconversion of an ester and its component acid and alcohol. The basis for this picture is that the ester is stable in both phases, and only hydrolyzed when it encounters a water molecule in the active site of the enzyme catalyst. The catalyst is fixed as a membrane component and is located on the interface. The ester is very soluble in the non-aqueous phase and relatively insoluble in the aqueous phase, and the acid and alcohol are soluble in the aqueous phase and relatively insoluble in the nonaqueous phase. In such a system one can visualize that the ester can be synthesized by a simple condensation reaction in which the "driving force" is the partitioning of the substrates and products between the two phases.

In any reacting system in which a component is fixed in space in a uniform geometry such that one can distinguish between the two sides of the component, the reaction becomes directional or vectorial. [This is in contrast to a system where all the components are free to move at random (in ideal solutions) or are fixed in random order. Although each individual molecular event in such a random system is also vectorial, the integrated directional component for the total system over a long period of time is zero. Such reactions are referred to as scalar.] There are many ways in which vectorial reactions can be rationalized in a membrane system, where the membrane components are fixed on the lipoprotein phase separating the "inside" and "outside." The situation of an enzyme on the membrane surface, at the lipid–water interface, illustrated in Figure 9.2 represents one case of a vectorial reaction. Another model is shown in Figure 9.3. Here the architectural arrangement of fixed molecules allows the approach and release of some molecules in only one direction. The representation in Figure 9.3 implies a certain structural rigidity in the arrangement of enzymes, which must not be taken too literally. This concept will be discussed

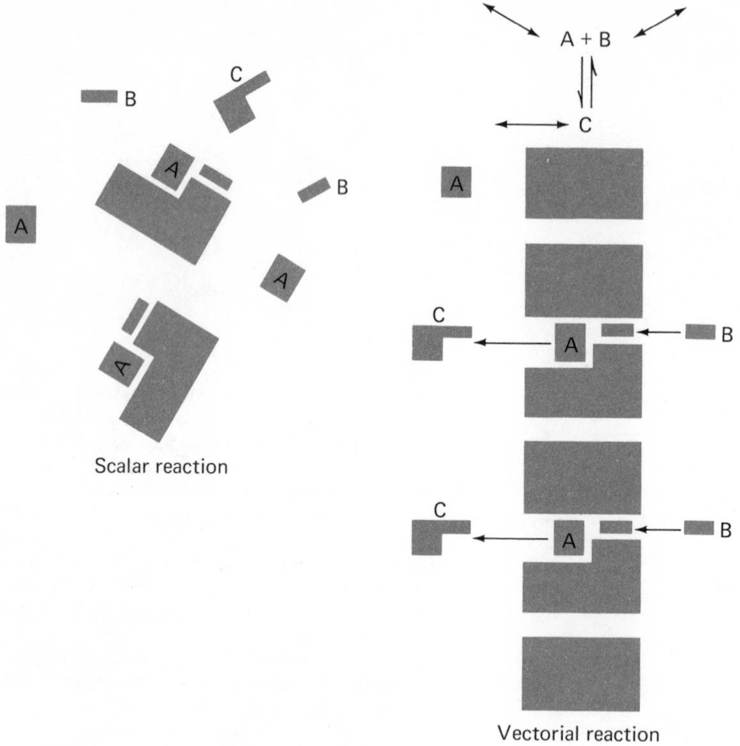

Figure 9.3 Vectorial and scalar reactions.

further in Chapter 10 in connection with mitochondria. As more refined methods of observation and fractionation are employed, it appears that more and more enzymes are found associated with the "particulate" membrane fractions of the cell. Thus, as was suggested in Chapters 1 and 6, DNA replication and protein synthesis may be associated with membranes.

Active Transport

Active transport is a typical "barrier phenomenon" and represents the oldest and most extensively studied functional property of membranes. All membranes show characteristic permeability properties for different solutes. As a general rule, water and small uncharged molecules pass readily through membranes, whereas charged compounds (ions) pass through only slowly; large molecules such as proteins are retained by the membrane. The natural tendency toward random distribution requires that all permeable components diffuse across the membrane to equilibrate the concentrations on both sides. This diffusion-driven flow of ions and other compounds is referred to as *passive transport*.

If one determines the concentration of a variety of components inside and outside the membrane, however, it is well known that in active, living systems, such an equilibrium very seldom exists.* Common examples of typical concentration differences are illustrated in Table 9.2. In all these cases the differential between inside and outside is maintained against the concentration gradient;

TABLE 9.2 MANIFESTATION OF ACTIVE TRANSPORT (CATION CONCENTRATION INSIDE AND OUTSIDE CELLS)

	Ion Concentrations (in Milliequivalents)/Liter			
	Na^+	K^+	Ca^{2+}	Mg^{2+}
Total body interstitial fluid (outside)	140	2	2	—
Total body intracellular level (inside)	10	153	0	32
Cerebrospinal fluid (outside)	141	2.5	2.5	—
Brain cells (assumed 80 percent of total; inside)	57	96	2	—
Plasma (outside)	140	4.5	5	2
Erythrocyte (inside)	20	140	—	3.5

this requires energy and the process thus is referred to as *active transport*. In studying the process at the molecular level, it appears that the mechanism can be explained in terms of specific *transport proteins*, which essentially are enzymes carrying out translocation instead of, or concurrent with, catalysis. If the only function of such a protein is transport, it is likely that the process can only be studied directly in the intact membrane. Assuming, however, that the transport involves a specific binding of the substrate to the transport protein, it may also be possible to work with isolated proteins and study their specific ligand binding. Both principles have been found to apply and have been used. In transport processes, too, the vectorial–reaction type is clearly an essential prerequisite in formulating molecular models.

Electrical Properties

Because of the slow passage of ions through the nonaqueous phase of the lipoprotein membrane, the intact membrane shows a very high electrical resistance (10^3 to 10^5 ohms/cm^2 compared to about 200 ohms/cm^2 for the soluble fraction of the cytoplasm). Also, because of the ion-concentration gradient across the membrane, a small potential difference can be measured across the membrane [10 to 100 mV (millivolts)]. Both of these properties have served as convenient handles in the study of membrane function. In any situation where the structural features or the functional status quo are changed so the ion-

* A "living" cell could be distinguished from a dead one on this basis: living means that a difference in concentration of a variety of compounds is maintained between the inside and outside; when the cell ceases to live, equilibrium between the inside and outside establishes rapidly.

concentration gradients are changed or the rate of ion transport is altered, significant changes in resistance and potential are observed. The characteristic drop in the "resting potential" when a membrane is stimulated or activated is referred to as the *action potential* and is a result of a rapid ion flux leading to a momentary elimination of the concentration gradient. The normal gradient and potential is rapidly restored when the stimulus is removed.

The biological activity of membranes can in this way be expressed in terms of chemical processes (catalysis and transport) or physical parameters (resistance and potential), and the approach to structure-function relationships is essentially the same as discussed for proteins and nucleic acids. Structural parameters are altered in a systematic and hopefully specific manner, and the resulting changes in functional properties are noted and correlated. The use of inhibitors with well-defined sites of action has been as important in the membrane field as it has been in the development of other areas of biochemistry.

9.4 ARTIFICIAL MEMBRANES

Artificial membranes have represented one of the most significant sources of information in the study of both structure and function of membranes. Because of the unique place of these model systems in membrane research, it is necessary to dwell briefly on their properties. It is actually difficult at present to decide whether the artificial membranes have been a blessing or a liability to the study of biological membranes; certainly they have been extremely useful in the general study of membrane phenomena. The difficulty with the artificial membranes is that they form spontaneously and appear to give typical bilayer membranes. This means that one has to suspect artificial membrane formation from normal membrane components during the isolation of natural membranes and also that models tend to be biased in favor of the commonly observed artificial structures. Artificial membranes are generally prepared by "painting" a mixture of hydrocarbon and phospholipid across a small orifice in a wall separating the two compartments of the experimental cell. A film forms across the entire orifice and it is stable for long periods of time. Table 9.3 shows some properties of such films in comparison to those of biological membranes. Depending on one's point of view, either the similarities or the differences may be emphasized. The interesting feature of the artificial membranes right now is that there are several reports of isolation of proteins that will confer "true" membrane properties (such as ion selectivity and electrical excitability) to lipid films, and it thus appears that these films may indeed serve as models for the study of biological membrane properties.

9.5 THEORETICAL ASPECTS

The discussion in Chapters 9 and 10 is rather qualitative and oriented toward the practical, experimental approach to the study of membrane properties. Lest

TABLE 9.3 SOME PROPERTIES OF NATURAL AND ARTIFICIAL MEMBRANES[a]

	Natural Membranes[b]	*Artificial (Bilayer) Membranes*[b]
Thickness in Å, by electron microscopy	40–130	60–90
Thickness in Å, by X-ray diffraction	40–85	—
Resistance (ohm/cm^2)	10^3–10^5	10^3–10^9
Potential difference (mV)	10–90	0–140
Interfacial tension (ergs/cm^2)	0.03–3.0	0.2–6.0
Water permeability (10^{-4} cm/sec)	0.25–58	2.3–24

[a] The data represent the results of many laboratories, taken from a review by H. T. Tien and A. L. Diana, *Chem. Phys. Lipids*, **2**, 55 (1968).

[b] Represent the range of values measured for different biological membranes and for artificial membranes with different composition.

this approach should leave a completely wrong impression, it is important for us to stress that membrane processes can also be described quantitatively on a sound theoretical basis. The kinetics and thermodynamics of both passive diffusion and active transport have been subjects of extensive theoretical treatment, the details of which are well beyond the scope of this book (consult the references at the end of this chapter). Perhaps one of the most significant aspects of these theoretical considerations has been the realization of the inadequacies of classical thermodynamics when applied to life processes. Based on the experiences with isolated systems studied in dilute salt solutions under fairly ideal conditions, it has become general practice not only to treat the experimental data as ideal but to formulate models and think about biochemical problems in terms of ideal-solution chemistry. The processes involving translocation are typical of living processes that cannot be treated in terms of the reversible cycle of classical thermodynamics, since at any time of observation of the living system, flow occurs in typically irreversible fashion. The application of irreversible thermodynamics to membrane transport has thus introduced a different approach to viewing all biochemical processes.

REFERENCES

Useful general reference books:

Chapman, D. (Ed.), *Biological Membranes*, Academic Press, New York, 1968.

Katchalsky, A., and P. F. Curran, *Nonequilibrium Thermodynamics in Biophysics*, Harvard University Press, Cambridge, Mass., 1965.

Peeters, H. (Ed.), *Protides of the Biological Fluids*, Elsevier, Amsterdam, 1968.

Smith, R. T., and R. A. Good (Eds.), *Cellular Recognition*, Appleton-Century-Crofts, New York, 1969.

Stein, W. D., *The Movement of Molecules Across Cell Membranes*, Academic Press, New York, 1967.

Recent review:

Korn, E. D., "Cell Membranes: Structure and Synthesis," *Ann. Rev. Biochem.*, **38**, 263 (1969).

TEN SOME MEMBRANE SYSTEMS

In this chapter some current facts, ideas, and molecular models of membrane structure and function are outlined. The mitochondrial membrane is discussed as a case of *functional coupling* of individual units in a membrane. The experimental approaches to the study of transport of metabolites across membranes are outlined and some models of specific transport systems are briefly considered.

10.1 THE MITOCHONDRION

The mitochondrion, as pointed out in Chapter 1, is a subcellular lipoprotein structure with an impressive list of biological functions. Both the structure and functions of mitochondria have been studied intensively for the last two or three decades, and a summary of some of the accumulated knowledge is given in Figure 10.1 and Table 10.1. Only one aspect will be considered here, namely the coupling of electron transport and oxidative phosphorylation, and the main emphasis will be on correlating these biological functions with what is known about the membrane structure. Figure 10.2 summarizes the main features of electron transport and oxidative phosphorylation. The metabolic significance of these processes is discussed in detail by Larner in this series.

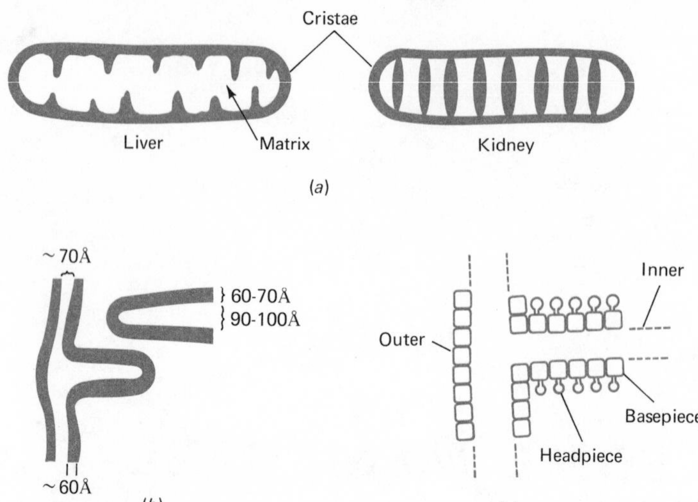

Figure 10.1 Diagrammatic representation of the structure of mitochondria at increasing magnification (see also Table 1.1). The characteristic convoluted membrane structure is perfect if a large membrane surface is desired within a certain space.

TABLE 10.1 DISTRIBUTION OF SOME ENZYMES IN MITOCHONDRIA

Inner Membrane	The components of the electron transport chain: NADH dehydrogenase, succinate dehydrogenase, and the cytochromes ATPase and "coupling factors" β-hydroxybutyrate dehydrogenase α-keto acid dehydrogenase
Matrix	All the tricarboxylic acid cycle enzymes except α-ketoglutarate dehydrogenase and succinate dehydrogenase Enzymes of the fatty acid β-oxidizing system Glutamate dehydrogenase
Outer Membrane	Enzymes involved in fatty acid biosynthesis Enzymes involved in phospholipid biosynthesis Monoamine oxidase Nucleoside diphosphate kinase

In trying to understand these processes in terms of the membrane "assembly-line concept," certain aspects will be emphasized, perhaps on completely false premises. Thus in the following, only two major phenomena are outlined: (1) the subunit division of both structure and function; and (2) the dissociation and reconstitution of functionally intact submitochondrial particles.

Subunits

Electron-transport particles can be obtained from mitochondria as intact lipoprotein-membrane subunits with a particle weight of about 3×10^6. The

ADP + P_i
NADH + H^+
FAD
NAD^+
(Fe)
$FADH_2$
ATP
QH_2
ATP
ATP
2 cyt b-Fe^{3+}
2 cyt c_1-Fe^{2+}
2 cyt c-Fe^{3+}
2 cyt a-Fe^{2+}
2 cyt a_3-Fe^{3+}
$O^=$
2 cyt b-Fe^{2+}
2 cyt c_1-Fe^{3+}
2 cyt c-Fe^{2+}
2 cyt a-Fe^{3+}
2 cyt a_3-Fe^{2+}
$\frac{1}{2}O_2$
Succinate
Q
FAD
(Fe)
Fumarate
$FADH_2$
ADP + P_i
ADP + P_i

(a)

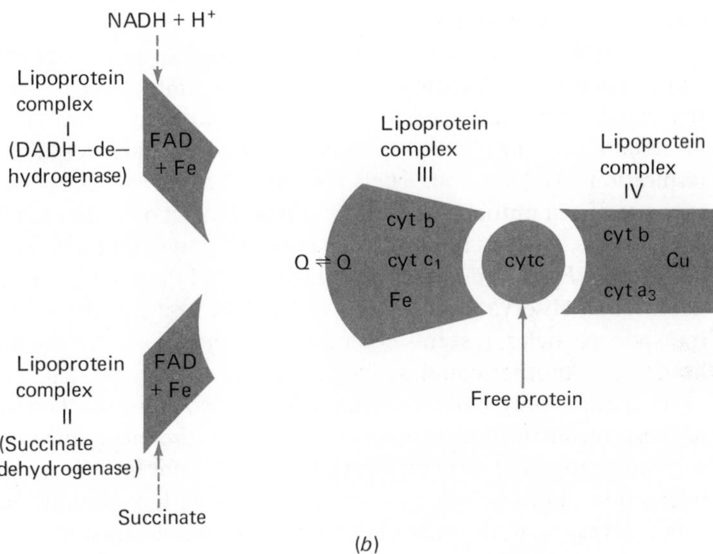

(b)

Figure 10.2 (*a*) The functionally-coupled series of reactions by which electron transport from the normal substrates (NADH and succinate) to O_2 with the concomitant synthesis of ATP (oxidative phosphorylation) is carried out. The reactions take place in intact mitochondria, but submitochondrial particles (ETP = electron-transport particles) which are capable of carrying out electron transport and oxidative phosphorylation have also been isolated. (*b*) The structural subparticles obtained from the electron transport chain when mitochondria or ETP are disrupted by sonication, treatment with mild detergents, organic solvents, or with salt solutions [see for example Fleischer, S. *et al., J. Biol. Chem.*, **236**, 2936 (1964)]; four smaller lipoprotein complexes and free cytochrome c are obtained. These complexes designated with roman numerals are shown schematically in the figure. Under proper conditions they can be recombined with the proper "coupling factors" to reconstitute active ETP with oxidative phosphorylation. [Taken in part from Green, D. E., in *Impact of Basic Sciences on Medicine*, B. Shapiro and M. Prywes (eds.), New York: Academic Press, 1966, p. 109.]

particle contains the entire electron-transport chain and can oxidize NADH with molecular oxygen but cannot phosphorylate ADP. This "elementary particle" can be further degraded by mild treatments to four complexes and cytochrome c, and again biological activity can be demonstrated in each of the complexes. This is most dramatically demonstrated by adding back one component at a time and demonstrating the gradual rebuilding of the electron-transport chain step by step. It is proposed that the intact unit is made up of "base-piece units" of the inner membrane (Figure 10.1), and that each complex contains a substantial quantity (50 percent) of structural protein. Since the base pieces have an estimated particle weight of about 400,000, the intact electron-transport particle (with a particle weight of 3×10^6) should contain some eight base pieces. It is tempting to propose that four of these base pieces correspond to the four complexes and the other four to structural-protein aggregates, but such a model is at present highly speculative. Nevertheless it appears that a model is emerging in which an observed structural continuity will bear direct correlation to a set of functionally coupled units.

The important biological role of electron transport is energy production, and one of the most exciting developments in "mitochondriology" is the recent reports of reconstitution of the process of oxidative phosphorylation by addition of certain *coupling factors* to submitochondrial particles that can carry out respiration. At least four such coupling factors are known and have been separated from mitochondria. They appear to act on different phosphorylation sites (Figure 10.2) but the specificity is not absolute. One of the coupling factors, designated F_1, is an unusual, cold-labile ATPase and, based on electron-microscope observations of both isolated F_1 and F_1 added back to electron-transport particles, it seems certain that it corresponds to the inner-membrane headpiece. Another coupling factor, F_4, may simply be structural protein, but again there is only limited evidence for this conclusion. The important point of these reconstitution experiments is again, the emergence of a model that correlates observed structure with functional coupling. It is clear that all the interactions, both between molecular species within subunits and between the subunits that form the intact functional membrane, must be quite specific.

Mechanism of Oxidative Phosphorylation

This fascinating area of research is discussed by Larner in this series, and will only briefly be summarized here to illustrate the probable role of the membrane structure. The mechanisms should account for three features of mitochondrial activity: (1) the molecular basis for the coupling of respiration (electron transport) to ATP production; (2) the transport of ions and metabolites across the mitochondrial membranes; and (3) the swelling and contraction of mitochondrion that accompanies the changes in ATP/ADP ratios.

The three models that have been proposed to explain oxidative phosphorylation are summarized in Figures 10.3 (*a*), (*b*) and (*c*). These three models clearly

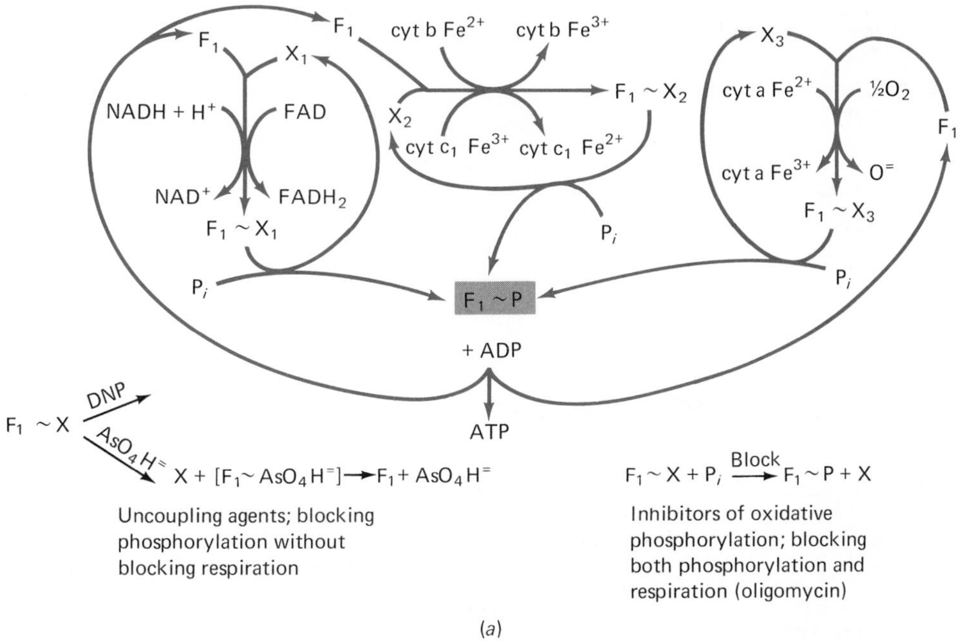

Figure 10.3(*a*) The covalent phosphate intermediate model of oxidative phosphorylation. The central point of this model is that it requires that some phosphorylated intermediate ($F_1 \sim P$ in the scheme above) is produced. The particular version presented above is completely hypothetical but represents an attempt to use the four different coupling factors, F_1, X_1, X_2, and X_3. The proposal of this model is that F_1, the ATPase, is common to all three phosphorylation sites, while X_1, X_2, and X_3 are specific, one for each site. Both F_1 and the proper X-factor are required for respiration to take place in the tightly-coupled mitochondria. In this way one can explain the fact that when ADP is depleted, respiration stops (F_1 and X are tied up). The effect of typical inhibitors is also rationalized on the same basis. Since no one has been able to find a phosphorylated intermediate, there is no direct evidence for this model.

differ only in mechanistic detail. They all have an "energized" or charged intermediate in common whether it is called a phosphorylated intermediate, a proton pressure, or a conformational change. The analogy to the wound spring is perhaps a useful mechanical model. Electron transport winds the spring; and in "coupled" mitochondria the catch is released with ADP + P_i, the spring unwinds against a frictional resistance, and work is done (ATP synthesis). In the uncoupled mitochondria the spring unwinds without any frictional resistance and no work is done (see also the model of the sodium pump in Figure 10.5). The first is the classical "covalent-intermediate model," which has become somewhat disreputable primarily because of the repeated failure over the years to actually demonstrate such an intermediate. Because of the negative nature of this argument, however, the model must still be considered a real possibility.

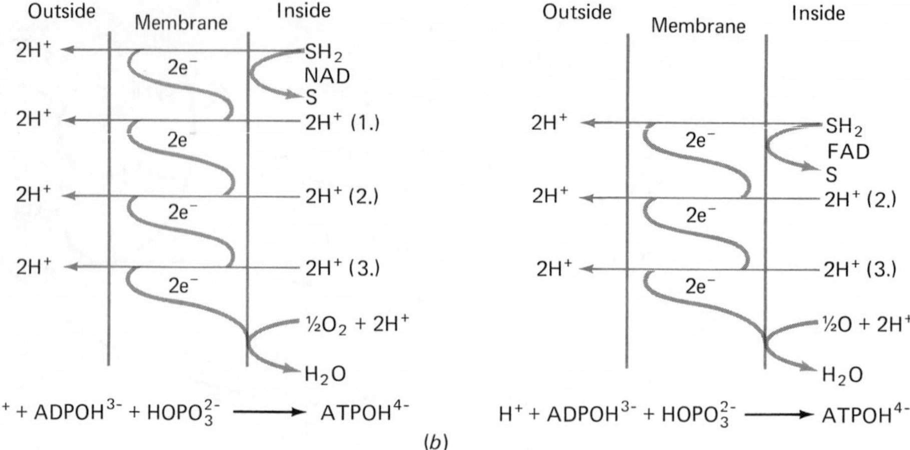

Figure 10.3(*b*) The chemiosmotic model of oxidative phosphorylation. [From Mitchell, P. and J. Moyle in Slater, E. C., Z. Kaninga, and L. Wojtczak (eds.), *Biochemistry of Mitochondria*, New York: Academic Press, 1967.] The basis for this model is the proposal that two protons are transported across the membrane at each phosphorylation step [three steps in the NADH oxidation (left) and two steps in the succinate oxidation (right)]. Since the membrane is impermeable to protons, the participation of the electrons is required in each transfer (the electron loops). Thus the net result of electron transport is to produce a "proton pressure" outside the mitochondria. Since the membrane is impermeable to protons, the proton pressure is relieved by ATP synthesis. The evidence for this model is primarily based on the observation of very rapid pH decreases in the medium when anaerobic and reduced electron-transport particles are subjected to short bursts of O_2.

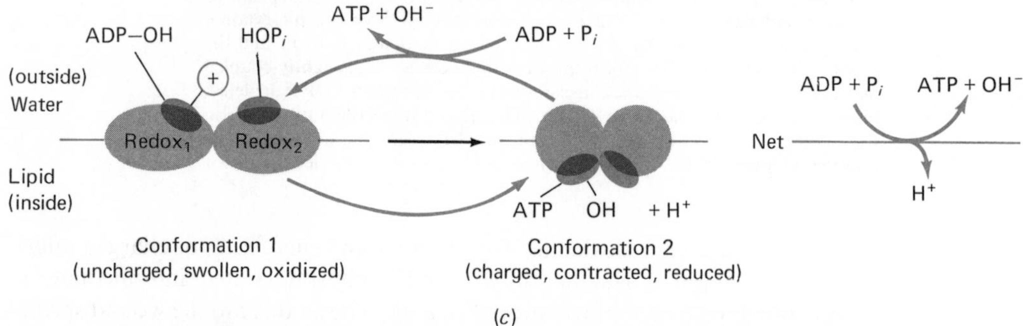

Figure 10.3(*c*) The conformational model of oxidative phosphorylation [freely interpreted from Boyer, P. D., and from Green, D. E., in King, T. E., H. S. Mason, and M. Morrison (eds.), *Oxidases and Related Redox Systems, Vol. 2*, New York: Wiley, 1965, pp. 994 and 1032]. The basic idea of this model is that the energy released from the redox reaction is translated into a conformational change in the membrane-bound redox systems associated with the coupling factors (shown in color). In the representation used here a hypothetical mechanism is illustrated which leads to synthesis of ATP with the net transport of one proton to the inside and one OH^- to the outside, in this way handling the "proton pressure" of the chemiosmotic model. The primary evidence for this model is the direct observation of different conformations of the mitochondrial crista in the electron microscope.

The two alternative models have a strong appeal as they provide an explanation of transport phenomena as well as of the phosphorylation process. The "chemiosmotic model" emphasizes the vectorial–reaction mechanism (see Figure 9.3) and thus takes advantage of the fixed orientation of the enzymes in the membrane complex to ensure a unidirectional flow of reactants and product. The main evidence in favor of this model is the apparent success in reversing the phosphorylation by applying a *proton pressure* or, in other words, shifting the equilibrium by adding protons. The "conformational model" also takes advantage of the membrane structure, but more along the lines of the interphase argument set forth in Chapter 9. Using the analogy of the wound spring: When all the ADP is used up and all the components are reduced, the springs are fully wound – the system is energized, and will stay that way until more ADP is provided (respiration in a tightly coupled mitochondrion can be blocked by removing ADP). Evidence for this model consists of essentially only electron micrographs of energized and nonenergized states with distinctly different appearances in agreement with the model's two conformational states. An interesting aspect of this model is the proposal that only one phosphorylation site exists and that all the individual "springs" of the electron-transport chain are coupled to this single site. This proposal would, in a general way, explain why it has been so difficult to reconstitute oxidative phosphorylation with anything less than the complete electron-transport chain.

The relationship in mitochondria between oxidative phosphorylation and ion transport and contraction is not well understood at present. It is clear that ion transport in the mitochondrion competes with ATP formation for the energy produced in electron transport. Thus if respiration in tightly coupled mitochondria is stopped by blocking phosphorylation [with the inhibitor (antibiotic) oligomycin], addition of Ca^{2+} or Mg^{2+} will restart the respiration and the cations will be taken up by the mitochondrion. Uptake of ions leads to swelling in most cases. If ATP or ADP is added to again start oxidative phosphorylation, the mitochondrion will contract again and water and ions are pushed out. It is highly relevant in this connection that there are recurring reports that mitochondria contain a component appearing in every way identical to actomyosin. If this is actomyosin, the contractile mechanism could be analogous to that of muscle. Clearly some very important questions about mitochondrial structure and function remain to be answered. It appears, however, that this functional unit illustrates better than any other all the unique and fascinating properties of membranes.

A couple of recent developments in the study of mitochondria should be added here for the sake of an up-to-date picture. First of all, recent work has given strong evidence that the mitochondrial DNA is self-replicating* and that the mitochondrion appears to have its own protein synthesizing apparatus;† this indicates that the mitochondrion is a fairly independent unit with both the blueprints and the synthesizing apparatus needed to produce at least some of

* B. Hudson, *et al.*, *Cold Spring Harbor Symp. Quant. Biol.*, **33**, 435 (1968).

† L. W. Wheeldon, and A. L. Lehninger, *Biochem.*, **5**, 3533 (1966).

its own protein components. Relevant to these findings but also greatly significant to the correlation of mitochondrial structure and function is the recent discovery of "promitochondria" in yeast.* Aerobically grown yeast has normal, functional mitochondria; however, when grown anaerobically, the same yeast has no recognizable mitochondria but does have some other subcellular particles (promitochondria) containing components characteristic of the normal mitochondria. The best description of the promitochondria appears to be that they are modified mitochondrial inner membranes containing ATPase and also mitochondrial DNA. Whether promitochondria represent an undifferentiated product of the total mitochondrial genome or the complete product of the mitochondrial DNA that must be complemented by missing components coded by the nuclear DNA (and expressed only under aerobic conditions) remains to be seen. The conversion of such "unfinished" lipoprotein particles into functional mitochondria by a systematic supplement of isolated mitochondrial components could be an invaluable tool in the study of mitochondrial structure and function.

10.2 TRANSPORT SYSTEMS

The plasma membrane is a key regulatory unit in the operation of the cell. The metabolic precursors must be transported in and the end products out through the semipermeable-membrane barrier. It is clear that virtually none of this transport is simply a passive diffusion process, and one of the major problems of membrane biochemistry is to establish the molecular mechanisms that regulate both the rate and the specificity of transport. The information available is encouraging but yet far from conclusive. Rather than to consider any one system in great detail, the following will treat some experimental approaches used in this important area of research. Most of the work has been done with microorganisms because of the very important advantages offered by comparing different mutants. The erythrocyte has also been a popular system for this type of study, especially the red-cell ghosts (essentially intact red-cell membranes with the hemoglobin removed), and different cultured mammalian cells have also been used to some extent.

It is perhaps useful to postulate from the start a general model for transport. It appears that there are general channels or pathways for transport through the membrane, that each type of channel is available to certain groups of compounds and not to others, and that each channel is linked to a group of highly specific transport agents; thus that every compound is subject to specific transport. When the transport agents are removed, the rate of transport is generally decreased and the specificity is lost. The transport agents may work by binding the substrate, carrying out translocation and then release, or by affecting a chemical alteration in the process so that a substrate derivative, rather than just the substrate, is released. It appears safe to predict that the transport agents are

* R. S. Criddle, and G. Schatz, *Biochem.*, **8**, 322 (1969).

proteins, often enzymes. All these rather tentative conclusions and predictions are based on the following types of experimental observations.

Experiments with Intact Cells

Most of the transport data have been obtained by following the transport of different simple compounds, properly labeled and supplied in the medium, into the cell. Amino acid uptake represents a good example of this approach. By carefully calibrated and controlled techniques, it has been possible to determine initial rates of amino acid transport, and it appears that this process can be treated completely analogously to enzyme kinetics according to the mechanism

$$S_{(\text{external})} + E \underset{k_{-1}}{\overset{k_1}{\rightleftharpoons}} ES_e \underset{k_{-2}}{\overset{k_2}{\rightleftharpoons}} ES_i \underset{k_{-3}}{\overset{k_3}{\rightleftharpoons}} E + S_{(\text{internal})}$$

In the simplest form based on initial rates, this would give the rate equation

$$V = \frac{V_e}{1 + K_e/[S_e]}$$

where $V_e = k_2 E_0$ and $K_e = k_{-1}/k_1$.

The more general form, allowing for flow in both directions, would be

$$V = \frac{V_e}{1 + K_e/[S_e]} - \frac{V_i}{1 + K_i/[S_i]}$$

where $V_i = k_{-2} E_0$ and $K_i = k_3/k_{-3}$.

Using this approach, it is possible to characterize the transport agent (E) in terms of $V_{\max}$ and K_m just as an enzyme. In addition, the competition between different amino acids can be studied, giving a good picture of the specificity of each process as well as of the interaction of several transport processes.*

The Isolation of Transport Agents (Proteins)

The method of isolating transport agents, specifically labeled in the intact membrane, has been used in connection with the lactose transport system of *E. coli* (see Chapter 8). The approach is best illustrated by an actual experiment [Figures 10.4 (*a*) and (*b*)]. After introducing the radioactive reagent into the active transport agent in question (M-protein), the radioactive label was found in an insoluble lipoprotein fraction, and this fraction was purified. It was found to be a protein, and, based on work with different mutants, it appeared to be the protein from the "permease" gene in the lac operon (see Chapter 8). It is interesting that the same approach will also work in cell-free systems, showing that in this particular case the galactoside permease retains its ability to bind

* H. N. Christensen, *Adv. Enzym.*, **32**, 1 (1969).

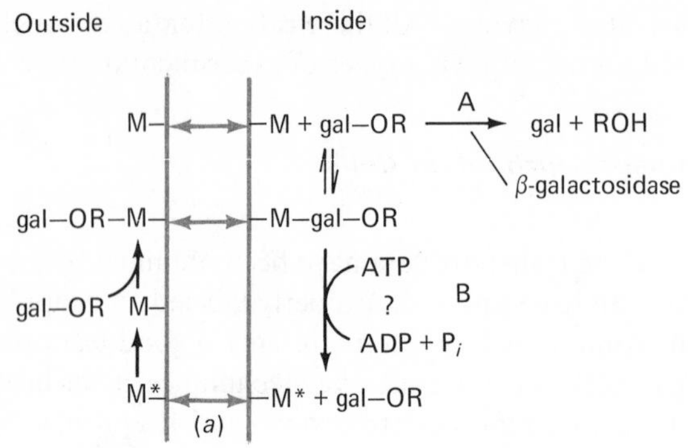

Figure 10.4(*a*) A model for galactoside transport in *E. coli* (see Chapter 8). The membrane-bound M-protein is responsible for galactoside (gal—OR) transport. (Solid colored arrows indicate transport across the membrane.) The affinity of M for gal—OR is very high, but, once inside, the dissociation could be favored by hydrolysis of gal—OR by β-galactosidase (A). Free galactose (and ROH) would thus accumulate. To explain the high cellular levels of gal—OR which are observed in *E. coli*, another mechanism must be considered (B). The proposal is that in an energy-requiring reaction (ATP above), M-protein undergoes a conformational change to M*, and that this new conformation has a low affinity for gal—OR, which is released. M* is next translocated and relaxes to the high-affinity conformation M which binds gal—OR and starts another transport cycle. Step A can be prevented by using galactosides which are not hydrolyzed by β-galactosidase, and step B can be blocked by "uncoupling agents" such as dinitrophenol (see Figure 10.3). (Redrawn from Kennedy, E. P., *Proc. Intern. Congress Biochem.*, Tokyo, 1967. *Intern. Union Biochem.*, **36**, 51.)

galactosides after the membrane has been disrupted. This general approach should be extremely useful for estimating the number of specific transport sites in each cell and for isolating individual transport agents.

Osmotic Shock Release of Membrane Components

It has recently been shown that bacterial protoplasts,* as well as intact cells, when exposed to a gentle osmotic shock, will release several proteins into the medium without disrupting the gross membrane structure.† The shocked protoplast generally shows greatly altered transport properties, in that it may become permeable to compounds not normally passing through its membrane and, more important, in that the rate of transport of other compounds has greatly decreased. The implication is that the specific transport agents have been lost from the membrane. When the released membrane proteins were characterized, several well-known enzymes were found in addition to some proteins that have no enzymatic activity but do have the ability to specifically

* The rigid cell wall can be digested by the enzyme lysozyme, which catalyzes the hydrolysis of glycosidic bonds in the glycopeptide component of cell walls (see Barker and Larner in this series), leaving a spherical protoplast with the membrane exposed.

† H. Neu, and L. A. Heppel, *J. Biol. Chem.*, **240**, 3685 (1965).

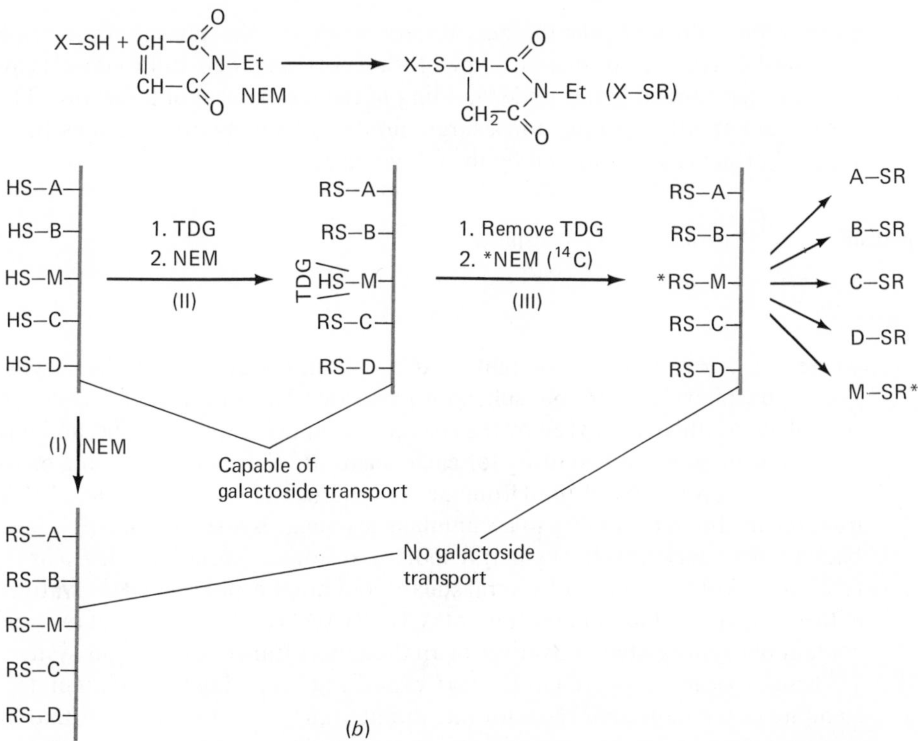

Figure 10.4(*b*) Differential tagging of M-protein (see general rationale in Figure 2.8). When intact *E. coli* cells are treated with the sulfhydryl reagent, N-ethyl maleimide (NEM) (reaction I) galactoside transport is irreversibly blocked. Attempts to purify a specific NEM-labeled protein show that a large number of SH-containing proteins reacted. Saturating the transport system with thiodigalactoside (TDG), complete protection from NEM inhibition was obtained, presumably without any effect on the reaction of NEM with the other SH-proteins (reaction II). Subsequent reaction of the product of reaction II (after removal of the transport substrate TDG) with radioactive NEM, led to a specific labeling of M-protein with the radioactive reagent. This then provided a sensitive experimental "handle" which made it possible to establish that M-protein is a part of the lipoprotein membrane and also to purify the protein after disruption of the membrane structure (from Fox, C. F., and E. P. Kennedy, *Proc. Natl. Acad. Sci.*, **54**, 891 (1965)].

bind ligands (amino acids, sulfate, and sugars have been found to be ligands).* It is reasonable to conclude that these are the specific transport agents for the different ligands. In the case of the shocked protoplasts, the method selects for the transport agents located on the surface of the membrane structure in a relatively loose interaction with the lipoprotein membrane core.

In connection with this approach, we should also consider the situation in which transport is associated with chemical alteration. This corresponds to vectorial enzyme reactions in which translocation is a part of the total enzyme function. It has been known for some time that sugars often are phosphorylated

*A. B. Pardee, *Science*, **162**, 632 (1968).

as they pass through membranes. Recent work on the phosphotransferase system of *E. coli** illustrates this type of transport very well and represents an exciting step toward better understanding of the mechanism of transport. The system apparently operates for a large number of sugars and contains three types of proteins as illustrated by the following scheme

$$\text{phosphoenol pyruvate} + X \overset{E_1}{\rightleftharpoons} \text{pyruvate} + X\text{—phosphate}$$

$$X\text{—phosphate} + \text{sugar} \underset{E_2}{\rightleftharpoons} X + \text{sugar phosphate}$$

where X is a low-molecular-weight protein that, in the presence of the enzyme E_1, is phosphorylated by phosphoenol pyruvate. The subsequent transfer of phosphate to sugar, catalyzed by the enzyme E_2, appears to be specific, and it is likely that there is a different E_2 for each sugar. By the shock technique mentioned above, X can be released from the membrane, and upon this removal of X from the medium, the ability to accumulate any sugar is lost. When X is added back to the shocked cell, sugar transport is restored. Mutations in E_1 or X result in loss of the uptake of several sugars, and no phosphorylated derivatives of these sugars are found in the mutant cells. Extracts of the wild type, however, contain phosphorylated derivatives of all the sugars transported by this system.

These experiments represent the very promising beginning toward an understanding of the molecular basis for membrane transport. It seems clear that in principle the process is very similar to enzyme action with the added aspect of translocation thrown in. The vectorial reaction mechanism is clearly essential to the translocation phenomenon. Based on our knowledge of enzymology, it seems reasonable to anticipate that regulatory mechanisms that apply to enzymes also will be found involved in the transport systems, and that the elucidation of transport thus will be of fundamental importance to the understanding of how the living cell operates.

10.3 A MOLECULAR MODEL OF ACTIVE TRANSPORT: Na-, K-ACTIVATED ATPase

Virtually all membranes studied to date have one enzyme component in common, namely an ATPase activated by Mg^{2+}, Na^{+}, and K^{+}.† The ATPase activity is always found strongly associated with the lipoprotein fraction of the membrane, and all attempts to obtain a lipid-free enzyme have resulted in partial loss of activity and total loss of the unique metal-ion–activation properties of the enzyme. Because of this fact, it is not certain that membrane Na-, K-activated ATPase is indeed a single enzyme, nor that the activity is expressed by the same protein or proteins in different membranes. In working with membranes, this

* W. Kundig, *et al.*, *J. Biol. Chem.*, **241**, 3243 (1966), and S. Tanaka, and E. C. C. Lin, *Proc. Nat. Acad. Sci.*, **57**, 913 (1967).

† R. W. Albers, *Ann. Revs. Biochem.*, **36**, 727 (1967).

is not a unique problem, however, and the significant feature for this discussion is that a Na-, K-activated ATPase appears to exist as a functional component of all membranes.

The relevant properties of this enzyme are outlined below. The enzyme catalyzes the overall reaction

$$\text{ATP} + \text{H}_2\text{O} \rightleftharpoons \text{ADP} + \text{P}_i$$

Under certain conditions, however, it also catalyzes a ^{32}P exchange:

$$\underset{(\text{ATP})}{\text{A-P-P-}^{32}\text{P}} + \underset{(\text{*ADP})}{\text{*A-P-P}} \rightleftharpoons \underset{(\text{ADP})}{\text{A-P-P}} + \underset{(\text{*ATP})}{\text{*A-P-P-}^{32}\text{P}}$$

This type of reaction implies an E—P intermediate

$$\text{ATP} + E \rightleftharpoons E\text{—P} + \text{ADP}$$

$$\text{*ADP} + E\text{—P} \rightleftharpoons \text{*ATP} + E$$

(* signifies a radioactive label in the adenosine of ADP and ATP) which has been confirmed by isolating the phosphorylated intermediate after short exposure of the enzyme to ATP labeled in the γ-phosphate with ^{32}P (A–P–P–^{32}P). The chemical properties of the enzyme-phosphate derivative strongly suggest that the phosphate is attached to a carboxyl group in the form of a mixed anhydride (the group involved appears to be a glutamate carboxyl). Under conditions different from those used in the phosphorylation, the enzyme-phosphate derivative rapidly hydrolyzes to free enzyme and inorganic phosphate. There are thus two distinct and separate parts of the ATPase activity that can be used to characterize the enzyme: (1) the ATP exchange (enzyme phosphorylation); and (2) the overall hydrolytic reaction.

The Role of Sodium and Potassium

The effect of metal activators is complex. There is apparently an absolute requirement for Mg^{2+}, but the Mg^{2+}-activated enzyme alone has low activity in both the exchange reaction and in the ATPase reaction. Addition of K^+ alone to the Mg^{2+}-activated enzyme has no effect on either reaction; addition of Na^+ alone has no effect on ATPase activity, but greatly enhances the exchange reaction. Addition of Na^+ and K^+ together gives a large enhancement of ATPase activity but eliminates the Na^+ stimulation of the exchange reaction. To complete the picture, the addition of K^+ to the phosphorylated enzyme intermediate leads to rapid release of inorganic phosphate. From this, it is clear that the first half of the overall ATPase reaction,

$$E + \text{ATP} \rightleftharpoons E\text{—P} + \text{ADP}$$

is activated by Na^+, while the second half,

$$E\text{—P} + \text{H}_2\text{O} \longrightarrow E + \text{P}_i$$

is stimulated by K^+, and the requirement for both ions in the overall reaction is thus explained. Several attempts have been made to study this enzyme by kinetic analysis, but it appears that with three activators, substrate, and a covalent enzyme-phosphate intermediate, the system is too complex to permit reasonably quantitative conclusions from kinetics. It is probably significant that the metal activation curves are sigmoidal, and show Hill coefficients of around 1.5, thus suggesting cooperative interactions between multiple binding sites.

Correlation of Sodium (and Potassium) Transport with the Properties of Na-, K-activated ATPase

One very general feature of the "sodium pump" (active transport of sodium) in all systems studied is that it is inhibited by the cardiac glycoside, ouabain. This compound is also a potent inhibitor of the Na^+- and K^+-activated ATPase activity. (Some activity, corresponding essentially to the level observed in the

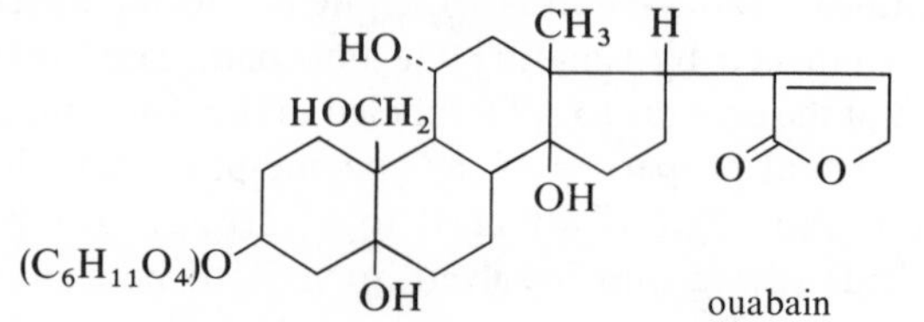

ouabain

absence of Na^+ and K^+, is not affected by ouabain.) This fact, together with the ubiquitous nature of Na, K–ATPase, represents the major basis for implicating this enzyme in the ion transport mechanism.

Based on these observations, it is possible to formulate a model for ion transport across a membrane, in which the energy released in the hydrolysis of ATP causes the conformational changes responsible for ion transport. The model is highly speculative and will not be easy to prove. The proposed mechanism is presented in Figure 10.5, in a rather formalistic manner. (It is probably usually true that the fancier the little round enzyme balls are drawn, the less is actually known about them.) The model has some interesting features, however. It encompasses the concepts of both conformational change and vectorial reactions of membrane-bound enzymes, and it does rationalize the correlation of ATP hydrolysis with flow of Na^+ out of the cell and K^+ into the cell. Since the first step of the reaction is activated by Na^+, it also allows for a regulatory mechanism: The pump is only turned on when the Na^+ concentration inside the cell is high enough to activate the enzyme. More importantly, the model does suggest further experimental tests, which appear practical. It should certainly be possible to demonstrate the existence of the two different conformers E and E^*, either one of which can be stabilized sufficiently by the absence and presence of Na^+, K^+, and ATP to be observed. And it may be possible to design proper experiments to look for the two different phosphorylated forms of the enzyme.

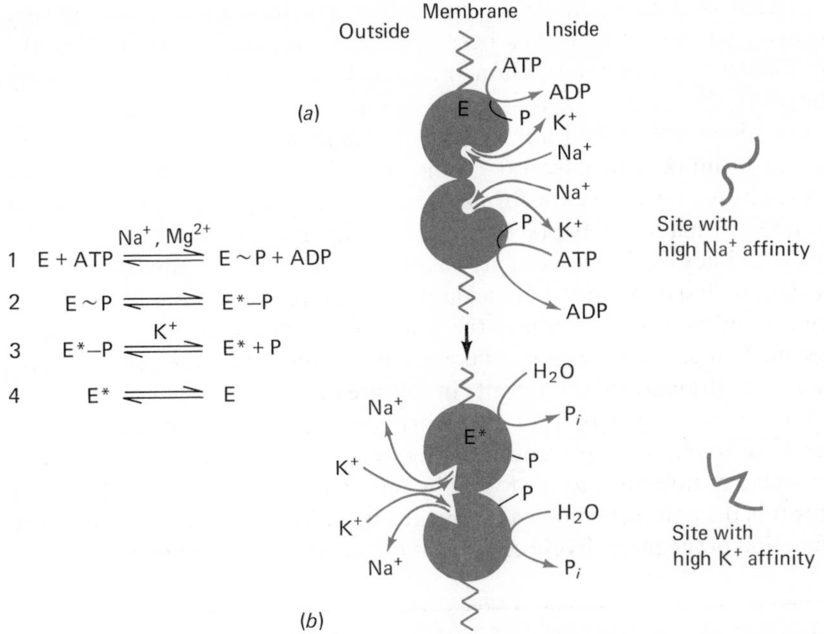

Figure 10.5 A hypothetical model of the "Na^+ (K^+) pump" (E is Na, K-activated ATPase). (*a*) High Na^+ affinity sites release K^+ and bind Na^+, and the enzyme catalyzes transfer of phosphate from ATP to form E ~ P (~ indicates high energy phosphate). In the conformational form E the E ~ P complex is stable (reaction 1 above). (*b*) The energy of E ~ P is used to transfer E to a higher energy state E* with a conformation that converts the high Na^+ affinity sites to high K^+ affinity sites (reaction 2 above). Na^+ is consequently released, K^+ bound, and E-P is hydrolyzed (reaction 3). The high energy E* spontaneously converts to the E conformer (reaction 4), and the process starts again.

There is no requirement that these must be different covalent derivatives, but if they are found to be, this would support the proposed model.

The analogy of this model to Kennedy's galactoside transport model (Figure 10.4) and the electron transport models (Figure 10.3) is obvious. Less obvious perhaps but highly significant is the analogy of these processes to other processes of translocation discussed in other chapters of this book : muscular contraction with ATP as energy source (Chapter 4) and ribosome-mRNA translocation with GTP as energy source (Chapter 7). It may be very fruitful indeed to consider all of these analogous systems together in the search for the mechanisms by which chemical energy is transformed into mechanical work (contraction or translocation).

10.4 CONCLUDING REMARKS

The few examples of membrane systems discussed here should illustrate the feasibility of studying these extremely complex systems at the molecular level.

The amount of data available on membrane structure and function is very impressive, but so far there has been very little success in coordinating the many different bits of information into a coherent picture. The qualitative treatment in this chapter is in part due to this variety of information and, as stated in Chapter 9, should not be taken to imply a disparity of quantitative studies in membrane biochemistry. It is possible that the membrane field is hampered by too restrictive and formal models. The analogy to the development of enzymology made in Chapter 9 can be used again to point out that once the first lock-and-key model has been proposed, it required a large volume of quantitative data to first confirm the model and then realize that the model needed to be refined and improved. Perhaps the study of membranes as functional units has reached the same stage now. There can be no doubt that the next few years will see some dramatic developments in this area.

The most exciting aspect of the work on biochemical characterization of membranes is that it represents a daring step away from the relative security of the isolated molecules toward the big and complex unknown of the intact cell, for it is the complete understanding at the molecular level of the structure and function of the intact living cell that is the ultimate goal of biochemistry.

REFERENCES

Useful books:

Green, D. E., and H. Baum, *Energy and the Mitochondrion*, Academic Press, New York, 1970.

Lehninger, A. L., *Bioenergetics*, W. A. Benjamin, Inc., New York, 1965.

Lehninger, A. L., *The Mitochondrion*, W. A. Benjamin Inc., New York, 1964.

Some reviews:

Albers, R. W., "Biochemical Aspects of Active Transport," *Ann. Rev. Biochem.*, **36**, 727 (1967).

Kaback, H. R., "Transport," *Ann. Rev. Biochem.*, **39**, 561 (1970).

Lardy, H. A., and S. M. Ferguson, "Oxidative Phosphorylation in Mitochondria," *Ann. Rev. Biochem.*, **38**, 991 (1969).

Skou, J. C., "Na^+- and K^+-activated ATPase," *Physiol. Rev.*, **45**, 596 (1965).

INDEX